EARTHQUAKE ENGINEERING

EARTHQUAKE ENGINEERING
Application to Design

CHARLES K. ERDEY
Northern Arizona University
Formerly Adjunct Professor, California State University Long Beach

JOHN WILEY & SONS, INC.

For general information about our other products and services, please contact our Customer Care Department within the United States at (800) 762-2974, outside the United States at (317) 572-3993 or fax (317) 572-4002.

Wiley also publishes its books in a variety of electronic formats. Some content that appears in print may not be available in electronic books. For more information about Wiley products, visit our web site at www.wiley.com.

Library of Congress Cataloging-in-Publication Data:

Erdey, Charles K., 1931–
    Earthquake engineering: application to design/Charles K. Erdey.
        p. cm.
    ISBN-13: 978-0-470-04843-6 (cloth)
    ISBN-10: 0-470-04843-3 (cloth)
    1. Earthquake engineering.   I. Title.
    TA654.6.E73 2006
    624.1′762—dc22

                                                    2006011329

Printed in the United States of America

10 9 8 7 6 5 4 3 2 1

CONTENTS

3   REINFORCED-CONCRETE STRUCTURES   36

4   SEISMIC STEEL DESIGN: SMRF   68

5 SEISMIC STEEL DESIGN: BRACED FRAMES   128

PREFACE

The primary motivation for writing this book is the causes of structural failures—*what went wrong*—during the earthquakes that hit the western states in the last decades.

In view of the relatively large number of steel moment-resisting frames damaged during the Northridge earthquake, the book expands on the evaluation and performance of structures of this type. The pre- and post-Northridge experimental research and new design strategies to improve moment connections for new buildings are also discussed, keeping in mind basic building code concepts to demonstrate the application of general strength-level load combinations.

Topics relevant to seismic design in other areas of engineering, such as concrete, masonry, and wood-framed buildings, are also included. An attempt has been made to maintain a practical approach. In lieu of problem-solving, single design issues, the book *walks* the reader through step-by-step design of actual projects in moderate-to-high seismicity areas in compliance with building regulations.

Chapter 12 introduces a new method of dynamic analysis and discusses the causes of joint failure in steel design. Subjects like matrices, differential equations, numerical analysis, and engineering applications are presented for completeness and ready reference for the reader.

It is hoped that the book will help practicing engineers not yet fully familiar with seismic design and graduating students to use the building codes in their seismic design practice.

ACKNOWLEDGMENTS

The author gratefully acknowledges the comments and suggestions of Tom King, professor of mathematics, California State Polytechnic University, Pomona, and John G. Shipp, S.E., senior member of SEAOC, for their suggestions, and to Victor F. Sanchez, P.E., for his valuable comments. Thanks are also due to Ernest Pappas, former plan examiner, Mohave County, for his contribution on some topics, Alex Zupanski, CSULB graduate student for facilitating the inputting of the author's mathematical formulas into the computer, and to Alice H. Zwiller for assisting with proofreading.

NOTATION

A = cross-sectional area (in.²)
A_B = ground-floor area
A_{gt} = gross area subjected to tension (in.²)
A_{gv} = gross area subjected to shear (in.²)
A_{nt} = net area subjected to tension
A_{nv} = net area subjected to shear (in.²)
A_w = effective area of weld (in.²)
A_x = torsional amplification factor (level x)
b_f = flange width of column
B = width of base plate (in.)
B_1, B_2 = factors used to determine M_u for combined bending and axial forces
C_a, C_v = seismic coefficients
C_e = height, exposure, and gust factor coefficient
C_q = pressure coefficient subject to function, geometry, and location of structure or element
D = dead load on structural element
d = depth of column
d_b = nominal bolt diameter (in.)
d_c = column depth (in.)
E, E_h, E_v = earthquake design components
E = modulus of elasticity of steel (29,000 ksi)
E_c = modulus of elasticity of concrete (ksi)
F_{EXX} = strength of weld metal (ksi)
F_i, F_n, F_x = design seismic forces applied to each level

f'_c = strength of concrete

F_p = design seismic force applied to part of structure

F_t = V portion of base shear on top of structure

F_u = specified minimum tensile strength of steel used (ksi)

F_y = specified minimum yield stress of steel used (ksi)

G = shear modulus of elasticity of steel (11,200 ksi)

g = acceleration due to gravity (386 in./s²)

H = average story height (above and below beam-to-column connection)

I = seismic importance factor related to Occupancy Category; moment of inertia (in.⁴)

I_w = importance factor subject to occupancy or function of building

K = coefficient for estimating natural frequency of beam (AISC *Specification*)

L = live load on a structural element; unbraced length (compression or bracing member)

L_b = laterally unbraced length

L_p = limiting laterally unbraced length

M = maximum moment magnitude

M_{max} = value of maximum moment in unbraced segment of beam (kip-in.)

M_A = absolute value of moment at quarter point of unbraced beam segment (kip-in.)

M_B = absolute value of moment at half point of unbraced beam segment (kip-in.)

M_C = absolute value of moment at three-quarter point of unbraced beam segment (kip-in.)

M_p = plastic moment of resistance of beam

M_u = required flexural strength (kip-in. or kip-ft)

N_a, N_v = near-source factors

P = design wind pressure; concentrated load (kips)

P_p = bearing load on concrete (kips)

P_u = required axial strength (kips)

Q_E = effect of horizontal seismic forces

q_s = basic wind pressure subject to basic wind speed

R = numerical coefficient applied to lateral-force-resisting systems

r = ratio used in determining ρ

R_I = response modification factor

R_p = component response modification factor

S_A, S_B, S_C, S_D, S_E, S_F = types of soil profiles

T = elastic fundamental period of vibration; tension force due to service loads (kips)

t_p = thickness of base plate (in.)

t_p = panel zone thickness including doubler plates (in.)
U = reduction coefficient
V = total design lateral force (of shear)
V_n = shear force component (kips)
W = total seismic dead load (UBC 1997, Section 1630.1.1)
w_z = width of panel zone between column flanges
Z = seismic zone factor; plastic section modulus (in.^{3})
α = fraction of member force transferred across a particular net section
Δ_M = Maximum Inelastic Response Displacement
Δ_S = Design-level Response Displacement
δ = deflection (in.)
ρ = Redundancy/Reliability Factor
ϕ = resistance factor
ϕ_b = resistance factor for flexure
ϕ_c = resistance factor for axially loaded composite columns
ϕ_c = resistance factor for compression
ϕ_t = resistance factor for tension
ϕ_v = resistance factor for shear
ϕ_w = resistance factor for welds
Ω_0 = Seismic Force Amplification Factor

EARTHQUAKE ENGINEERING

CHAPTER 1

OVERVIEW

As usual in our vibrant, free society, it is up to us to decide whether to face the reality of the seismic threat and embrace the availability of the solutions or to continue to lie helpless before quakes which can flatten our houses, destroy our employers, damage our national economy and national defense, and wipe out the financial equity of a lifetime in a mere thirty seconds of groundshaking.
John J. Nance, *On Shaky Ground,* Morrow & Co., New York, 1988

1.1. INTRODUCTION

During an earthquake an individual could be thrown out of bed at night, be unable to stand upright and be forced to kneel on the ground, fall down stairs, or even be tossed out of the swimming pool by the violent sloshing of the water. In the aftermath of the 6.7-Richter-magnitude Northridge earthquake of January 17, 1994 (Figure 1.1), the author spoke with the resident of a two-story house who had been through a similar experience during the 1971 San Fernando earthquake. At that time she was repeatedly knocked down while attempting to reach her baby daughter downstairs. Twenty-three years later, living in a different location but still near the epicenter, the violent quaking of Northridge prevented her once again from reaching the ground floor. Both seismic events happened in the early morning.

The author collected these and other personal accounts in the course of inspections of nearly 100 homes damaged by the Northridge earthquake.

There are ways of making structures safer than the current ones. Researchers and the engineering community have mobilized to achieve that goal, working on removing shortcomings in the design of structures that have not

Figure 1.1 Parking structure that collapsed during the 1994 Northridge earthquake, California State University, Northridge Campus.

performed well in seismic events and coming up with improved versions capable of standing up to a certain level of earthquakes. (See Figures 1.2 and 1.3.)

One option is to build or retrofit on seismic isolators or structural dampers. An example is the Los Angeles City Hall, retrofitted with a *viscous-device* type of supplemental damping to improve seismic response. However, placing such a massive stone building and historic landmark on an earthquake damage control system comes at a cost that not all areas can afford.

1.2 CONCEPTS, TERMINOLOGY, AND SOURCE OF EARTHQUAKES

Specific Gravity

The specific gravity of a substance refers to how much heavier than water a unit volume of the substance is. Some specific gravities related to earthquake engineering are as follows:

Earth's crust	2.7–3.0
Mantle (inner periphery)	5.7
Core (periphery)	9.7
Center	12.3

The earth's crust *floats* on the surface of the mantle (Figure 1.4), which possesses a viscoelastic character. This equilibrium is called *isostasy*.

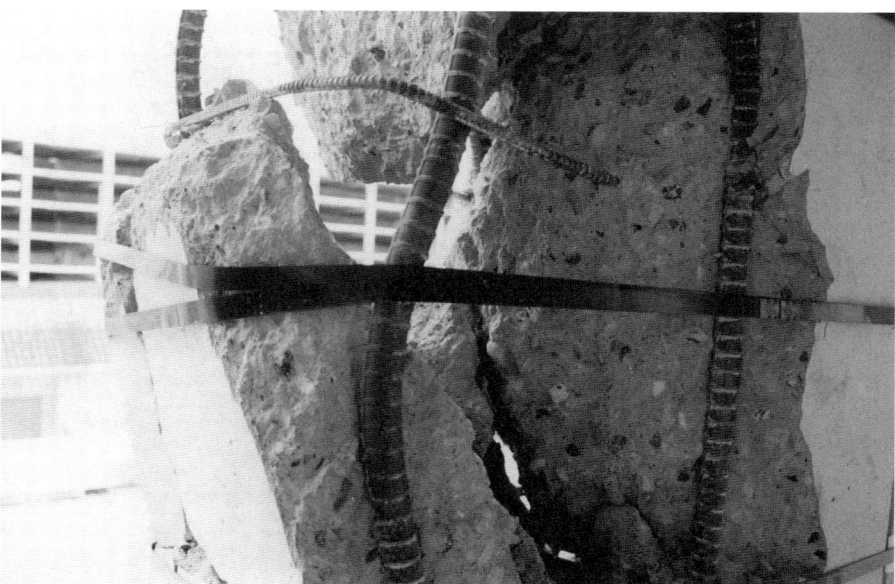

Figure 1.2 Straps are holding the crumbled lightweight concrete in a multistory residential building in Santa Monica, California, severely damaged by the Northridge earthquake.

Orogenic Movements and Crust Convection (Conveyor Belt)

Orogenic movements and crust convection are mainly responsible for mountain building and valley forming—in other words, the constant changes affecting the surface of the earth.

In the first half of the twentieth century Alfred Wegener asserted that at one time continents such as Africa and South America were connected and then drifted away from each other. Wegener, who was ridiculed at the time for his *continental drift* concept, perished on an expedition to the North Pole. Since then, fossil and geological evidence has substantiated the fact that these continents were once one massive piece. High-technology developments of the 1960s and deep-diving submarines have produced interesting findings about ocean floor fissures and left-and-right movements that, like a giant conveyor belt, have the power of moving continents that float on the viscous mantle. A similar movement at Lake Victoria in Africa is slowly splitting the African continent.

Subduction Zones

As the ocean floor exerts pressure on the coastline of the continent, the leading edge of the ocean floor is pushed under the continent, carrying down sea deposits, including the remains of organisms (Figure 1.5). The matter reaches

Figure 1.3 Detail of parking structure that collapsed during the Whittier Narrows earthquake of 1987.

intensive heat under the continent and produces geothermal irregularities— gases and molten matter that tend to rise to the surface. This *subduction* process can be seen in the series of active volcanoes along the Pacific shore- line of the American continent from Alaska to Chile, and is responsible for the earthquakes that affected Chile, Colombia, California, and Washington State.

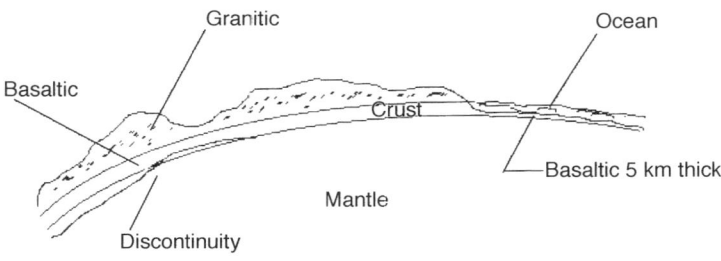

Figure 1.4 Discontinuity of seismic waves, named the *Mohorovicic discontinuity.*

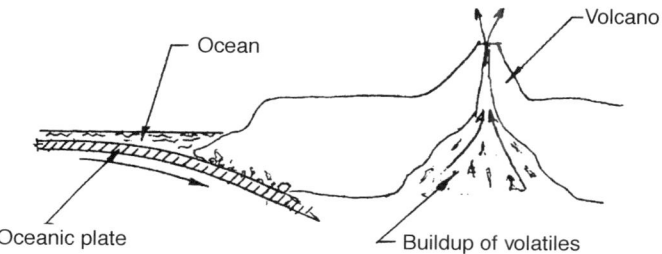

Figure 1.5 Subduction process.

Volcanoes

Around 900 years ago the Sunset Crater volcano eruption coupled with strong ground motion caused panic among the native population in what is today Flagstaff and the surrounding areas of Arizona. We can still see the geologically fresh lava flow. In northern California, Mount Shasta bombarded the neighboring region with boulders that scattered for miles, some weighing about 5 tons. The upheaval was accompanied by severe ground motion.

Land Erosion

The 1812 New Madrid earthquake in Missouri is considered the largest earthquake in what is regarded a *low-seismicity* area. What could have caused such an event was the mighty Mississippi constantly eroding the land mass and making the earth's crust lighter. Since the earth's crust cannot adjust immediately to the river's action, from time to time it springs up.

Summary of Main Sources of Earthquakes

1. Orogenic movements such as mountain building
2. Subduction and plate convection followed by geothermal and mechanical disturbances
3. Volcanic activity
4. Land erosion

1.3 WAVE PROPAGATION AND VELOCITIES

Wave Propagation

The focal point of an earthquake under the surface of the earth is called the *hypocenter* and its corresponding point on the surface the *epicenter*. It is customary to refer to earthquakes with relation to the epicenter.

When an earthquake hits the hypocenter, it sends out shock waves. There are two types of shock waves:

Push waves—denoted *p*
Shock waves or *shear* waves that produce transverse vibration with respect to the direction of travel, also named *s* waves

The *p* waves are faster than *s* waves and arrive first, produce a relatively mild vibration, and cause less damage. They are messengers of the severe ground shaking that will follow. The moment the *s* waves arrive, seismographic diagrams start recording the magnitude of ground shaking (Figure 1.6).

If the distance from a given observation point to the hypocenter is *s*, the propagation velocity of the transverse waves is v_s and the propagation velocity of the longitudinal (push waves) is v_p. Then *T*, the time difference between the arrival of *p* and *s* waves, is given as

$$T = \frac{s}{v_s} - \frac{s}{v_p} = s\left(\frac{1}{v_s} - \frac{1}{v_p}\right)$$

where the distance $s = (1/v_s - 1/v_p)^{-1} T$ from simple arithmetic. We need three observation points to use triangulation *and* the geology of the ground, which determines v_s and v_p by measurement. The 1997 UBC gives some rough values for v_s and v_p. Nonetheless, it is advisable to have a good geotechnical report for accuracy.

Wave Velocities (Body Waves)

$$v_p\sqrt{\frac{E(1 - \sigma)}{\rho(1 + \sigma)(1 - 2\sigma)}}$$

where E = Young's modulus
σ = Poisson's ratio (usually 0.25)
ρ = density

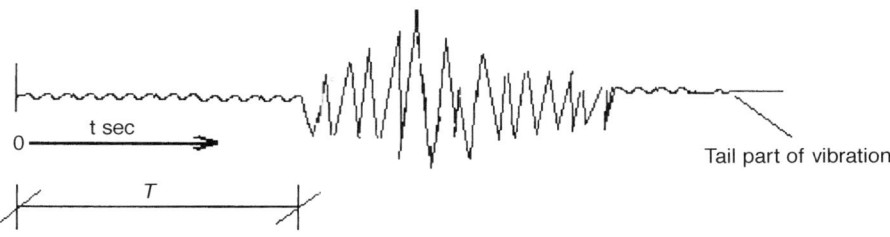

Figure 1.6 Seismograph reading of ground vibration caused by an earthquake.

and

$$v_s \sqrt{\frac{E}{2\rho(1 + \sigma)}}$$

Velocities v_s of typical transverse waves to propagate through the ground for selected materials are as follows (in meters per second):

Sand	60
Reclaimed sand	100
Clay	250
Gravel	600
Tertiary rock	1000 and up

1.4 MAGNITUDE OF EARTHQUAKES

To compare earthquakes, we need some yardstick or scale such as the one created by C. F. Richter. Using a standard horizontal Wood–Anderson seismograph, the magnitude

$$M = \log_{10} A$$

where A denotes the *trace* amplitude in micrometers for an epicentral distance of 100 km. When the distance from the epicenter is other than 100 km,

$$M = M_\Delta - \frac{1.73 \log_{10} 100}{\Delta}$$

where M_Δ is the magnitude at a distance Δ calculated from the basic Richter formula. The magnitudes of significant earthquakes in the United States are given in Table 1.1.

1.5 BUILDING DAMAGE

As a measure of the magnitude of destruction caused by the 1994 Northridge earthquake, following is a summary of the structural damaged suffered in a densely populated area, the City of Los Angeles:

Total Number of Buildings Damaged 93,200 (1900 *red,* 8800 *yellow,* 82,500 *green*). Of these, 3000 buildings suffered moderate to major damage.

TABLE 1.1 Significant U.S. Earthquakes

Location	Date	Magnitude
Cascadia Subduction Zone[a]	1700	~9.0
New Madrid, Missouri	1811, December 16	8.1
	1812, February 7	~8.0
Ventura, California	1812	7.1
Fort Tejon, California	1857	7.9
Ka'u District, Island of Hawaii	1868	7.0
Lanai, Hawaii	1871	6.8
Owens Valley, California	1872	7.4
California/Oregon coast	1873	7.3
Denver, Colorado	1882	6.2
Charleston, South Carolina	1886	7.3
Imperial Valley, California	1892	7.8
Cape Yakataga, Alaska	1899	7.9
Yakutat Bay, Alaska	1899	8.0
Eureka, California	1899	7.0
San Andreas, California	1906	8.3
San Francisco, California	1906, April 18	7.8
Oregon	1910	6.8
Pleasant Valley, Nevada	1915	7.1
Eureka, California	1922	7.3
Humboldt County, California	1923	7.2
Santa Barbara, California	1925	6.3
Clarkston Valley, Montana	1925	6.6
Lompoc, California	1927	7.1
Fox Islands, Aleutians, Alaska	1929	7.8
Valentine, Texas	1931	5.8
Cedar Mountain, Nevada	1932	7.3
Long Beach, California	1933	6.4
Excelsior Mountain, Nevada	1934	6.5
Hansel Valley, Utah	1934	6.6
Helena, Montana	1935	6.25
Central Alaska	1937	7.3
East of Shumagin Islands, Alaska	1938	8.2
Imperial (El Centro), California	1940	7.1
Ossipee Lake, New Hampshire	1940	5.5
Skwenta, Alaska	1943	7.4
Unimak Islands, Alaska	1946	8.1
Wood River, Alaska	1947	7.2
Southwest Montana	1947	6.25
Manix, California	1947	6.4
Seattle, Washington	1949	7.1
White Wolf, California	1952	7.7
Kern County, California	1952	7.3
Near Islands, Alaska	1953	7.1
Rainbow Mountain, Nevada	1954, August	6.8
Fairview Peak, Nevada	1954, December	7.1
Andreanof Islands, Alaska	1957	8.6

TABLE 1.1 (*Continued*)

Location	Date	Magnitude
Fairweather, Alaska	1958	8.0
Wyoming	1959	6.5
Hebgen Lake, Montana	1959	7.3
Prince William Sound, Alaska	1964, March 27	9.2
Rat Islands, Alaska	1965	8.7
Puget Sound, Washington	1965	6.5
Rat Islands, Alaska	1966	7.61
San Fernando, California	1971	6.7
Sitka, Alaska	1972	7.6
Near Islands, Alaska	1975	7.6
Eastern Idaho	1975	6.1
Kalapana, Hawaii	1975	7.2
Mount St Elias, Alaska	1979	7.6
Imperial Valley, California/Mexico border	1979	6.4
Eureka, California	1980	7.4
Coalinga, California	1983	6.5
Borah Peaks, Idaho	1983	7.0
Andreanof Islands, Alaska	1986	8.0
Chalfant Valley, California	1986	6.4
Whittier, California	1987	6.0
Gulf of Alaska	1987	7.9
Gulf of Alaska	1988	7.8
Loma Prieta, California	1989	6.9
Crescent City (offshore) California	1991	7.0
Sierra Madre, California	1991	5.8
Joshua Tree, California	1992	6.2
Big Bear, California	1992	6.5
Cape Mendocino, California	1992	7.2
Landers, California	1992	7.3
Bishop. California	1993	6.2
Northridge, California	1994	6.7
Cape Mendocino, California	1994	7.0
Northern California (off coast)	1994	7.1
Andreanof Islands, Alaska	1996	7.9
Hector Mine, California	1999	7.2
Seattle, Washington	2001	6.8
Denali Fault, Alaska	2002	7.9
Rat Islands, Aleutians, Alaska	2003	7.8
Offshore Oregon	2003	6.2
San Simeon, California	2003	6.5
Central California	2004	6.0
Northern California (off coast)	2005	7.2
Gulf of California	2006, January 4	6.6

[a] A 600-mile-long region that covers northern California, Oregon, Washington, and southern British Columbia. The earthquake triggered off a tsunami that reached Japan. Written records place the earthquake in the evening of January 26, 1700.
Source: U.S. Geological Survey.

Failures by Building Class
   Wood-framed homes: 1650
   Wood-framed apartments, condominiums, and hotels: 630
   Tilt-ups, masonry: 350
   Unreinforced masonry retrofitted: 213
   Structural steel buildings: 100

1.6 STRUCTURAL FAILURES: OVERALL FAILURE

Structural failures may be categorized as *overall failure* and *component failure*. Overall failure involves collapse or overturning of the entire structure. The choice of the type of structure is instrumental and often a predetermining factor for failure.

1. *Moment Frame in Longitudinal Direction, Shear Walls at Each End in Short Direction*   This structural system was approved as a major structural category to resist earthquakes by Structural Engineers Association of California (SEAOC), which influenced the Uniform Building Code (UBC) and the 2000 and 2003 International Building Codes (IBCs). This type of structure, however, fared badly in the January 17, 1994, Northridge earthquake. Figure 1.7 shows the collapsed wing of an office building in the 1994 Northridge earthquake. The longitudinal moment frame underwent large lateral oscillation imposed by the longitudinal component of the quake.

The moment frame pushed over and destroyed the shear wall, leaving the structure defenseless against the lateral force component in the short direction, which caused the building to collapse. Each new UBC edition and its IBC successors adopted this type of building system, granting it a *prequalified status* design category. Such structures, they reasoned, would be able to resist earthquakes, along with a few recognized structural types, such as moment frames, braced frames, shear wall structures, and hybrid combinations of these.

The basic concept behind the moment frame–shear wall combination advocated by SEAOC is that the end shear walls will take care of the earth force component in the short direction while the moment frame will resist the longitudinal component acting in the longitudinal direction of the structure (Figure 1.8). As explained below, the laws of nature challenge such an assumption.

When an earthquake hits, the structure undergoes lateral oscillations that amplify in the longitudinal direction. The springlike response of the moment frame and the large floor mass contribute to the excitation. Measured lateral floor displacements—*story drift*—can be on the order of 10 in. This fact has been verified by the author while performing a dynamic analysis on

Figure 1.7 Northridge earthquake, 1994. Collapsed end shear wall, Kaiser Permanente Office Building, Granada Hills, CA. (Photo courtesy of the University of California Library at Berkeley.)

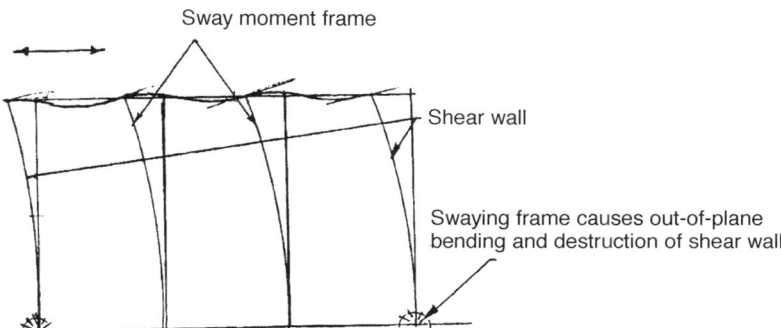

Figure 1.8 Moment frame and end shear wall, a bad combination to counter earthquakes.

earthquake-generated ground acceleration data at the base of an actual structure. Rapid and violent oscillations of the moment frame exceeding 2 Hz (cycles per second) will cause the end shear wall to bend back and forth until it breaks at the base. This was the case of the Kaiser Office Building in Granada Hills during the 1994 Northridge earthquake. Once the end shear walls are destroyed, the building is susceptible to catastrophic failure. It is estimated that 80% of high-rise hospital structures in California have this type of construction.

Champagne Towers, an upscale high-rise apartment building overlooking the Santa Monica Bay in Southern California was built utilizing a similar system, that is, end shear walls in the short direction and reinforced-concrete moment frames acting in the long direction of the structure. When the 1994 Northridge earthquake hit, residents of the towers woke up to a frightening sound and violent ground shaking accompanied by lateral sway. Initially not all that significant, within seconds the sway turned violent and uncontrollable followed by the sound of an explosion when the concrete columns broke up (Figure 1.9). The alarmed residents fled the building. The first daylight revealed severe damage to the main load-bearing columns with diagonal cracks up to $\frac{1}{2}$ in. Within hours the building was declared dangerous, uninhabitable, and condemned.

Mechanism of Destruction of Moment Frame–End Shear Wall Construction. Shear walls are sensitive to out-of-plane bending. Normally just a

Figure 1.9 Reinforced-concrete columns severely damaged in the Northridge earthquake, Champagne Towers, Santa Monica, CA.

single layer of vertical reinforcement is provided for the wall and it is tradi-
tionally placed at the theoretical elastic centerline of the wall, thereby offering
poor resistance to out-of-plane bending (Figure 1.10).

When the moment frame undergoes lateral sway in the longitudinal direction, it causes serious out-of-plane flexural deformation to the shear wall. As
the shear wall begins to sway back and forth, it crushes the outer and inner
fibers of the concrete until there is no concrete left to support the rebars. Once
the rebars are deprived of the confining effect of the concrete, they buckle
under the load imposed by the shear wall. The process develops very fast,
the sway occuring at 2–3 Hz. The deformation of the wall can be very large
and the lateral floor displacement or story drift can reach a magnitude of 10
in. or more.

Exceedingly large sway of frame structures occurred during the San Fernando earthquake of 1971 in California. The stair/elevator towers of the Vet-
erans Hospital swayed out, acting like a sledgehammer against the wings of
the main building and destroying it.

By now it should be clear to the reader that moment frame–end shear wall
construction is a dangerous combination based on static force considerations
that ignore the dynamic response of the entire structure. Unfortunately this
type of construction was (and is) stated as accepted practice in successive
editions of the UBC and its successors, the IBC 2000 series.

Figure 1.10 It is estimated that 80% of hospital structures in California are of the
moment frame–shear wall type of construction, a bad design arrangement to resist a
strong earthquake.

2. *Multistory Buildings with First-Story Shear Walls Not Aligned with Upper Story Shear Walls* When first-story walls are not aligned with upper story shear walls, they create an unnatural configuration that severs the continuity of the vital lateral force resisting system. Figures 1.11–1.13 show a cross section and a side elevation of a building with the type of design forces normally—but erroneously—applied to the main cantilever structure corbels or cantilevered beams. These forces include gravity loads but ignore dynamic impact by seismic forces.

While numerous engineering seminars teach that the floor of a structure transfers the seismic shear to the nonaligned shear wall below, reality contradicts this concept. In addition to horizontal shear, the cantilever floor transfers the overturning moment, which is greatly amplified by the rocking motions induced by the *dynamic impact* of the earthquake.

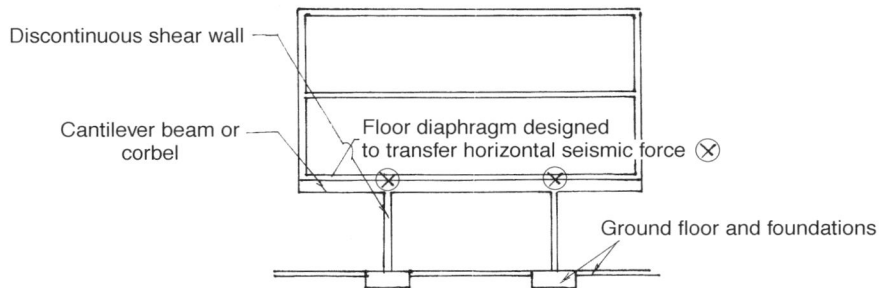

Figure 1.11 Cross section of building with discontinuous shear walls.

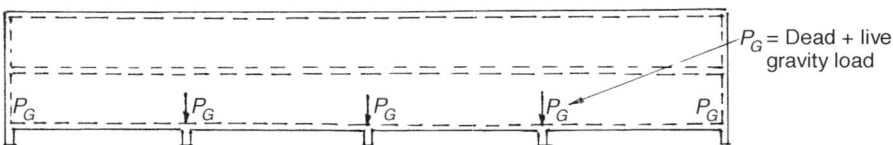

Figure 1.12 Side elevation showing incomplete cantilever design forces. Dynamic force caused by rocking motion can exceed P_G gravity load for which the building is normally designed.

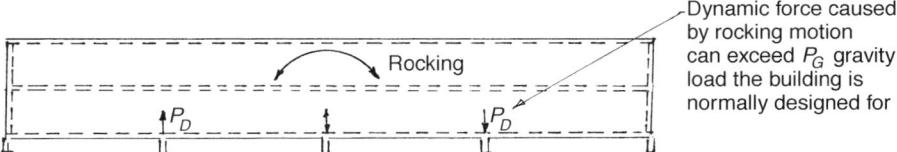

Figure 1.13 Side elevation showing seismic reaction forces on cantilever ends.

The cantilever element, not designed for dynamic forces that may well exceed dead- and live-load gravity, will be underdesigned and prone to failure. Such was the case at the chemistry-physics building of the Santa Monica Community College (SMCC) in Santa Monica, California, destroyed by the 1994 Northridge earthquake. The damage was so severe that the building collapsed in a heap of rubble and no attempts were ever made to repair and/or rehabilitate it. Instead, the building was demolished before the restoration of the majority of the other damaged buildings even began.

Numerous high-rise hospitals and hotel structures built using this system are in danger of collapsing if a significant earthquake hits them. An example is the Santa Ana Tustin Hospital in Santa Ana, California, where shops occupy the first story space provided by the offset shear wall system.

3. *Dry-Jointed Connections in Precast Parking Structures without Monolithic Connections* An example of this type of construction was the California State University precast parking structure located near the epicenter of the 1994 Northridge earthquake (Figure 1.14). Being dry jointed, that is, the precast beams were supported on corbels of the equally precast concrete columns without monolithic connection, the structure relied on friction of the support reaction created by gravity forces (mostly dead weight) across a few inches of seating.

Figure 1.14 California State University parking structure. A corner of the cast-in-place framework was pulled in by the collapsed interior of the multistory structure during the 1994 Northridge earthquake.

During an earthquake of relatively large magnitude, such as the 1994 Northridge earthquake, the several-feet-long lateral movement of the ground tends to pull the beams off the inadequate support. Worse, the vertical ground acceleration makes the structure weightless, thus overcoming any attempt to rely on friction connection.

Regrettably, a moratorium has never been declared on these potentially dangerous structures. Paradoxically, just a few months after the 1994 Northridge earthquake another precast parking structure was built at a sister campus, the *California State University Long Beach,* using the same system as the collapsed Northridge parking structure.

Some California *freeway bridges* are perfect examples of ill-fated uses of dry-jointed connections. An expansion joint is provided for the otherwise continuous bridge deck, separating a main span into a short cantilever (the supporting portion) and a long span (the supported portion). The main span rests on a few-inch-wide seating. When an earthquake hits, it causes several feet of measured lateral movement between adjacent or dry-joint connected structural components. As the long span springs up, its support, the initially continuous system breaks and suddenly converts into one short and one large span cantilever element no longer capable of supporting their own weight. The I-14/I-5 interchange near the town of San Fernando, California, crushed a motorist to death when it collapsed in the 1971 San Fernando earthquake. Yet the bridge was rebuilt using the same structural system, only to kill a highway patrol officer that was on the bridge when the Northridge earthquake hit.

It is estimated that more than 90% of the major California freeway bridges are built using the same faulty concept and construction method. Following the Loma Prieta and Northridge earthquakes, CALTRANS attempted to retrofit the joints by locking the two bridge segments with high-strength steel tendons. The tendons were inserted through holes drilled into the supporting joints. However, the soft concrete matrix has a Brinell hardness of about 10 with the high-strength steel at about 100 on the Brinell hardness scale. Thus, when a strong earthquake occurs, there is likely to be a cheese-cutter effect that will cause the cable to cut through the concrete and separate the joints. Once good construction is compromised, it seems virtually impossible to reverse the inevitable course of events, such as earthquakes exposing inherent structural weaknesses.

Is there any other way to create expansion joints in continuous structures? Perhaps doubling the columns would create safer expansion gaps between separable bridge components.

4. *When a Structure Is Too Strong to Break Up* A solid, several-story apartment structure built as a monolithic box of cast-in-place concrete tilted without structural damage in the 1964 Niigata earthquake in Japan. Evidently the rocking motion of the building, generated by the earthquake, created a pumping action to the partially saturated soil, increasing its potential for liquefaction.

1.7 COMPONENT OR JOINT FAILURE

Component failure refers to failure of one or more structural elements, mostly joints, due to a type of damage that makes the structural component or joint unusable. Such a condition necessitates repair or replacement. Because the failure mechanism differs according to the choice of material and type of structure, it seems best to categorize the structure by the construction materials used, especially *steel* and *concrete,* and then create subcategories.

Steel Structures

Apart from hybrids, *two major lateral force-resisting systems exist:* (1) *moment-resisting frames* and (2) *braced frames.*

1. *Moment-Resisting Frames* The most frequently observed damage to these structures is beam–column joint failure. To ensure continuity and moment transfer, the solution by the construction industry has been to butt weld the beam flanges to the column. Shear transfer from beam to column (and column to beam!) is provided by the shear tab, a vertical steel plate welded to the column. The shear tab and the beam web provided with boltholes allow prompt and easy erection. By tightening the bolts between the beam web and the plate, the beam is kept in place until the beam flanges are welded to the column flanges (or webs).

Several California earthquakes have proven that such a construction method is defective. The lateral oscillations caused by ground motion on a highly elastic steel frame create large internal forces that correspond to the mass times acceleration of the massive concrete floor acting as the driving force.

The amplification of dynamic lateral displacements often goes out of control, overtaxing the joint. During the rapid cycle, reversed stresses observed by the author far exceeded the nominal yield strength of the steel. In fact, often even the nominal ultimate strength (F_u) of the metal was exceeded. Under such conditions, the structure had to depend upon additional reserves.

Such reserves are provided by the moment of resistance of the shear tab by utilizing resistance of the bolts. When the *friction grip bolts* slip to form a couple, the bolts begin exerting an excessive force on the shear tab. Not meant for such extreme use (or misuse), the hardened high-strength bolts split the shear tab. The author observed such damage during the postearthquake retrofitting of the Anthony Building, head office and control building for the Los Angeles County Water and Power utility (Figure 1.15).

A fully welded web joint, the shear tab was welded to the beam web instead of being bolted. It could have performed better if it had utilized the full flexural resistance of the entire beam section consisting of flange and web. Yet this method was compromised in favor of a fast and easy erection.

Figure 1.15 Left: crack in beam web during the Northridge earthquake, Anthony Building, Department of Water and Power, Los Angeles County.

Needless to say, tearing up the shear tab represents the last phase of the integrity of a structure.

When the destroyed shear tabs can no longer carry the vertical reaction, the floor will collapse on the floor below, which could create a catastrophic chain reaction of progressive failure (*pancaking*) that could eventually wipe out the entire building. Fortunately, in the case of the Anthony Building, the January 1994 Northridge earthquake was of relatively short duration. Had it lasted longer, it would have caused a floor collapse and possibly catastrophic failure of the entire control building.

2. *Braced Frames* Concentric braced frames that proved their value in situations involving static loads have a rather poor performance in an earthquake. Being a rather rigid structure, its shock absorption under dynamic impact is almost negligible.

The damage to the concentric braced frames of the Oviatt Library Annex in the California State Northridge Campus, California, is a clear demonstration of the lack of shock-absorbing properties of this type of structure (Figures 1.16 and 1.17).

The 4-in.-thick base plates connecting the Oviatt Library Annex structure to its foundation, behaving like glass, cracked and failed in brittle fracture. In addition to splitting the plates horizontally, as shown in Figure 1.16, a punching shear failure started to develop around the perimeter of the frame

Figure 1.16 Cracks in 4-in.-thick base plates during the Northridge earthquake, Oviatt Library Annex Building.

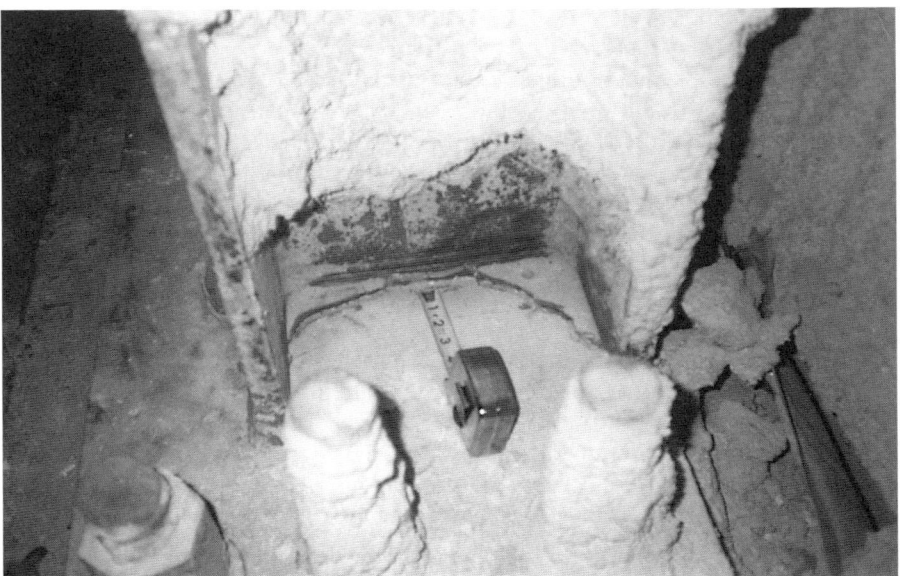

Figure 1.17 Shear failure at perimeter of frame leg–base plate solidly welded connection, Oviatt Library Annex Building. Note the crack in the center and the initial stage of shear failure on the left.

leg–base plate solidly welded connection. Torn from its footings, this rather slender structure would have overturned had the earthquake lasted a bit longer.

An alternative to the concentric braced frame which offers improved shock absorption properties is the hybrid eccentric braced frame, which has members eccentrically connected at the joints. However, there is a trade-off for improved shock absorption. During the time-dependent impacts of the earthquake, the beam providing eccentric connection for the braces will bend and undergo permanent deformation. Should the deformation be of significant magnitude, it will affect the entire structure and leave an out-of-alignment building that is difficult and expensive to repair or a permanently damaged building that is impossible to repair. Developed alternatives for added safety that do not compromise serviceability are presented at the end of this book.

Although there has been much debate about this structure, the improved shock absorption is at the expense of sizable and permanent deformations that make the entire building unacceptable for future use.

Reinforced-Concrete Structures: Moment-Resisting Frames

Of all structures, perhaps the *reinforced-concrete moment-resisting frame* is the most vulnerable to earthquakes. Initially named *ductile moment-resisting frame* by SEAOC and in several UBC editions, it was renamed *special moment-resisting frame* in the 1997 edition of the Code. It was the preferred structural design category and enjoyed a special low lateral coefficient as compared to other structures. The problem is that there is nothing *ductile* about this type of structure.

An example of the nonductile performance of the reinforced-concrete moment-resisting frame was a fashion center parking structure in the Whittier Narrows earthquake of 1987 in California (Figures 1.18–1.20). The author established that, during each aftershock, the large, rapidly reversed horizontal shear forces produced a grinding action at the beam–column joint that pulverized the concrete until it totally disappeared. Once the concrete was gone, the slender rebars, lacking lateral confinement, could no longer support the weight of the massive concrete floor structure and buckled.

The mechanism of beam–column joint failure can be described in the following manner. As the significant mass of the floor (or roof) deck starts swaying back and forth, the frame columns attempt to resist the movement. This causes the acceleration and dynamic forces to be imparted by the floor beam to the beam–column joint. The dynamic response characteristics of a deck are normally different from the frame column. The reversal of dynamic forces, with their back-and-forth movement, grinds the concrete between beam and column until it entirely disappears from the joint, leaving the column rebars exposed. The rebars, no longer confined, behave like slender columns and buckle under a large vertical beam reaction that they were not meant to support. A progressive failure mechanism then results in collapse of the entire structure.

Figure 1.18 Initial stages of column joint degradation in the progressive failure of a fashion center parking structure in the Whittier Narrows earthquake, 1987.

What went wrong with the Whittier fashion center parking structure and the numerous other earthquake-damaged reinforced-concrete moment frames that complied with UBC and SEAOC *Blue Book* requirements? According to the SEAOC–UBC concept, the column joint was supposed to yield under the sway action of the frame. Since moment frames are referred to as *rigid frames* in most textbooks, it seems odd to adopt them in earthquake areas, thereby endowing them with qualities they do not possess. Such attributes are joint yielding, plastic joint rotation beyond 3 rad, and excessive strength reserve under the dynamic load of successive strain reversals caused by an earthquake.

Another major problem is underestimating earthquake-generated forces in the UBC and IBC codes, as discussed next.

1.8 CODE DESIGN FORCES: RESERVE STRENGTH TO COUNTER EXTREME FORCES

The author has analyzed actual forces acting upon a newly built, Code-complying structure damaged by the Northridge earthquake. By reconstructing time deformations, frequency of internal forces, and characteristics of

Figure 1.19 Exterior columns that supported the second floor broke at the second-floor beam–column intersection, fashion center parking structure.

structural vibration caused by the earthquake, it was determined that the *large internal forces exceeded several times the UBC-predicted forces.*

At this point it is important to evaluate and compare the UBC-recommended design forces with actual earthquake forces as measured at the site. The 1979 UBC *lateral coefficient for base shear* (the maximum lateral force coefficient) was 0.094 *g* at *working stress level,* or 13% at *strength level,* using the 1.4 UBC 1997 load factor for conversion. The 1988 UBC lateral coefficient for base shear was a mere 11.3% *g* at strength level, to be increased again to 13% by the 1997 UBC.

How do these predictions compare to actual field measurements? The *lateral and vertical* earthquake force was 100% *g* during the 1971 San Fernando earthquake, measured at the Pacoima Dam. It was nearly 200% *g* near the epicenter of the 1994 Northridge earthquake, at a nursery north of the California State University Northridge Campus.

The effects on buildings with a disproportion between projected or *design forces* and actual forces are clearly obvious. In addition, building code regulations are prescriptive. As such, design professionals are expected to follow somewhat rigid design rules based on the *law of man* rather than the *law of*

Figure 1.20   Collapsed upper floor and a large portion of the collapsed parking structure, later demolished, Whittier Narrows earthquake, 1987.

physics. The latter is inherent in the nature of earthquakes, based on proportions of predictable forces and actual structural resistance.

If the gap between actual and predicted design forces were not too large, it could be assumed that a structure would remain safe by applying a bit of additional resistance as an adjustment. However, if the actual forces were in excess of approximately 15 times the UBC's predicted design forces, something drastic is destined to happen. Fortunately the duration of California earthquakes has been short as compared to other U.S. regions such as Alaska. Time plays an essential role: The longer the duration of an earthquake, the more damage it will cause, such as the 1964 Great Alaska earthquake, which lasted more than 3 min.

The UBC lateral coefficient, applied horizontally to a ductile moment frame, was between 12 and 18% *g* depending on height, geometry, and other factors. This is a markedly underestimated value as compared to the 100% g lateral ground acceleration measured at the Pacoima Dam during the 1971 San Fernando earthquake, that is, more than five times the UBC-estimated equivalent static force that overlooks the *magnifying dynamic impact factor due to structural response.*

As mentioned earlier, the horizontal and vertical ground accelerations measured almost 200% g—the strongest ground movement recorded in the Northridge earthquake—at the Cedar Hills nursery in Tarzana,* near the epicenter. Such readings substantiate the fact that structures engineered following UBC regulations in force at the time were about 10 times underdesigned.

*The Tarzana Shake. The strong-motion accelerograph in this location recorded 1.82 g vertical accelerations for approximately 8 seconds *after* the Northridge earthquake. The puzzling readings and unrelenting shaking intrigued seismologists around the world and attracted them from Africa, England, Japan, and New Zealand. It was the strongest measurement recorded in seismic history. Equally inexplicable was the fact that Tarzana houses did not suffer significant damage and people in the community were fine.

The accelerograph was implanted into a rock close to the ground, on top of a hill in a ranch once owned by author Edgar Rice Burroughs, creator of Tarzan, now the grounds of Cedar Hills Nursery. The fact that the measuring instrument was on shallow rock proves that the readings were not augmented. Site and instrument evaluations done afterward also confirmed the validity of the readings.

CHAPTER 2

SEISMIC DESIGN REGULATIONS

2.1 BUILDING CODES

The 1997 UBC, the 75th and last UBC issued, was replaced by the IBC in 2000. Revised twice as we write these pages, the 2006 IBC has been published. The UBC underwent several modifications, some changes undoubtedly influenced by significant seismic events such as the 1933 Long Beach earthquake. Here we discuss some of the most significant provisions that affected seismic design.

The 1960 UBC provided the following equation in addressing the total lateral force acting at the base of the structure:

$$V = KCW \qquad \text{for base shear}$$

This approach was soon modified. Subsequent editions of the UBC included zone factor Z, which depends on the expected severity of earthquakes in various regions of the United States; coefficient C, which represents the vibration characteristic based on the fundamental period T of the structure; horizontal force factor K, which measures the strength of the structure against earthquake impact; and the total dead-load weight W of the structure:

$$V = ZKCW \qquad \text{expression for the base shear.}$$

The coefficients I and S were added to the formula in the 1970s, where I represents the importance of the structure (such as hospitals and fire and police stations) and S the soil structure, a "site structure" resonance factor in UBC 1982:

$$V = ZIKCSW$$

The 1988 UBC provided a modified expression,

$$V = \frac{ZIC}{R_W} W$$

where C incorporated the soil structure response factor S into the expression for the fundamental period of the structure and R_W assumed the role of the former K factor for basic braced frame, special moment-resisting space frame (SMRSF), and other types of structures. Contrary to expectations, the new formula produced virtually the same relatively low value for base shear as its predecessors.

Unlike the 1997 UBC, previous codes did not indicate the steps to be taken by the design professional. The implementation of code provisions was left to the interpretation of individuals, namely engineering seminar instructors, which inevitably led to a disparity of code interpretations and uncertainty.

The uncertainty over code interpretation was acutely felt by design professionals taking courses for the structural licensing examination. A palpable example is the discontinuity of seismic forces through shear walls from upper stories down to the first floor where the shear wall is offset. Seminar instructors maintained that the second-floor slab would pick up the horizontal reaction of the upper shear wall and transmit it to the offset lower shear wall. This flawed concept ignored the existence of other severe internal forces, evident in the collapse of the Santa Monica Community College Chemistry Building and other structures in the 1994 Northridge earthquake.

2.2   UBC 1997: A MODEL CODE

The 1997 UBC can be considered the model code for years to come. The structure and methodology will be practically the same for the IBC editions that followed, with the exception of setting up earthquake forces. The 1997 UBC currently in use in California and other western states led to the IBC. A quantum leap from previous seismic provisions, the 1997 UBC was drafted after the 1994 Northridge earthquake and benefits from the lessons learned from a seismic event that proved the need for revision and significant improvement of the design provisions contained in the building code. It resulted in significant change in structural configurations, member sizes, and types of beam-to-column connections.

Among the new features introduced by the 1997 UBC is the modification of internal design forces such as column loads and forces in braces and connections in an attempt to increase design values. Regrettably, the design values stem from unrealistically low seismic design forces applied to the main structure. While the method to increase some internal design forces might be arbitrary, such as to multiply values of basic code seismic analysis by 2.8 or 3.0, 1997 UBC effectively guides the designers as to how to modify values in their analysis.

Two design examples based on the complex, elaborate 1997 UBC regulations will be presented later, one using load and resistance factor design (LRFD) and the other based on allowable stress design (ASD) analysis. The examples are analytical with regard to 1997 UBC prescriptive provisions and are illustrated by a step-by-step design of all structural elements—except foundation and base-plate design—in accordance with 1997 UBC requirements.

The design seismic forces in the 1997 UBC are based on earthquake zones with due consideration given to existing faults. The 2000 and 2003 IBC earthquake forces are based on statistics of ground velocities and accelerations. A question arises: Where do such nationwide data come from? When IBC 2000 was issued, there were U.S. seismograph networks present in three states: California, Oregon, and Washington. It is critical to have accurate data from the rest of the nation to accomplish the ambitious seismic design proposed by the IBC.

2.3   INTERACTION OF BUILDING CODES AND OTHER STANDARDS

The IBC, as well as its predecessor the UBC, is the product of a joint effort by engineering bodies and design professionals. Both reference other building standards, such as the American Concrete Institute (ACI), the American Society of Civil Engineers (ASCE) 7, the American Institute of Steel Construction (AISC) *Manual,* the American Society for Testing and Materials (ASTM), and others in whole or in part, with or without modification. Thus those standards become part of the code. Practicing professionals and most students at the graduate level are aware of the system and how it works. The 2005 third edition of the AISC *Manual* (p. vi) warns the reader in this respect: "Caution must be exercised when relying upon other specifications and codes developed by other bodies and incorporated by reference herein since such material may be modified or amended from time to time subsequent to the printing of this edition."

As a general rule, the UBC and its successor the IBC adopt AISC standards for structural steel design, ACI 318 for structural concrete design, and ASCE 7 for design loads for buildings and other structures, all with due consideration to ASTM standards with regard to material properties and performance as well as American Welding Society (AWS) tests and standards and other engineering and construction entities. A significant example of this interaction is IBC 2006, Section 1613, "Earthquake Loads." What was several pages of definitions, examples, and design philosophy in IBC 2000 has been reduced substantially by adopting the ASCE 7 provisions for structures and nonstructural permanent components to resist the effects of earthquakes. Following are examples of close interaction and cross-referencing among codes and adopted standards from other entities.

Section 2205 of IBC 2003 adopts AISC provisions when addressing structural steel design:

> 2205.1 General. The design, fabrication and erection of structural steel for buildings and structures shall be in accordance with the AISC LRFD, AISC 335 or AISC-HSS. Where required, the seismic design of steel structures shall be in accordance with the additional provisions of Section 2205.2.

The 2003 IBC also includes ACI guidelines in its seismic requirements:

> 2205.3 Seismic requirements for composite construction. The design, construction and quality of composite steel and concrete components that resist seismic forces shall conform to the requirements of the AISC LRFD and ACI 318. An R factor as set forth in Section 1617.6 for the appropriate composite steel and concrete system is permitted where the structure is designed and detailed in accordance with the provisions of AISC 341, Part II. In Seismic Design Category B or above, the design of such systems shall conform to the requirements of AISC 341, Part II.

In turn, Section 1617.6 adopts ASCE 7 provisions stating that for seismic-force-resisting systems the provisions given in Section 9.5.2.2 of ASCE 7 should be used except as modified in Section 1617.6.1. *Exception:* "For structures designed using the simplified analysis procedure in Section 1617.5, the provisions of Section 1617.6.2 shall be used."

In turn AISC, issues *Seismic Provisions for Structural Steel Buildings,* now in its 2005 edition, published as American National Standards Institute (ANSI)/AISC 341-05.

It is worth noting that the latest revision of the AISC *Seismic Provisions* implemented a number of changes to reflect the Federal Emergency Management Agency (FEMA)/SAC recommendations. As we know, such recommendations were part of the massive effort undertaken by the engineering community to improve seismic design of structures after the 1994 Northridge earthquake.

The AISC revisions to the seismic provisions also include modifications "to be consistent with the ASCE 7-02 document, *Minimum Design Loads for Buildings and Other Structures.* This allows these provisions to be incorporated by reference into both the 2003 IBC and NFPA 5000 building codes that use ASCE 7-02 as their basis for design loadings."

In the area of masonry design, IBC codes adopt provisions and design practice and philosophy of entities such as The Masonry Society (TMS), the Masonry Standards Joint Committee, the ACI, and the ASCE. CHAPTER 21 ("Masonry"), Section 2103, of IBC 2003, which addresses masonry construction materials, relies heavily on ASTM standards. The chapter includes several tables for mortar proportions and properties, compressive strength of clay masonry and reinforced masonry, and a table that summarizes specific requirements for masonry fireplaces and chimneys. Section 2106, "Seismic Design," of IBC 2003 adopts the provisions of ACI 530/ASCE 5/TMS 402 and specific sections of each of those entities depending on the seismic design category of the structure.

There are other examples of the strong relationship between the main building code and engineering bodies. In the field of design, manufacturing, and use of open web steel joists, the IBC adopts the Steel Joist Institute (SJI) specifications. Section 2206 of IBC 2003 establishes that the design, manufacturing, and use of open web steel joists and joist girders should be in accordance with one of the following SJI specifications:

1. Standard Specifications for Open Web Steel Joists, K Series
2. Standard Specifications for Longspan Steel Joists, LH Series and Deep Longspan Steel Joists, DLH Series
3. Standard Specifications for Joist Girders

In addition, where required, the seismic design of buildings should be in accordance with the additional provisions of Section 2205.2 or 2211.

Section 2211 addresses the design of cold-formed steel light-framed shear walls in great detail, taking into account wind and seismic issues, and provides tables for the nominal shear values for wind forces and seismic forces for shear walls framed with cold-formed steel studs. The 2003 IBC is also concerned with the shear resistance adjustment factor C_o and provides a table for maximum opening height ratio and height.

In general, IBC 2003 refers to AISI-NASPEC for the design of cold-formed carbon and low-alloy steel structural members but also includes ASCE provisions. These are given in Section 2209, "Cold-Formed Steel," stating that the design of cold-formed carbon and low-alloy steel structural members should be in accordance with the *North American Specification for the Design of Cold-Formed Steel Structural Members* (AISI-NASPEC). However, the design of cold-formed stainless steel structural members should be in accordance with ASCE 8. Cold-formed steel light-framed construction should comply with IBC 2003, Section 2210.

2.4 IBC 2006

At the time of this writing the International Code Council (ICC) launched the third edition of the IBC, IBC 2006.

Like its predecessors, the latest edition strives to provide an up-to-date building code that addresses the design and construction of building systems with emphasis on performance. The aim is also to offer a forum for building professionals to evaluate and discuss the performance and prescriptive requirements of IBC 2006 at an international level.

The preface of the 2006 edition "presents the code as originally issued, with changes reflected in the 2003 edition and further changes approved through the ICC Code Development Process through 2005."

Significant Changes from 2000 and 2003 Editions

CHAPTER 16. STRUCTURAL DESIGN

Table 1604.5 Occupancy Category of Buildings and Other Structures

The 2006 IBC has attempted, and achieved, simplification. One of the significant changes in structural design is in Table 1604.5. Initially issued as "Classification of Buildings and Other Structures for Importance Factors," the table is now presented as "Occupancy Category of Buildings and Other Structures." The *seismic factor, snow factor,* and *wind factor* have been eliminated. The first column is now "Occupancy Category" and, as IBC 2003, includes categories I–IV.

From our structural and seismic point of view, other changes in Chapter 16 worth noting are mentioned next.

Section 1602 Definitions and Notations

Several items have been deleted from the list, such as basic seismic-force-resisting systems and boundary members. Dead loads include additional items such as plumbing stacks and risers; electrical feeders; heating, ventilating, and air-conditioning systems; and fire sprinkler systems. The concept "element" has been deleted. The "frame" definition and listing have been eliminated in their entirety.

The definition of *occupancy category* as a category used to determine structural requirements based on occupancy has been added to reflect the changes in Table 1604.5 as noted above.

Section 1603 Construction Documents

Minor modifications have been made to this section.

1603.1.5 Earthquake Design Data The 2006 IBC shows the evolution of concepts and philosophy since the IBC 2000 issue. The IBC 2000 listed: (1) Seismic use group, (2) Spectral response coefficients S_{DS} and S_{D1}, (3) Site class, (4) Basic seismic-force-resisting system, (5) Design base shear, and (6) Analysis procedure. In IBC 2003 the list was modified to add other information:

1. Seismic importance factor I_E and seismic use group
2. Mapped spectral response accelerations S_S and S_1
3. Site class
4. Spectral response coefficients S_{DS} and S_{D1}

5. Seismic design category
6. Basic seismic-force-resisting system(s)
7. Design base shear
8. Seismic response coefficient(s) C_S
9. Response modification factor(s) R
10. Analysis procedure used

Consistent with the revisions discussed earlier, IBC 2006 modifies item 1 to read "Seismic importance factor, I, and **occupancy category**."

Section 1604 General Design Requirements

1604.3.3 Steel has been modified once again in IBC 2006, this time to show a more concise and specific list: "The deflection of steel structural members shall not exceed that permitted by AISC 360, AISI-NAS, AISI-General, AISI-Truss, ASCE 3, ASCE 8, SJI JG-1.1, SJI K-1.1 **or** SJI LH/DLH-1.1, as applicable."

Table 1604.5 Was discussed above in the introduction to IBC 2006.

Two subsections have been added to **1604,** "General Design Requirements," of IBC 2006, and it is interesting to note that they address resistance to earthquake and wind, including the role of seismic isolation in structures:

1604.9 Counteracting Structural Actions Prescribes that structural members, components, *and cladding* must be designed to resist earthquake and wind forces, taking into consideration *overturning, sliding*, and *uplift*. Continuous load paths should be provided to transmit those forces to the foundation. The 2006 IBC specifically mandates that the force should take into account the "effects of friction between sliding elements" when sliding is used to isolate the components.

1604.10 Wind and Seismic Detailing Prescribes that lateral-force-resisting systems should meet the provisions of the Code and ASCE 7, excluding Chapter 14 and Appendix 11A.

1605.2.1 Basic Load Combinations Modifications are made to equations where strength design or LRFD is used:

(Equation 16-1): $1.4(D + F)$
(Equation 16-2): $1.2(D + F + T) + 1.6(L + H) + 0.5(L_r$ or S or $R)$
(Equation 16-6): $0.9D + 1.6W + 1.6H$
(Equation 16-7): $0.9D + 1.0E + 1.6H$

1605.3.1 Basic Load Combinations Where ASD (working stress design) is used, this section has also undergone modifications to the equations, not listed here for simplification.

Section 1606 Dead Loads

The wording is changed in **1606.1** to note that dead loads are those defined in Section 1602.1 and "dead loads shall be considered permanent loads." Although not marked in the margin, **1606.2** has also been modified: "**Design dead load.** For purposes of design, the actual weights of materials of construction and fixed service equipment shall be used. In the absence of definite information, values used shall be subject to the approval of the building official."

Table 1607.1 Minimum Uniformly Distributed Live Loads and Minimum Concentrated Live Loads   Item 30, "roofs," of the table has undergone substantial changes, not included here for simplification.

Section 1609 Wind Loads

1609.2 Definitions   A number of definitions have been deleted from IBC 2003. Two are left in IBC 2006: **hurricane-prone regions**—areas vulnerable to hurricanes—and **wind-borne debris region**—portions of hurricane-prone regions that are within 1 mile of the coastal mean high water line where the basic wind speed is 110 mph or greater or portions of hurricane-prone regions where the basic wind is 120 mph or greater; or Hawaii.

1609.3.1, 1609.4, 1609.4.1, 1609.4.2, and 1609.4.3   These sections have undergone significant changes; not listed here for simplification.

Section 1610 Soil Lateral Loads

No changes were made to this section from 2003.

Section 1613 Earthquake Loads

General Note   Again, overall great simplification in IBC 2006 as compared to 2003, namely:

- IBC 2006 has reduced a number of definitions from the rather extensive list included in IBC 2003.
- Sections 1614–1623 of IBC 2003 have been eliminated, including the tables for *design coefficients for basic seismic-force-resisting systems.*

1613.1 Scope   In 1613.1 IBC 2006 prescribes that every structure and portions thereof, including nonstructural components that are permanently attached to structures and their supports and attachments, must be designed and built according to ASCE 7, with exclusion of Chapter 14 and Appendix 11A. The seismic design category can be determined according to either Section 1613 of IBC 2006 or ASCE 7. Exceptions are certain family dwellings, some wood-frame buildings, agricultural storages and structures that *require special considerations that are not addressed by* IBC *or* ASCE 7.

1613.2 Definitions   IBC 2006 has included here some of the earthquake nomenclature that was listed in Section 1613 of IBC 2003.

Table 1613.5.2 Site Class Definitions   Replaces Table 1615.1.1 in IBC 2003 with no changes.

Table 1613.5.3(1) Values of Site Coefficient F_a   Replaces Table 1615.1.2(1) of IBC 2003 with a change in caption b that now reads: "Values shall be determined in accordance with Section 11.4.7 of ASCE 7."

Table 1613.5.3(2), Values of Site Coefficient F_v   Replaces Table 1615.1.2(2) of IBC 2003 with a change in caption b that now reads: "Values shall be determined in accordance with Section 11.4.7 of ASCE 7."

1613.5.6 Determination of Seismic Design Category   The subject of this subsection was under 1616.3 in IBC 2003 that was based on the seismic design use group and has been substantially modified. Essentially, "Occupancy Category I, II or III structures located where the mapped spectral response acceleration parameter at 1-second period, S_l is greater than or equal to 0.75 shall be assigned to Seismic Design Category E." Occupancy category IV structures under seismic conditions should be assigned to seismic design category F. All others should be assigned to a category based on their occupancy category and the design spectral response acceleration coefficients S_{DS} and S_{DI} according to Section 1613.5.4 or the site-specific ASCE 7 procedures.

Tables 1613.5.6(1) and 1613.5.6(2),   which were 1616.3(1) and 1616.3(2), respectively, in IBC 2003, are presented with a modification to the title to include occupancy category I or II in the first column.

1613.6.2 Additional Seismic-Force-Resisting Systems for Seismically Isolated Structures   IBC 2006 incorporates this subsection (the last in the structural design chapter) to add an exception to the end of ASCE 7, Section 17.5.4.2. The exception is "for isolated structures designed in accordance with this standard [ASCE 7] the Structural System Limitations and the Building Height Limitations in Table 12.2-1 for ordinary steel concentrically braced frames (OCBFs) . . . and ordinary moment frames (OMFs) . . . are permitted to be taken as 160 feet . . . for structures assigned to Seismic Design Category D, E or F, provided that . . . 1. The value of R_l as defined in Chapter 17 is taken as 1; 2. For OMFs and OCBFs, design is in accordance with AISC 341."

CHAPTER 19. CONCRETE

Section 1908

Like previous IBC editions, IBC 2006 adopts ACI 318 provisions with significant modifications included in 1908.1–1908.1.16. **Section 1908.1.3** modifies and adds ACI 318 definitions. One of them is "design displacement" as "total lateral displacement expected for the design-basis earthquake, *as specified by Section 12.8.6 of* ASCE 7."

CHAPTER 21. MASONRY

We are mostly concerned with **Section 2106, Seismic Design.** The 2006 IBC introduces a modification to 2106.1, "Seismic Design Requirements for Masonry," stating that masonry structures and components should comply with the requirements in "Section 1.14.2.2 and Section 1.14.3, 1.14.4, 1.14.5. 1.14.6 or 1.14.7 of ACI 530/ASCE 5/TMS 402 depending on the structure's seismic design category as determined in Section 1613." The same sections of the ACI, ASCE, and TMS standards are referred to in Section 2106.1.1, "**Basic seismic-force-resisting system**"; 2106.1.1.2, "**Intermediate prestressed masonry shear walls**"; 2106.1.1.3, "**Special prestressed masonry shear walls**"; and others that show minimal changes from IBC 2003.

CHAPTER 22. STEEL

The 1997 UBC devoted about 36 pages to steel design, which were then reduced to 6 pages in IBC 2000 and 8 in IBC 2003. Inasmuch as IBC 2006 adopts chiefly AISC 360 for the design, fabrication, and erection of structural steel for buildings and structures, the chapter on steel has been reduced to two and a half pages. "Seismic Design Categories A, B, or C" in Section 2205.2.1, which were the subject of modifications in IBC 2003, are changed once again in IBC 2006.

The *R* Factor

In IBC 2003 an *R* factor was set forth in 1617.6 for the appropriate steel system and was permitted where the structure was designed and detailed according to the provisions of AISC, Parts I and III. The 2006 IBC refers to an *R* factor as set forth in Section 12.2.1 of ASCE 7 for the appropriate steel system and is permitted where the structure is designed and detailed conforming to the provisions of AISC 341, Part I. For structural steel systems not specifically detailed for seismic resistance, the designer shall use the *R* factor in ASCE 7, Section 12.2.1.

Steel Joists

It is interesting to note that IBC 2006 devotes almost one full page to Section 2206, which addresses the design, manufacture, and use of open web steel joists, as compared to about 10 lines in IBC 2003. In both cases, though, the codes refer to the additional provisions of Section 2205.2 or 2210.5. In general, the design regulations rest with one of the SJI specifications, that is, SJI K-1.1, SJI LH/DLH-1.1, or SJI JG-1.1, already mentioned above in reference to Chapter 16, "Structural Design."

Cold-Formed Steel, Light-Framed Construction

The 2006 IBC devotes Sections 2209 and 2210 to cold-formed steel and light-framed steel construction, respectively. In general, the subject is referred to AISI-NAS and to ASCE 8 for cold-formed stainless steel structural members. Section 2211 included in IBC 2003 has been eliminated.

Wind and Seismic Loads

2210.5 Lateral Design   Modified to prescribe that the design of light-framed cold-formed steel walls and diaphragms to resist wind and seismic loads should be in accordance with AISI-Lateral.

CHAPTER 23. WOOD

The 2006 IBC devotes virtually the same number of pages to the design of wood-framed structures as its predecessor, IBC 2003.

Section 2307 Load and Resistance Factor Design LRFD,

which was in accordance with ASCE 16* in IBC 2003, is now modified to state that the structural analysis and construction of wood elements and structures using LRFD design should be in accordance with the American Forest and Paper Association (AF&PA) National Design Specification (NDS).

Table 2308.12.4

With regard to seismic design, there is a significant change in this table, "Wall Bracing in Seismic Design Categories D and E," with minimum length of wall bracing per each 25 linear feet of braced wall line. Once again IBC 2006 achieves simplification. Story location in the table has been changed to "Condition" and reduced to *one* story.

---

* ASCE 16, "Load and Resistance Factor Design Standard for Engineered Wood Construction."

CHAPTER 3

REINFORCED-CONCRETE STRUCTURES

3.1 INTRODUCTION

We mentioned the inadequate performance of *reinforced-concrete moment-resisting frames* in Chapter 1, which addressed structural failures during an earthquake. However, properly engineered reinforced-concrete structures normally fare well in an earthquake.

Concrete design procedures are thoroughly described in current textbooks, namely *Design of Reinforced Concrete* by J. C. McCormac (Wiley, New York, 2005), among others. The procedures adhere to the principles set up by ACI 318, now in its 2005 edition and adopted by the IBCs. This ensures compliance with the latest seismic regulatory standards.

Concrete as Construction Material

Concrete has high compressive strength but comparatively low tensile strength. To compensate, *reinforced concrete* was invented by placing steel reinforcement in anticipated zones of tension of members such as beams and columns. The outcome of this blend was a successful and highly popular material that, if well designed and properly constructed, offers a number of advantages:

Economical   Can be constructed of local materials. Stone and sand aggregates are found near almost any job site.

Not Susceptible to Buckling   Unlike steel, the structural components of concrete are relatively robust.

Easy to Form The continuity of components such as beam-to-column connection facilitates construction. Cast-in-place concrete is monolithic, provided we ensure continuity of reinforcement. Both concrete and steel must be used to form the composite known as reinforced concrete.

Steel Reinforcement Grade refers to the specified yield strength. Per ASTM A 615:

Grade 40: $f_y = 40$ ksi
Grade 60: $f_y = 60$ ksi

Grade 40 is used mostly for smaller projects, whereas grade 60 is used for major jobs such as large-span beams and columns for multistory buildings. However, we must not draw an oversimplified conclusion about the usefulness of grade 40. Despite its lower strength, grade 40 is more ductile than its stronger counterpart and extremely well suited for ties and stirrups, where corners of relatively tight curvature must be formed without risk of cracking the reinforcement. As a general rule, the stronger the grade, the least ductile is the reinforcing steel.

Aggregates To cast a strong concrete, we must have coarse and fine aggregates. Coarse aggregate may be gravel such as from a river deposit or crushed stone. Fine aggregate is sand that must not contain any sizable amount of silt, organic matter, and so on. The ratio of coarse to fine aggregate is determined by trial and error using test mixes for the local aggregate. If concrete is made of stone, we call it *stone concrete* or *normal-weight concrete.* *Lightweight concrete* is made of aggregates that are the byproduct of industrial processes.

Cement In addition to strong and durable aggregates, we must *cement* the aggregates and steel together to form a monolithic artificial stone. The amount of water added to make the chemical composition work is controlled by the water–cement ratio. The ratio of water by weight to the amount of cement is critical for the strength of concrete. Too little water might not be sufficient to trigger the chemical reaction and might result in a nonworkable mixture. On the other hand, too much water will dilute the effectiveness of cement and result in weak concrete. A suitable water–cement ratio is just a bit over 0.4 for concrete without plasticizers and between 0.30 and 0.35 with plasticizers. Making good concrete, casting it, and leaving it at the mercy of the elements does not mean the job is done. The fresh concrete should be protected from dehydration, excessive heat, and frost.

Reinforcing Bars Reinforcing steel bars may be *plain* or *deformed.* Deformed bars are preferred over plain bars because they provide a better bonding ("grip") between concrete and steel. Bars are identified by numbers that refer to the diameter expressed in one-eighth of an inch. For instance, the diameter of a #10 bar is $10 \times \frac{1}{8}$ in. $= 1.25$ in.

The most commonly used deformed bars are #3–#11. The #14 bar is considered too large for buildings, but is suitable for bridge construction.

The bar carries a force of $1.27 \times 60 = 76.2$ k *nominal* axial force. This is a very large force that has to be transmitted to the body of concrete through development length at the ends of the beam, by lap length.

The development of bars has become a major issue in recent years.

The #4 and #5 bars are mostly used for slab reinforcement, ties, and stirrups for columns and beams. The #6 bars and larger are normally used for main flexural reinforcement of beams and tension–compression reinforcement of columns.

Mechanical Properties of Concrete

Modulus of Elasticity   The modulus of elasticity is the ratio of stress to strain and is applicable to elastic materials. A member will have smaller deformation if the value is higher. The general formula for all concrete, normal or lightweight, is

$$E_c = w^{1.5}(33)(f'_c)^{0.5}$$

where w is the unit weight expressed in pounds per cubic and f'_c the strength of concrete in pounds per square inch for a 28-day cylinder. For the normal-weight or *stone* concrete

$$E_c = 57,000 \ (f'_c)^{0.5}$$

where both E_c and f'_c are in pounds per square inch.

Tensile Strength   As a rule, ACI 318 does not include the tensile strength of concrete in any strength calculation, except under special conditions. Apart from prestressed concrete (transfer), ACI 318 is silent about the tensile strength of concrete. Concrete tensile strength values are given in ACI 9.5.

Modulus of Rupture   A cracked concrete member possesses less stiffness to counter deflection as compared to an integral member.

The tensile resistance of normal-weight concrete in ACI 9.5.2.3 is given as the modulus of rupture

$$f_r = 7.5(f'_c)^{0.5}$$

as measured in destructive bending tests on plain concrete beams. The modulus of rupture is the upper bound of the measured tensile resistance. Its value is higher than the values obtained by more accurate test measurements obtained in Europe in tensile tests on *dogbone* coupons.

Following are the tensile strength values for grades of concrete according to the modulus-of-rupture tests:

$$f'_c = 3000 \text{ psi} \qquad f_r = 410 \text{ psi}$$

$$f'_c = 4000 \text{ psi} \qquad f_r = 475 \text{ psi}$$

$$f_r = 5000 \text{ psi} \qquad f_r = 530 \text{ psi}$$

where

$f'_c = 4000$ psi is generally considered the most popular for buildings in the construction industry, except for very tall multistory structures and bridges

$f'_c = 5000$ psi is chosen where there is a supply of local, adequately strong aggregate available

Weight of Concrete

Normal weight: 150 pcf
Light weight: 100–120 pcf

ASTM Standard Bar Sizes and Areas

#3	0.11 in.^{2}	#9	1.00 in.^{2}
#4	0.20 in.^{2}	#10	1.27 in.^{2}
#5	0.31 in.^{2}	#11	1.56 in.^{2}
#6	0.44 in.^{2}	#14[a]	2.25 in.^{2}
#7	0.60 in.^{2}	#18[a]	4.00 in.^{2}
#8	0.79 in.^{2}		

[a] Use with discretion.

Design for Bending

For years the only method used was *working stress design* (WSD). It was based on actual or *working* loads with a load factor of 1. Following European practice, this method was gradually replaced by *ultimate strength design* (USD), now referred to as *strength design* or *limit-state design*. The major steps in this transition are:

The 1956 ACI code included *ultimate strength design* in the Appendix.
The 1963 ACI code awarded the two methods equal standing.
The 1971 ACI code became almost totally a *strength* code.

According to the rules of strength design, there are two options to design for flexure:

(a) Using two of the three basic static equations $\Sigma M = 0$ and $\Sigma H = 0$, and allowing an idealized plastic response to take place for both concrete and steel
(b) Using the stress–strain compatibility relationship to determine internal stresses as a function of the external moment

The latter method, however, is mandatory for some special cases, such as combined bending and axial compression and doubly reinforced beams.

Assume that, after vertical flexural, a crack has propagated to the extreme compression fiber at the top face of a simply supported beam. Then the compression zone has to bear on a single line A of an infinitesimal depth. If the compression zone has enough strength, it will plasticize and build up an internal force C required to resist the external moment.

This method is known as the *rectangular compression block method*. It was used in Europe for years before being incorporated into the ACI code. The reader will note that the results were easily obtained: Let the compression block develop a sufficient fully plasticized depth to build up the resistance to counter the factored external moment. For this we assumed an *ideal* plasticizing process that would result in a uniform plastic stress distribution.

In reality the edges of the compression stress diagram are rounded off, especially at the *neutral* axis, where transition takes place. The idealized rectangular stress block sets up an equivalent, simplified, uniformly distributed stress block. The block in turn, with adjusting parameters such as $0.85f'_c$, would produce the same results as the complex and more laborious stress–strain compatibility method.

3.2   SHEARING RESISTANCE OF RC BEAMS

The shearing resistance of RC beams should be renamed *combined bending and shearing resistance* of reinforced concrete beams because the two concepts cannot be separated.

In the 1950s the flexural resistance/ultimate flexural strength was already in use very much as we are using it today. At that time Europe and Russia gradually introduced the concept of *ultimate strength design* into their building codes, as did the United States about 10 years later. What was named *ultimate strength design* or *strength design* was only applicable to *pure* bending, that is, for a single section along a beam. Combined bending and shear were present everywhere. Building codes were still using ASD formulas for shear strength design, which did not make much sense. The ASD shear formulas were unclear and did not fit the physics of beam behavior in the ultimate strength state. This author led the way to solve the problem by *combining shear and bending* as it naturally occurs in a reinforced-concrete beam and developed a full *ultimate strength design method for combined shear and bending moment* (Figures 3.1–3.6).

Looking at the free-body diagram of Figure 3.1 we note that both sums of the horizontal forces must be equal to zero,

$$\mathbf{C - T = 0}$$

but also the sum of the vertical forces must be in equilibrium and equal to zero; otherwise the two pieces A and B would move vertically with respect to each other and the beam would collapse in *shear failure.*

The basic derivation without building code restrictions is as follows: We design the beam for bending, that is,

$$M_u = 40 \times 2 = 80 \text{ k-ft} = 960 \text{ k-in.}$$

We are concerned with the 40 k vertical force acting on **A**. If the shearing resistance of the compressed zone counteracts the entire 40 k shear, we would have nothing to worry about, but in most cases this does not work out. In our example, **C** is about 48 k and the combined *frictional* and *intrinsic* shear resistance of the compressed zone is about 22.5 k. That leaves an unbalanced shear force of 40 k − 22.5 = 17.5 k that must be counteracted by some means.

Among others, the most practical solution is the use of stirrups. We must place enough stirrups along $0.577(d - a) = 11.0$ in. because efficiency dictates that all stirrups that accounted for resisting the unbalanced shear *must intersect the inclined crack* with a horizontal projection of $\tan 30° (d - a) = 11.0$ in. Let us try #3 U-shaped stirrups.

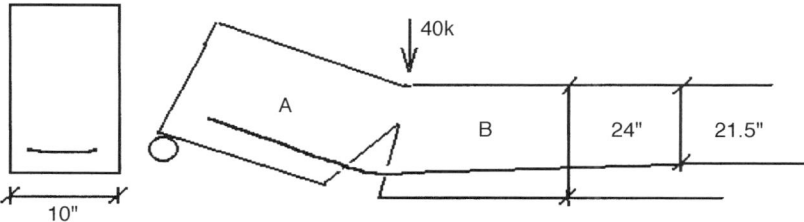

Figure 3.1 Diagrammatic representation of beam under combined shear and bending forces.

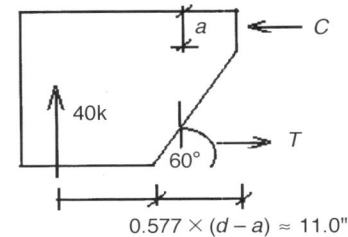

Figure 3.2 Free-body diagram showing summation of forces.

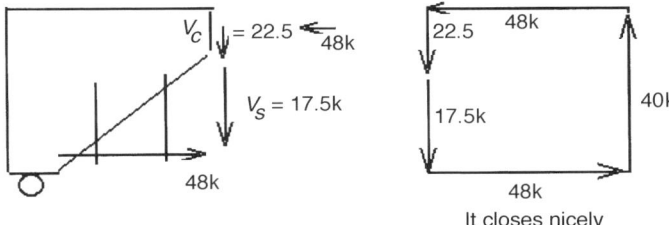

Figure 3.3 Equilibrium of the free-body diagram.

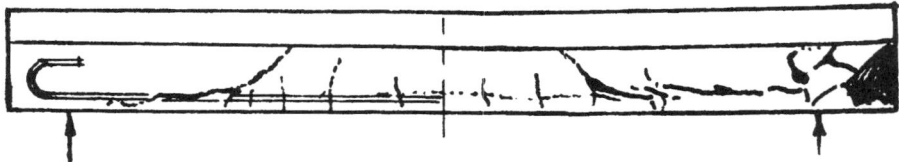

Figure 3.4 Horizontal splitting at the level of main reinforcement of a test beam before final destruction of bond and anchorage occurred.

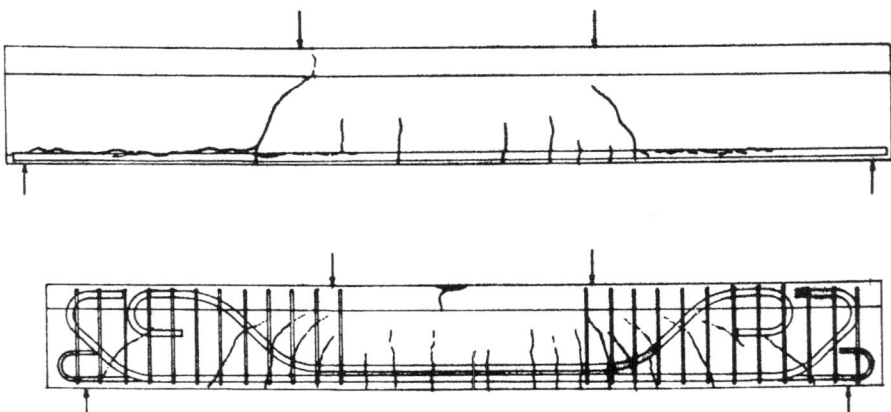

Figure 3.5 RC beam shear and bond failure patterns.

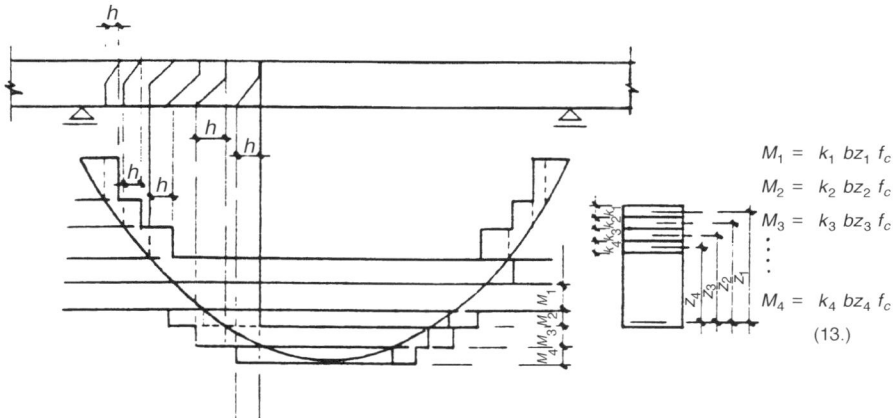

$$M_1 = k_1\,bz_1\,f_c$$
$$M_2 = k_2\,bz_2\,f_c$$
$$M_3 = k_3\,bz_3\,f_c$$
$$\vdots$$
$$M_4 = k_4\,bz_4\,f_c$$
(13.)

Figure 3.6 Shift in the moment diagram by horizontal projection h of the inclined crack. Reproduced from the author's paper presented at the European Concrete Committee Symposium on Shear, Wiesbaden, Germany, 1963.

Every U-shaped stirrup has a cross-sectional area of $2 \times 0.11 = 0.22$ in. and a vertical resistance with $f_y = 40$ ksi $= 0.22 \times 40 = 8.8$ k. We need $17.5/8.8 = 1.99$ stirrups, say 2.0, to counteract the unbalanced 17.5 vertical force. Thus we have an ultimate design without the added sophistications and restrictions imposed by the ACI code, for example, the strength reduction factor.

Notice the *shift* of force in the tensile reinforcement. It leaves only $24 - 11 = 13$ in. development length. No wonder bars failed in the past, pulling out of the support area of the concrete. The author's findings mentioned above were that the stirrups not only resist in bending but also have a horizontal lever arm of $\tan 30° (d - a)/2 = 5.5$ in. There is a shift in the location of the horizontal tensile component T.

After our introduction we discuss the design for shear in the ACI regulations.

ACI Regulations

In the course of research on shear the author found that *cracks follow the stress trajectories or patterns of the principal stresses of a homogeneous material.* Reinforced concrete, just as plain concrete, tends to behave like a homogeneous material until it cracks along the path of principal tensile stresses. The role of stirrups is to add to the shearing resistance of concrete in a combined effort to resist V_u. The basic equation is

$$V_u = \phi V_{sn} + \phi V_{cn}$$

where V_{sn} is the shearing resistance of the stirrups and V_{cn} is the shearing resistance of the concrete. We added the subscript n to indicate nominal resistance, even though this is not included in the ACI 318.

Another key equation is

$$V_s = A_v f_y n$$

where $n = d/s$
$\quad d$ = effective depth
$\quad s$ = spacing of stirrups

Thus

$$V_s = A_v f_y \frac{d}{s} \quad \text{or} \quad s = \frac{A_v f_y d}{V_s}$$

where V_s must be defined from the first basic equation,

$$V_u = \phi V_s + \phi V_c \qquad V_s = \frac{V_u - \phi V_c}{\phi}$$

For the above operation we need V_c, which is defined as

$$\boxed{b_w d 2 \sqrt{f_c'} = V_c}$$

where b_w is the width of the *beam web*. There are other, more complicated expressions for V_c in the ACI 318, such as

$$V_c = \left(1.9\sqrt{f_c'} + 2500\rho_w \frac{V_u d}{M_u}\right) b_w d \leq 3.5\sqrt{f_c'}b_w d$$

The three ACI limits to stirrup spacing are as follows:

1. Calculate

$$\phi V_c = 2\phi b_w d \sqrt{f_c'}$$

If $V_u < \frac{1}{2}\phi V_c$ or $\phi b_w d \sqrt{f_c'}$
Use nominal stirrups like
$50 b_w s = A_v f_y$ or $s = A_v f_y / 50 b_w$

If $V_u > \frac{1}{2}\phi V_c = \phi b_w d \sqrt{f_c'}$
Calculate amount of stirrups

$$\boxed{s \leq d/2}$$

2. Between

$$\phi b_w d \sqrt{f_c'} \quad \text{and} \quad V_s \leqq 4\sqrt{f_c'} b_w d$$

stirrup spacing **must be less than or equal to** $d/2$.

3. If $V_s > 4 b_w d \sqrt{f_c'}$ but $V_s < 8 b_w d \sqrt{f_c'}$ the maximum stirrup spacing is $d/4$

The absolute limit is

$$V_s = V_{s,\text{max}} = 8 b_w d \sqrt{f_c'}$$

Practical Note. Stirrup sizes are normally #3, #4, and #5. Anything larger is for a megaproject. As mentioned earlier, do not use it unless you are designing a bridge.

The *best* f_y *for stirrups* is 40,000 psi. At this strength, stirrups bend well. For harder steel, f_y = 60,000 or higher, stirrups tend to be brittle and might crack around hooks. Then you have development length for the stirrups, which might fail. *Minimum spacing: 3 in.*

3.3 DEVELOPMENT LENGTH

By now it should be obvious to the reader that there is a potential danger of bars pulling out. The force in the bars can be much larger and closer to the support of simply supported beams than anticipated by a conventional engineering approach—a shift in the bending moment diagram.

When the 1963 ACI code made strength design the preferred method of analysis for bending, there was no ultimate strength methodology—*limit strength* as it became known later—to calculate shear strength or development strength. Both are essential parts of beam design. At that time the provisions of the code and the technical literature were only good for pure bending, that is, for analysis of a section subjected to bending. It did not take into consideration the presence of other forces such as shear and compression that are inevitably combined with the external moment.

In a simply supported beam loaded with a uniformly distributed load there is only one cross section that receives a pure moment, while mathematically speaking, there are infinite cross sections subjected to both bending and shear.

What methods were then used to evaluate the limit-state effects of these forces? For shear, a parabolic stress distribution matched with a uniform stress distribution extending from a working stress evaluated the neutral axis to the centerline of the tensile reinforcement. Still, these stresses where applied to a nonexisting part of the cross section split by a crack. That is, stresses were applied over a void and results evaluated on a hypothesis that contradicted the laws of physics. Evaluation of resistance to bond failure of bars embedded in concrete was equally not adapted to ultimate strength or *strength* requirements.

A homogeneous material—reinforced concrete is anything but homogeneous—was assumed subjected to differential tensile forces at the location of the rebar. This means a continuum of infinitely small elements was assumed capable of forming a continuous chain with the tendency to shear off along a plain parallel to the axis of the beam. Again, splitting of concrete by numerous diagonal cracks was not considered. These cracks actually reacted against slippage under the high bearing pressure created by lugs of rebars that would either deform or pull out.

In 1963 the author presented the unique findings of research about the relationship of those hitherto ignored forces that affected the limit-state response and strength of reinforced concrete to the Wiesbaden Symposium on Shear of the European Concrete Committee. The paper contained the first mathematical model and rational method that showed by actual design examples how to combine shear and moment in one ultimate strength operation, that is, to determine the amount and disposition of rebars for shear and flexural resistance. It was also the first time that it was noted that the stirrups take part in the resisting moment.

The author also derived mathematically the amount of moment sharing between stirrups and main tensile reinforcement. He named it *shift* in the moment diagram. The concept was quoted by others without crediting the original source and was eventually incorporated into ACI 318.

Application to Design of Structural Members

Consider a continuous beam with a uniformly distributed load. The following steps will determine the total shear reinforcement:

1. Divide the length over which shear reinforcement is required into equal increments; then calculate each cross section separately for the given static conditions.
2. Provide each zone with the reinforcement required for the worst case in that zone, that is, the biggest area of steel at the smallest pitch obtained from the two results of the calculations made on the two boundaries of the zone—1-1 and 2-2 of zone II in Figure 3.7.

Thus a typical system of shear reinforcement will be as in Figure 3.7, where heavy lines represent stronger shear reinforcement and thinner lines lighter

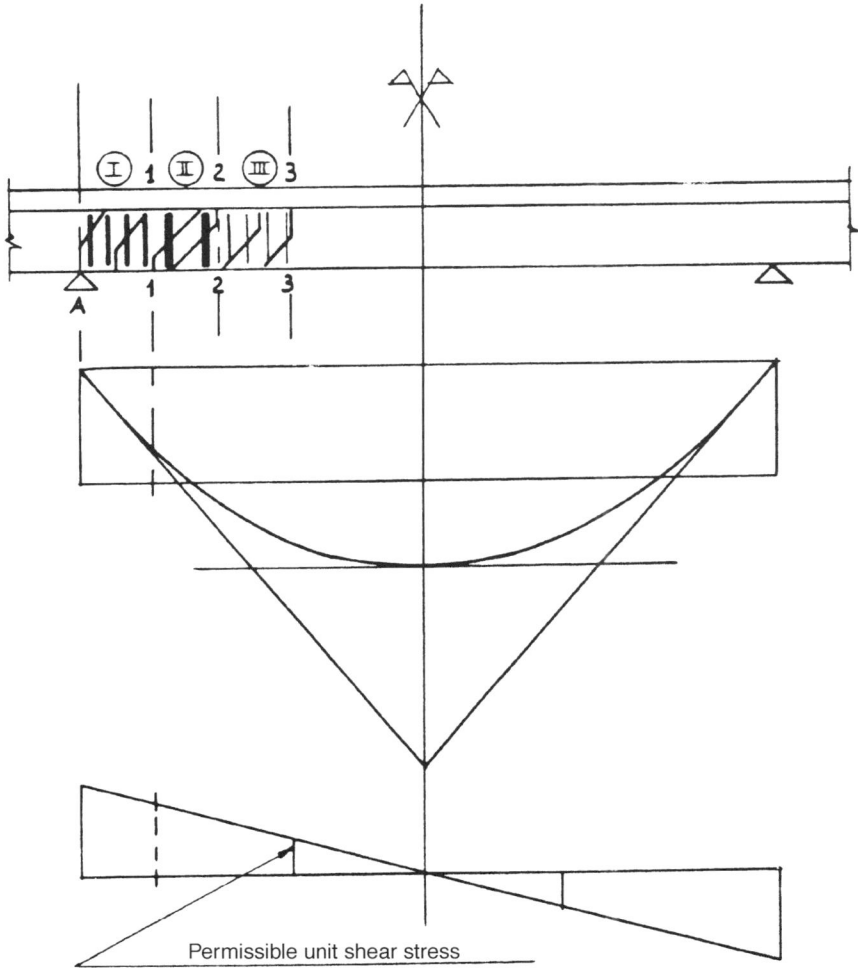

Figure 3.7 Typical system of shear reinforcement. Reproduced from the author's paper presented at the European Concrete Committee Symposium on Shear.

reinforcement. The area of each pair of stirrups is large in zone II because this coincides with the point of zero bending moment. No shear resistance is provided by the bending moment and the intensity of the shear force is still high.

The pitch of the shear reinforcement is quite large, though, because the inclination of the crack is 45°. In zone III the area of stirrups is the smallest. This is due to the decrease of the shear forces and the increased shear resistance of the compression zone caused by the increasing bending moment.

The inclination of the cracks varies with the type of load, which makes it difficult to give actual figures for the spacing of stirrups. In beams reinforced with stirrups large secondary cracks tend to form at the final stages of loading, which also weakens the structure. It follows that it is more practical to combine stirrups with bent-up bars.

We will now present how the ACI addresses the *development length* requirement. Given the ACI equation (12-1) with $K_{tr} = 0$,

$$l_d = \left(\frac{3}{40} \frac{f_y}{\sqrt{f'_c}} \frac{\alpha\beta\gamma\lambda}{[(c + K_{tr})/d_b]} \right) d_b \tag{12-1}$$

we have

$$\boxed{\frac{l_d}{d_b} = \frac{3}{40} \frac{f_y}{\sqrt{f'_c}} \frac{\alpha\beta\gamma\lambda}{[(c + K_{tr})/d_b]} = \frac{3}{40} \frac{f_y}{\sqrt{f'_c}} \frac{\alpha\beta\gamma\lambda}{1.5/d_b}}$$

We now develop a #7 *bottom bar* for $f'_c = 3.0$ ksi and $f_y = 40$ ksi:

$$\frac{l_d}{d_b} = \frac{3 \times 40000}{40 \times 54.77} \frac{(1.0)(1.0)(1.0)(1.0)}{1.5/(7/8)} = 32 \text{ in.} \qquad \text{42-in. top bar, } \tfrac{7}{8} \text{ in.}$$

That is, the development length $l_d = 32$ in. for a $\tfrac{7}{8}$-in. diameter bottom bar.

To develop a #8 bottom bar for $f'_c = 3.0$ ksi and $f_y = 40$ ksi,

$$\frac{l_d}{d_b} = \frac{3 \times 40000}{40 \times 54.77} \frac{(1.0)(1.0)(1.0)(1.0)}{1.5} = 36.5 \approx 36 \text{ in.} \qquad \text{47-in. top bar, #8}$$

To develop a #6 bottom bar for $f'_c = 3.0$ ksi and $f_y = 40$ ksi,

$$\frac{l_d}{d_b} = 54.77 \frac{(1.0)\text{Bott}(1.0)\text{Epox}(0.8)(1.0)\text{Ltw}}{(1.5 + 0)/0.75}$$

$$= 27.38 \approx 28 \text{ in. for #6 bottom bar}$$

That is, 36-in. top bar, #6.

Redo the above for $f_y = 60$ ksi ($f'_c = 3.0$ ksi):

#7 Bottom bar	48 in.	62-in. Top bar
#8 Bottom bar	55 in.	71-in. Top bar
#6 Bottom bar	41 in.	54-in. Top bar

For $f_y = 60$ ksi ($f'_c = 4.0$ ksi):

#7 Bottom bar	42 in.	54-in. Top bar
#8 Bottom bar	48 in.	62-in. Top bar
#6 Bottom bar	36 in.	46-in. Top bar

Building Code Provisions

As a rule, all IBC editions adopted the ACI 318 regulations for concrete design of structures and components to resist seismic forces with some added requirements of ASCE 7. In addition, IBC 2006 has modified existing definitions and added the following definitions to ACI 318, Section 21.1:

Design Displacement   Total lateral displacement expected for the design basis earthquake, as specified by *Section 12.8.6 of* ASCE 7.

Detailed Plain Concrete Structural Wall   A wall complying with the requirements of *Chapter 22, including 22.6.7.*

Ordinary Precast Structural Wall   A precast wall that complies with the requirements of *Chapters 1–18.*

Ordinary Reinforced Concrete Structural Wall   A cast-in-place wall complying with the requirements of *Chapters 1–18.*

Ordinary Structural Plain Concrete Wall   A wall complying with the requirements of *Chapter 22, excluding 22.6.7.*

3.4   NORTHRIDGE EXPERIENCE

Reinforced-concrete *parking structures* and *tilt-up concrete walls* were areas of major concern for the City of Los Angeles after the Northridge earthquake. The Parking Structure Subcommittee of the Department of Building and Safety/SEAOSC Task Force investigated 20 heavily damaged parking structures including 8 that had suffered partial or total collapse. The failures had occurred primarily as result of:

1. Excessive drift in lateral resisting concrete frames or shear walls
2. Lack of ductility in interior, nonlateral resisting concrete frames
3. Lack of strength and proper detailing of diaphragm boundary members

A series of emergency measures followed to prevent similar shortcomings in the design of future parking garages.

   The City of Los Angeles Building Bureau addressed the deficiencies found in *tilt-up wall buildings,* namely that the wall anchoring system and continuity ties had performed poorly in over 350 buildings, of which about 200 had partial roof collapses. A number of the structures were of reinforced masonry, although the damage was not as extensive as for tilt-up panels.

3.5   CASE 1: REINFORCED-CONCRETE PARKING GARAGE

Seismic: High-Seismicity Area

The parking structure analyzed next suffered extensive damage during the Northridge earthquake. Nevertheless it was not until two years later that the facility was red-tagged by county officials, after it was determined that the damage to the structural system was more severe than initially observed.

DESCRIPTION OF FACILITY

The structure is an underground four-level, approximately 200,000-ft^2 reinforced concrete parking garage in downtown Los Angeles. It contains four square configured levels, A to D. The parking facility was designed in 1968 under the UBC then in effect. The vertical loads are resisted by precast, prestressed T beams and poured-in-place reinforced beams which are framed into concrete columns. It is not a moment-resisting frame. The lateral loads are resisted by four shear walls in the north–south direction, two shear loads at level D, and three shear walls from level A to level C in the east–west direction. Floor slabs are poured-in-place, post-tensioned concrete.

   The floor-to-wall connection is by dowel bars of moderate reinforcement. The concrete diaphragm—floors and ramps—are connected to the exterior walls by #4 dowels at 24 in. on center (OC) with 24-in embedment in the slab and the walls. Conventional isolated concrete pad footings and continuous spread footings serve as foundations supporting reinforced columns and walls. Design bearing pressures were 12,000 psf dead load (DL) plus live load (LL) allowing 50% overstress for a combination of earthquake and gravity (DL + LL). Ramps are built of 7 in. concrete cast in place on top of precast, prestressed beams resting on corbels. Concrete shear walls serve as the seismic-resisting elements. The drive ramps are supported by post-tensioned precast, prestressed lightweight concrete T beams of approximately 64 ft span, spaced 19 ft on center. Four-foot-wide pilasters cast monolithically with the concrete walls support the precast beams on 2-ft-wide, 2.75-ft-deep concrete corbels. The ramp floor is a 7-in.-thick concrete slab post-tensioned in both directions. Cast-in-place 5-ft × 5-ft reinforced concrete girders and a 10-in.-thick slab form the roof of the parking structure. As noted below, the roof carries about a 5-ft depth of earth fill and landscaping.

Above the concrete parking structure is a street-level landscaped and paved "flag" plaza. During the years 1988–1992 the Metropolitan Transit Authority built an elevator and escalator/stair entrance to the subterranean Civic Center Red Line Station north of the western entrance to the parking garage stairs. Long-span girders and columns support the landscaped plaza. A short tunnel leads to a three-bay double-loaded parking level with the center bay ramping down to three additional levels.

EARTHQUAKE-CAUSED AND OTHER DAMAGE

The cast-in-place concrete walls, floor, roof slabs, and precast lightweight concrete beams that supported the parking structure ramps cracked profusely during the January 17, 1994, Northridge earthquake (Figure 3.8). The damaged areas that caused most concern were the support ends of the precast T beams and the beams that supported the fan room, which were shattered by numerous cracks (Figure 3.9). The cracks showed a *limit-state pattern* and prompted the closure of the facility.

Previous earthquakes such as Whittier Narrows and Sierra Madre had already left a mark in the structure, which worsened under the seismic impact of Northridge and became unsafe. However, seismic activity was not the only culprit: Excessive tensile stresses caused by shrinkage in relatively large, shrinkage-sensitive floor slab areas contributed to the deterioration of the structural members.

Another source of the damage is corrosion. Rusty reinforcing bars precipitated the fragmentation of the concrete beams in an outdoor canopy structure. Seepage from the landscaped plaza irrigation had gradually corroded the reinforcements and deteriorated the concrete in a number of places (Figure 3.10).

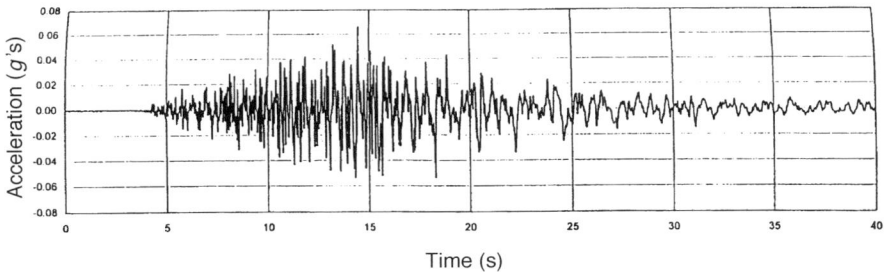

Figure 3.8   Seismograph readings of the 1994 Northridge earthquake ground motion, vertical acceleration taken in Los Angeles Civic Center location.

Figure 3.9   Earthquake-caused cracks in cast-in-place concrete beams.

Figure 3.10   Water damage in roof structure due to seepage from the plaza above.

The author was part of the team that inspected the facilities and was appointed to report on the causes and extent of damage:

Damaged T-4 and supporting corbels at D-3 (half level)
Several T beams and supporting corbels
The bearing area between T beams and corbel is $9 \times 9 = 81$ in.^{2} of drypack. T beams are placed at 19.0 ft OC.

Load on T Beam

Service loads:

DL 7-in. concrete slab	$0.088 \ (19.0) = 1.66$ k/ft
Precast beam	$1.0(2.67)(0.15) = 0.40$ k/ft
LL during earthquake (EQ), estimated	$19.0(0.01) = \underline{0.19}$ k/ft
	2.25 k/ft

Factored load:

$$w_u = 1.4(2.06) + 1.7(0.19) = 3.21 \text{ k/ft, say } 3.2 \text{ k/ft}$$

The reaction due to $w_u = 3.2$ k/ft is

$$R_A = R_B = \tfrac{1}{2}(63.83)(3.2) = 102.35 \text{ k}$$

The actual bearing pressure (*static*) is

$$p = \frac{102.35}{91.0} = 1.125 \text{ ksi} < 0.85(4.0) = 3.4 \text{ ksi}$$

The actual bearing pressure during the January 17, 1994, earthquake must have been *3 times larger* than the static bearing pressure.

Shearing Resistance of Corbel Supporting T-1 Precast

The available A_{vf} for bars crossing the potential crack at the face of the support is

$$A_{vf} = 3(0.6) = 1.8 \text{ in.}^2$$

$$V_u = \phi A_{vf} f_y \mu = 0.85(1.8)(60)1.4 = 128.5 \text{ k} > 102.35 \text{ k}$$

However, a minimum of $0.2V_u$ tensile force due to shrinkage must be applied (ACI 11.9.3.4) resulting in a loss of steel area

$$A_n = \frac{0.2(128.5)}{60} = 0.428 \text{ in.}^2$$

The effective reinforcement available for shear friction is determined as

$$1.8 - 0.43 = 1.37 \text{ in.}^2$$

$$V_{u,\text{corrected(ACI 11.9.3.4)}} = 0.85(1.37)(60)(1.4) = 98.0 \text{ k} \approx 102.35 \text{ k applied}$$

Not much extra safety factor in corbel.   OK for static load.

The fact that some of the corbels cracked leads to the conclusion that the earthquake produced much larger reactions by dynamic impact than expected from the design static load.

CAST-IN-PLACE CONCRETE BEAMS SUPPORTING ELECTRICAL EQUIPMENT ROOM

The first step was to calculate the load-carrying capacity of the critically damaged AB-14A beam *that provided support to heavy equipment in the fan room above* (Figure 3.11). The structural assembly consists of a 7-in reinforced-concrete floor slab supported by beam AB-64, 18 × 30 in. spanning north–south. The entire structural system, in turn, rests on two beams of approximately 64 ft span: AB-14A on the north side and AB-16A on the south side. Of the two, AB-14A was the weakest even though it carried about the same load:

Concrete strength: $f'_c = 4000$ psi
Reinforcing steel: Grade 60, $f_y = 60,000$ psi

We used ACI and UBC recommendations:

Tributary width for each AB beam	$\dfrac{28.84}{2} = 14.42$ ft
Weight of 7-in. concrete slab	$\dfrac{7/12(150)(14.42)}{1000} = 1.26$ k/ft
Weight of AB-64 beam	$(1.5)\left(\dfrac{23}{12}\right)(14.42)(150)\left(\dfrac{1}{15.5}\right)\dfrac{1}{1000} = 0.40$ k/ft
Weight of AB-14A beam	$\left(\dfrac{26}{12}\right)\left(\dfrac{35}{12}\right)\dfrac{150}{1000} = \underline{0.95}$ k/ft
	$= 2.61$ k/ft

Figure 3.11 Cracks in AB-14A beam that supported the fan room and other electrical equipment on the floor above.

Housekeeping concrete pads to mount fans, 4 in. high, acting on only half of the beam span at B

$$\frac{2(10 \times 6)(0.33)(150)}{1000} \left(\frac{1}{32}\right) = 0.56 \text{ k/ft}$$

Converting the 0.56 k/ft into a uniformly distributed load over the span yields

$$w_{\text{pad}} = \frac{0.56(32.0 \text{ ft})}{64.0 \text{ ft}} = 0.28 \text{ k/ft}$$

The factored dead load is

$$w_D = 1.4(2.61 + 0.28 + 0.05 \text{ (metal studs, lath, and plaster)}$$

$$= 4.116 \text{ k/ft} \quad \text{say } \underline{4.12 \text{ k/ft}}$$

The shearing resistance of the beam at the support is determined as follows:

#4 stirrups @ 6 in. OC for 6 ft, 0 in.

$$V_c = \frac{2\sqrt{4000}(26)(42 - 5.4)}{1000} = 120.37 \text{ k}$$

$$V_s = \frac{(0.4)(60)36.6}{6} = \underline{146.40 \text{ k}}$$

$$266.77 \text{ k}$$

The shearing resistance at the support (span 63.83 ft) is

$$0.85(266.77) = 226.75 \text{ k}$$

The equipment only occupies the west side of BM (from the centerline to B).

The weight of the equipment as obtained from the architect's report is as follows:

Fans	5.3 k each	
Duct work		
350 lb + 130 + 90/100 = 285 lb = 0.3 k each	*Factored loads*	
5.6 k each	1.7 × 5.6 = 9.52 k	
Electric transformer 4.0 k	1.7(4.0) = 6.8 k = 6.80 k	
Floor mounted 5-kV load interruptor		
switch 1.35 k	1.7(1.35) = 2.3 k = 2.30 k	
Floor-mounted switchboard 1.20 k	1.7(1.20) = 2.00 k = 2.0 k	

$$V_{B_{\text{equip.}}} = \frac{2.3(38.83) + 6.8(43.25) + 2.0(43.83) + 9.52(51.83 + 57.83)}{63.83}$$

$$= 23.73 \text{ k}$$

$$V_{B_{\text{D+equip.}}} = \left(\frac{63.83}{2}\right)(4.12) + 23.73 = \underline{155 \text{ k}}$$

Reaction of factored DL at B

DL + equipment + 75 psf

LL (UBC Table 23-A, item 29, light)

The factored live load is

$$w_L = 0.075 \ (28.84/2)(1.7) = 1.84 \ \text{k/ft}$$

The maximum shear at B due to $w_D + w_{\text{equip}} + w_L$ is

$$V_{\text{Bmax}} = (4.12 + 1.84) \ 63.83/2 + 23.73 = \underline{214 \ \text{k}} < \underline{226.75}$$

Original shear resistance OK for static load.

The AB-14A beam that supported the fan room and equipment on the floor above had pulled away from the concrete pilaster support at B5. A wide crack approximately 85% of the entire height and the full width of the beam separated the beam from the pilaster. Only four #11 top reinforcement held the beam in shear friction (Figures 3.12 and 3.13). There was widespread moisture throughout the structure and rusting of rebars in the two major beams AB-14A and AB-16A. To determine the shear friction resistance of AB-14A we assumed that 50% of the cross section was impaired by rust and 50% of the reinforcement area was still providing some resistance.

$$f'_c = 4000 \ \text{psi, grade 60}$$

$$A_{vf} = (0.50)(4)(1.56) = 3.12 \ \text{in.}^2$$

From the shear friction equation UBC (11.26)

$$V_n = A_{vf} f_y \mu = 3.12(60)(1.0) = 187.2 \ \text{k} < 214 \ \text{k}$$

$$\mu = 1.4\lambda, \qquad \lambda = 1.0 \ \text{normal weight}$$

factored reaction of equipment + DL + LL.

where

$$\mu = 1.4 \ \lambda = 1.4(\text{I}) \ \text{for monolithically cast stone concrete.}$$

Therefore there is *not enough resistance* based on the embedment along the top of the beam.

Actual Bond Provided for Hook of #11 Bars into B4

Required:

$$L_{nb} = \frac{1200 d_b}{\sqrt{4000}} = 26.75 \ \text{in.}$$

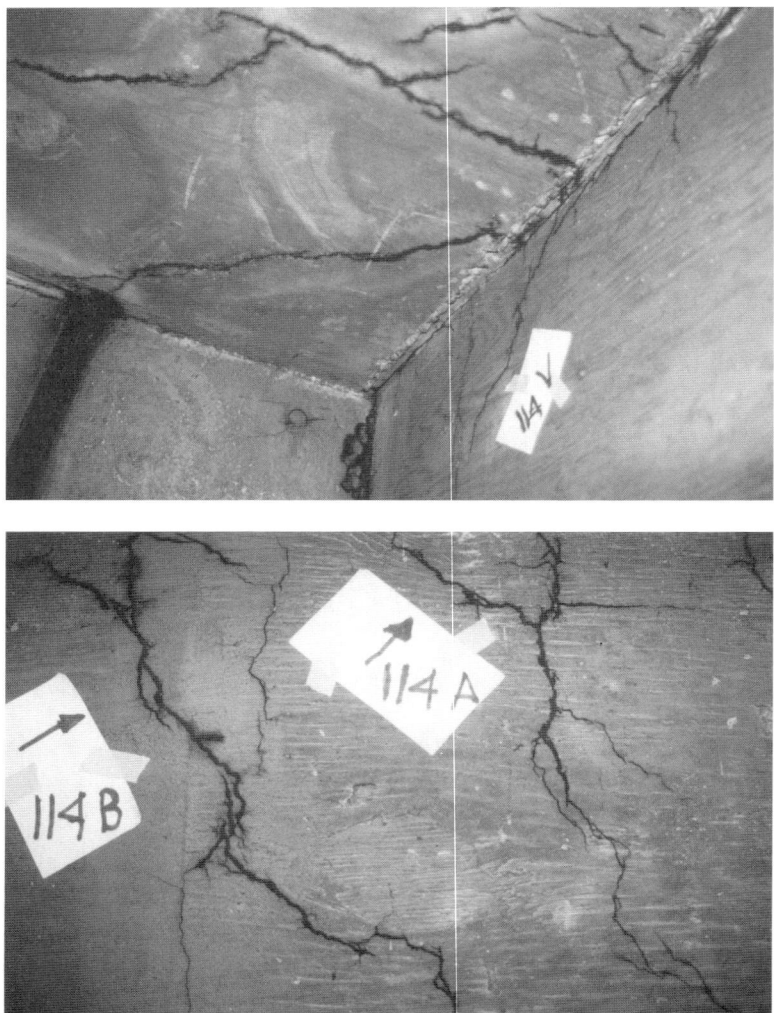

Figure 3.12   Close-up of earthquake-caused cracks in RC beams. One of the beams separated from the concrete pilaster support and was held in friction by four #11 reinforcement.

Provided: 17 in. (per original engineering drawings, March 1968).

17 in. < 26.75 in. This is not enough.

The resistance of the hook end of the bars is

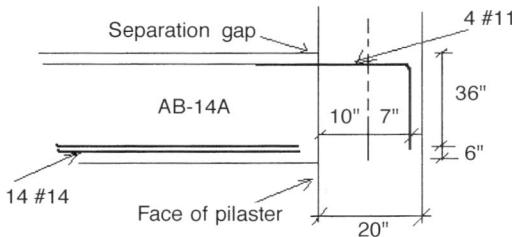

Figure 3.13 AB-14A beam and pilaster configuration.

$$17/26.75 \ (187.2) = 119.0 \text{ k} \qquad \textbf{GOVERNS}$$

just slightly over the unfactored maximum support shear of its own weight and equipment load (110.7 k).

There is not enough safety in shear.

Precast Lightweight Concrete Beam, 63.83-ft Span

We first determined the maximum factored shear and the shearing capacity of the beam at supports:

$$f'_c = 4000 \text{ psi} \qquad \text{Beams } 19.0 \text{ ft apart}$$

$$f_y = 60,000 \text{ psi}$$

Two bottom bars are available for shear friction. *Pretensioned tendons would not help in shear friction since they do not extend to provide support and do not build up adequate bond and compressive strength near the end.*

The factored DL + LL on the beam is as follows:

19-ft-wide, 7-in.-thick lightweight concrete slab	$(7/12)0.11(19.0) = 1.22$ k/ft
Weight of 12 in. $\times$ 2.5 ft lightweight concrete beam	$(1.0)2.5(0.11) = \underline{0.275}$ k/ft
	1.495 k/ft

$$w_D = 1.495 \times 1.4 = 2.10 \text{ k/ft}$$

The live load using UBC is determined as follows:

$$R = 23.1\left(1 + \frac{D}{L}\right) = 23.1\left(1 + \frac{1.495}{19(0.05)}\right) = 59.45\% \qquad \textbf{GOVERNS}$$

or

$$R = r(A - 150) = \left[\left(\frac{19.0}{63.83}\right) - 150\right] \times 0.08 = 85\%$$

Reduction formulas

$$w_L = \frac{0.5945(50) \times 19.0}{1000} = 0.5648 \text{ k/ft}$$

$$w_{L,\text{ult}} = 0.96 \text{ k/ft} \qquad \text{Factored live load}$$

$$w_{Du} + w_{Lu} = 2.10 + 0.96 = 3.06 \text{ k/ft}$$

The maximum factored shear at B (Figure 3.14) is

$$3.06(63.83/2 - 2.7) = 90.0 \text{ k applied static shear}$$

We then apply UBC shear friction formula (11-27) assuming a 20% failure plane with the vertical.

Contribution of the two #7 horizontal reinforcements:

$$V_n = A_{vf}f_y\,[(1.4\lambda \sin \alpha_1) + \cos \alpha_1]$$
$$= 2(0.6)60[1.4(0.85) + 0.34] = 33.12 \text{ k}$$

Contribution of #4 at 9-in.-OC vertical stirrups:

$$V_{n,\text{stirrups}} = 3(2)(0.2)(60)(1.19 \times 0.34 + 0.94) = 96.8 \text{ k}$$

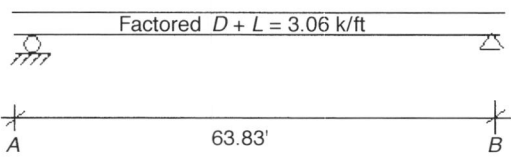

Figure 3.14

Combined shear resistance:

$$V_u = \phi V_n = 0.85(33.12 + 96.8) = 110.4 \text{ k} > 90.0 \text{ k}$$

or 13% over **OK**

Corbel Supporting T-1 Precast Beam

Now we calculate the shearing resistance of the corbel. The available A_{vf} for bars crossing the potential crack at the face of the support is

$$A_{vf} = 3(0.6) = 1.81 \text{ in.}^2$$

$$V_u = \phi A_{vf} f_y \mu = 0.85(1.8)(60)1.4 = 128.5 \text{ k} > 102.35 \text{ k}$$

However, a minimum of $0.2V_u$ tensile force due to shrinkage must be applied (ACI 11.9.34), resulting in a loss in steel area:

$$A_n = \frac{0.2(128.5)}{60} = 0.428 \text{ in.}^2$$

The effective reinforcement available for shear friction is determined as

$$1.8 - 0.43 = 1.37 \text{ in.}^2$$

$$V_{u,\text{corrected (ACI)}} = 0.85(1.37)(60)(1.4) = 98.0 \text{ k} \approx 102.35 \text{ k applied}$$

No significant extra safety in corbel.   **OK for static load.**

The fact that some of the corbels cracked leads to the conclusion that the *dynamic impact* of the earthquake produced much larger reactions than estimated by purely static analysis.

CAUSES OF FAILURE OF STRUCTURAL SYSTEMS

(a) *General* Numerous cracks were observed on cast-in-place concrete walls, floor and roof slabs, prestressed, precast lightweight concrete T beams supporting the parking structure floors and ramps, and the cast-in-place normal-weight concrete beams. Some of the damage suffered by the structural elements during the Northridge earthquake was due to the impact of earlier earthquakes; other causes were choice of materials and insufficient protection to prevent corrosion of reinforcing elements.

(b) *Fan Room Structure, Beam AB-14A*   Only four #11 top bars were opposing shear failure at the support in lieu of the previously solid concrete section. The embedment of the four top bars into the wall was short, 17 in. versus 26.75 in. as required by the UBC current at the time. Therefore the shearing resistance was only (17/26.5) 187.2 = 119 k versus the unfactored shear of its own weight and equipment weight of 110.7 k. *Thus the beam was left without safety factors.*

(c) *Precast, Prestressed Lightweight Concrete Beams*   Many of the T beams in the parking structure displayed the pattern of *ultimate limit state,* that is, formation of a plastic mechanism when the reinforcement yields to form plastic hinges at enough sections to *make the structure unstable.*[1,2] Cracks displayed a shear failure pattern as would be caused by excessive vertical and horizontal forces that occur during earthquakes. The *lightweight aggregate, lacking the strength of normal-weight aggregate, exacerbated the cracks.* Lightweight concrete exhibited weak performance in other structures impacted by the Northridge earthquake, such as the Champagne Towers discussed in Chapter 1 of this book.

(d) Originally cast as a monolithic mass, the concrete in the prestressed, precast T beams was fragmented into smaller segments held together by reinforcement, stirrups, and pretensioned tendons that were *not protected by sleeves or sheathing* (see Figure 3.15).

3.6   CASE 2: REINFORCED-CONCRETE RETAINING WALL SYSTEM

Seismic: Moderate-Seismicity Area

PROJECT DESCRIPTION

The system of retaining walls designed for this project was on sloping ground at the back of the property. (See Figures 3.17 and 3.18). The steep hill and nature of the soil required an extensive soil investigation on which we based our calculations.

Soil profile S_C. Allowable bearing pressure was 2000 psf per geotechnical exploration and report. The test pits at the site encountered dense sand with gravels and a few boulders below the existing ground surface. The soils at the site were considered suitable for use as struc-

Figure 3.15 A fragment of lightweight concrete that broke off a precast T beam and landed on the parking level below about two weeks after the Northridge earthquake. This was a typical occurrence in the process of deterioration that followed the earthquake.

tural backfill for the proposed retaining wall system if larger material were screened out.

The design parameters followed the recommendations contained in the geotechnical report, which in this case prevailed over seismic considerations:

(a) Soil pressure for cantilevered retaining walls 40 pcf
(b) Coefficient of lateral pressure K_a: 0.4
(c) Allowable bearing pressure 2000 psf

CASE 1

12.0 ft Maximum Height of Earth Retained

A 50-psf surcharge with a 30% load reduction will be included.

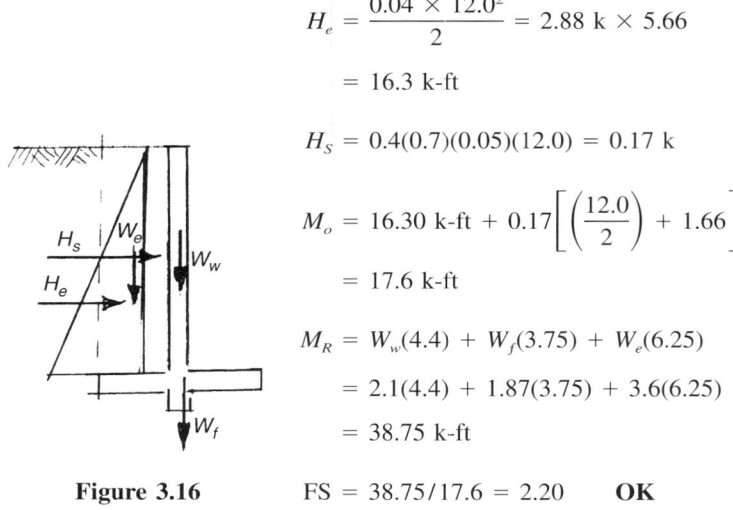

$$H_e = \frac{0.04 \times 12.0^2}{2} = 2.88 \text{ k} \times 5.66$$

$$= 16.3 \text{ k-ft}$$

$$H_S = 0.4(0.7)(0.05)(12.0) = 0.17 \text{ k}$$

$$M_o = 16.30 \text{ k-ft} + 0.17\left[\left(\frac{12.0}{2}\right) + 1.66\right]$$

$$= 17.6 \text{ k-ft}$$

$$M_R = W_w(4.4) + W_f(3.75) + W_e(6.25)$$

$$= 2.1(4.4) + 1.87(3.75) + 3.6(6.25)$$

$$= 38.75 \text{ k-ft}$$

Figure 3.16   FS $= 38.75/17.6 = 2.20$ **OK**

The location of the resultant of the vertical forces affected by lateral earth pressure is

$$\frac{38.75 - 17.60}{7.57} = 2.80 \text{ from the toe}$$

The maximum toe pressure is

$$7.57 \times \tfrac{2}{3} \times 2.80 = 1.80 \text{ ksf} < 2.0 \text{ ksft}   \textbf{OK}$$

Design against Sliding

$$H = H_e + H_S = 3.05 \text{ k}$$

The passive resistance is $2.0(0.3) + 6.33 \times 0.4 = 3.13 \text{ k} > 3.05$; nevertheless, a 16-in.-wide, 1.0-ft-deep keel will be provided under the stem wall to enhance resistance against sliding.

Design of 16-in.-Thick RC Stem Wall

6.0-ft high from top of footing:

$$M_{u,\text{max}} = 1.6 \times [\tfrac{1}{3}(2.88 \times 12.0) + \tfrac{1}{2}(0.17 \times 12.0)] = 20.00 \text{ k-ft}$$

$$= 240 \text{ k-in.}$$

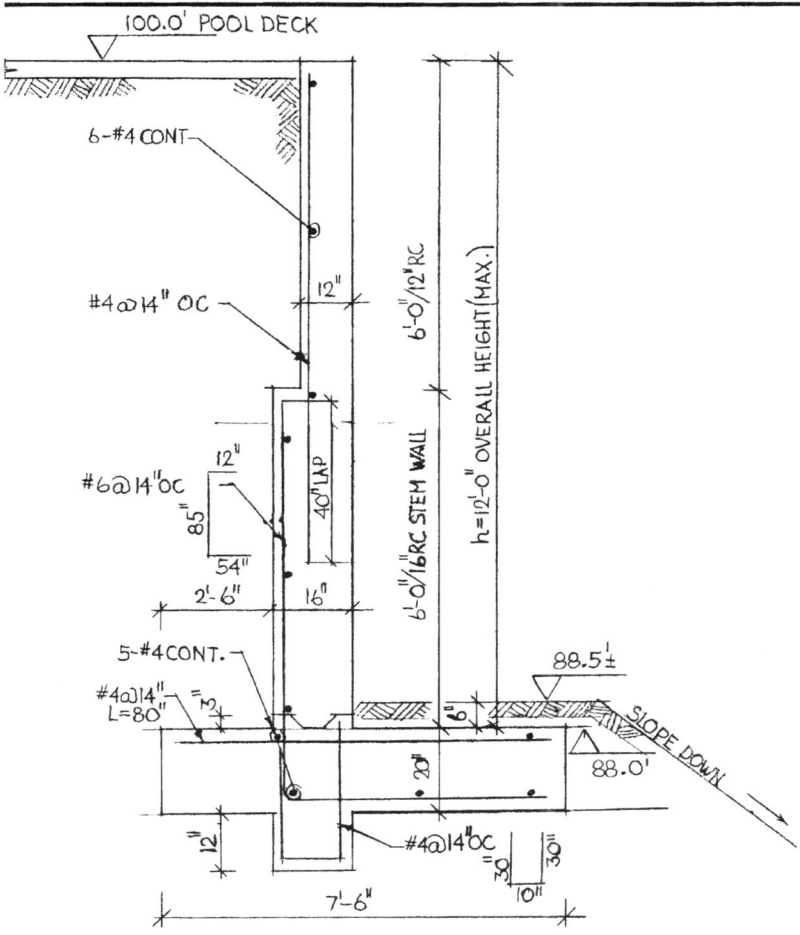

Figure 3.17   Retaining wall range 10 ft $> h \le$ 12 ft. Scale: $\frac{1}{2}$ in. to ft.

$$A_{st} = \frac{240}{0.9 \times 0.88 \times 13.62 \times 60} = 0.37 \text{ in.}^2 < 0.377 \text{ in.}^2$$

Provided by #6 @ 14 in. OC vertical.

Design of 12-in. RC Stem Wall

Attached to top of 16-in. stem wall:

$$M_{u,\max} = 1.6[\tfrac{1}{6}(0.04 \times 6.0^3) + \tfrac{1}{2}(0.014 \times 6.0^2)]$$

$$= 2.70 \text{ k-ft} = 32.5 \text{ k-in.}$$

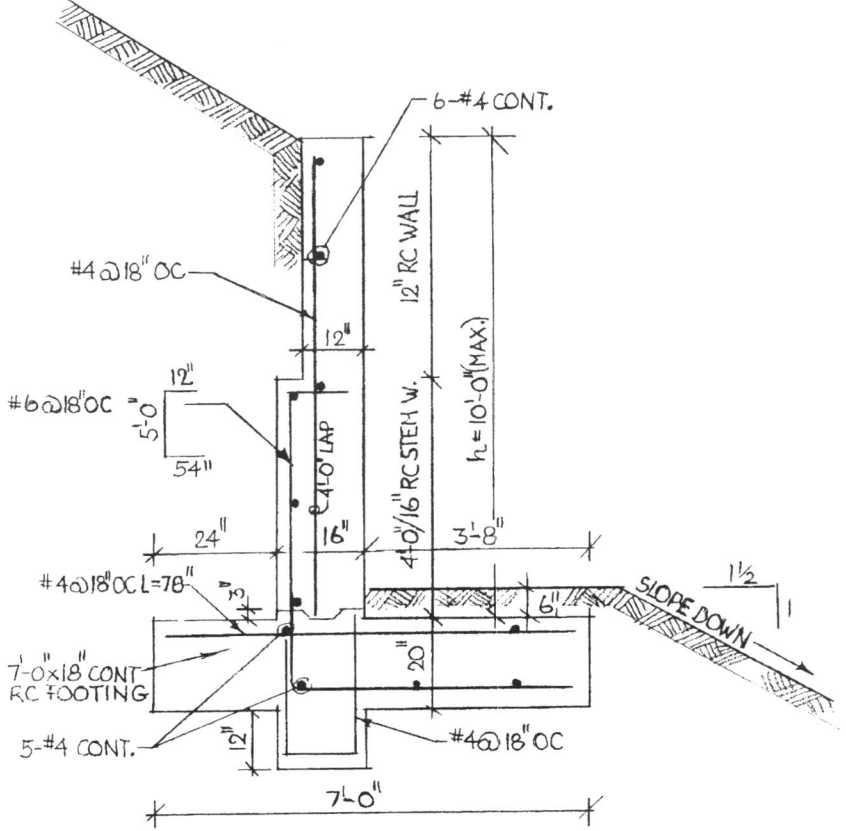

Figure 3.18 Retaining wall range 8 ft $\geq h <$ 10 ft. Scale: $\frac{1}{2}$ in to ft.

$$A_{st} = \frac{32.5}{0.9 \times 0.8 \times 9.625 \times 60} = 0.08 \text{ in.}^2 < 0.17 \text{ in.}^2$$

Provided by #4 @ 14 in. OC.

CASE 2

10.0 ft. maximum height of earth retained a 2:1 slope. The equivalent fluid pressure is 60 pcf:

$$H = \tfrac{1}{2}(0.06 \times 10.0^2) = 3.0 \text{ k} \times (3.33 + 1.5) = 14.49 \text{ k-ft}$$

Because both overturning and horizontal force are almost identical to case 1, the overall concrete dimensions for case 2 will remain the same *except* footing 1 width at 7.0 ft.

Design of 16-in. RC Stem Wall

$$M_{u,\text{max}} = 1.6[\tfrac{1}{6}(0.06 \times 10.0^3) + \tfrac{1}{2}(0.014 \times 10.0^2)]$$

$$= 17.12 \text{ k-ft} = 205 \text{ k-in.}$$

$$A_{st} = \frac{205}{9 \times 0.95 \times 13.69 \times 60} = 0.292 \text{ in.}^2 < 0.293 \text{ in.}^2$$

Provided by #6 @ 18 in. OC vertical.   OK.

REFERENCES

1. McCormac, J. C. 1993. *Design of Reinforced Concrete,* 3rd ed. HarperCollins. New York.
2. Erdey, C. K. 1963. "Ultimate Resistance of Reinforced Concrete Beams Subjected to Shear and Bending." Paper presented at the European Concrete Committee Symposium on Shear, Wiesbaden.

CHAPTER 4

SEISMIC STEEL DESIGN: SMRF

4.1 DESIGN OF SMRF STRUCTURE: LRFD METHOD

Knowledge of the design principles and methodology established by 1997 UBC makes it a relatively easy transition to the IBCs. The seismic design procedure discussed in this chapter will demonstrate seismic design following the basic 1997 UBC requirements. It guides the reader through the full design of an actual structure originally engineered to the 1994 UBC provisions and redesigned in compliance with the stringent seismic requirements of the 1997 UBC.

The 2000 IBC and its successive editions were issued to replace the three existing model codes: the National Building Code (NBC), Standard Building Code, and UBC. Under the IBC the entire country is considered subjected to varied degrees of seismicity. Apart from setting up earthquake forces from seismic maps, the IBC follows the same seismic design philosophy and methodology set forth by 1997 UBC.

One of the objectives of this chapter is to show the impact on seismic design philosophy and methodology brought forth by the Northridge experience reflected in 1997 UBC. The design procedure presented here is not merely an exercise in structural analysis; we also endeavor to provide interpretation and explanation of the complex, prescriptive provisions of 1997 UBC.

It is hoped that this chapter will be useful to civil and structural engineers not yet exposed to seismic design as well as to those familiar with earthquake regulations. It offers the double benefit of a step-by-step seismic design as well as guiding the reader through the basics of the Load and Resistance

Factor Design (LRFD). A method still new to many design professionals, LRFD is considered the preferred design alternative by the steel industry.

While every effort has been made to show detailed analysis and design, some steps of standard textbook application—such as distribution of torsional story shear—have been deliberately omitted to help the reader concentrate on seismic design procedures rather than repetitive textbook solutions.

4.2 DESIGN STEPS

Before starting the seismic design establish a methodology and a plan consisting of six major steps:

1. Study the 1997 UBC provisions thoroughly to acquaint yourself with the new regulations and set up the basis for your design. You might find it useful to keep a log of notes.
2. Determine external loads, which are of three types:
   - Gravity
   - Earthquake
   - Wind

   Special provisions in CHAP. 22, DIV. IV, address specific seismic design and overstrength requirements that call for added strength for connections and specific structural elements.

   Advice. Prepare two *independent sets of load cases* that correspond to **(a)** CHAP. 16, DIV. I, formulas (12-1)–(12-6), for the general design and **(b)** CHAP. 22, DIV IV, formulas (3-1)–(3-8) and (6-1) and (6-2) for the detailed systems design requirements mandated by Section 1633.
3. *Before* starting the analysis check the provisions of the ρ value in (30-3), a function of r_{max}, to avoid member overstress:

$$\rho = 2 - \frac{20}{r_{max} \sqrt{A_B}}$$

   If this step is done too late and the value of ρ is too high, you will have to thoroughly revise your structural layout and may have wasted considerable effort.
4. *Before* you do the structural analysis, determine lateral displacements and compare these with the 1997 UBC story-drift limitations (interstory drift, SEAOC *Blue Book,* September 1999, seventh edition). They are very restrictive; if not checked, your previous work may have been wasted because some of the structural members selected may not fit the story-drift criteria.

5. Establish member sizes.
6. Do a separate structural design for specific elements and connections to comply with CHAP. 22, DIV. IV.

Note. Steps 5 and 6 interact and must be done simultaneously.

4.3   PROJECT DESCRIPTION: FOUR-STORY OFFICE BUILDING

The structure is a four-story office building in UBC seismic zone 4. It was designed by the 1994 UBC regulations. Our task is to redesign it in accordance with the 1997 UBC provisions. Our exercise will be enlightening, as it will show the basic differences between the 1994 and 1997 designs. You will notice major changes in structural layout and member sizes.

4.4   PROJECT LAYOUT AND TYPICAL SMRF PER UBC 1994

Figure 4.1 shows the typical floor layout. Figure 4.2 is a typical SMRF as it was designed for the 1994 project. Our task will be as follows:

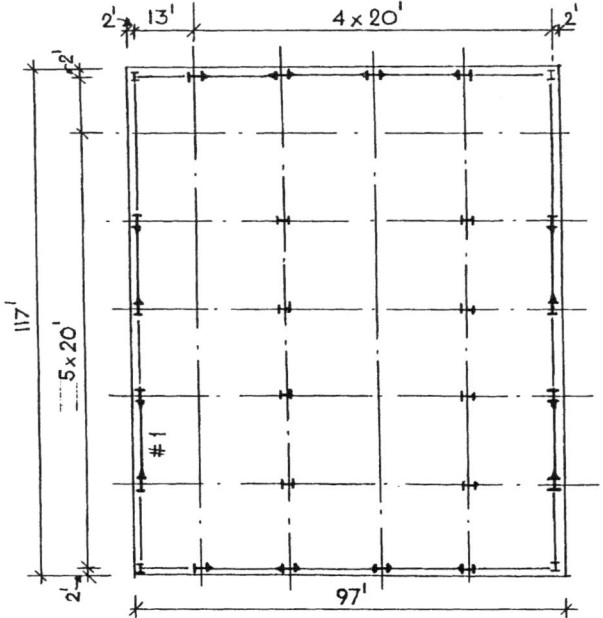

Figure 4.1   Typical floor layout of four-story SMRF structure.

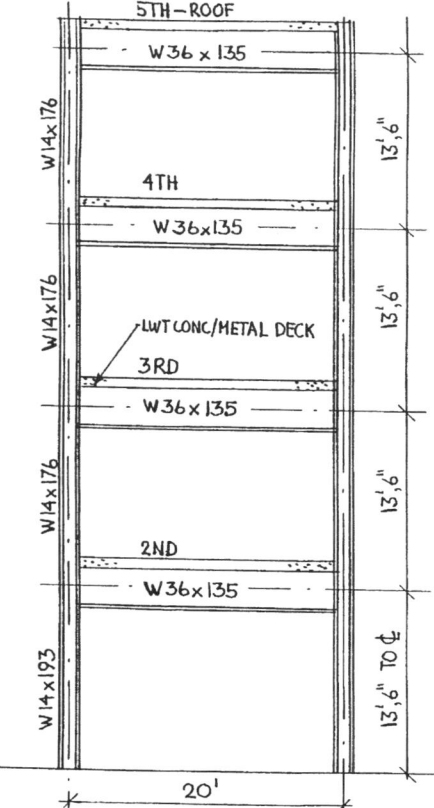

Figure 4.2 Special moment-resisting frame, UBC 1994 regulations.

(a) Go through the plans and a new design process following the provisions of UBC 1997.

(b) Analyze the structure for gravity loads (DL, LL), earthquake loads, and wind loads.

(c) Carry out the engineering design on selected structural members and connections as though the building had not yet been constructed.

4.5 1994 DESIGN

Four identical SMRFs were provided to resist lateral forces—wind and earthquake—in the north–south direction. Figure 4.2 shows one of the identical frames. The SMRFs act in conjunction with the gravity force supporting structural steel elements—girders, beams, columns—not shown in the figure for simplification. We will carry out our design on frame 1.

4.6 WIND ANALYSIS: 1997 UBC, CHAPTER 16, DIV III

The 1997 UBC states:

Design wind pressures for buildings and structures and elements therein shall be determined for any height in accordance with the following formula

$$p = C_e \, C_q \, q_s \, I_w \qquad (20\text{-}1)$$

where

C_e = height, exposure, and gust factor coefficient	Table 16-G
C_q = pressure coefficient subject to function, geometry, and location of structure or element	Table 16-H
q_s = basic wind pressure subject to basic wind speed	Table 16-F
I_w = importance factor subject to occupancy or function of building	Table 16-K

The 1997 UBC gives two options for determining wind loads:

Method 1: Normal-Force Method (1621.2)   The wind pressure is applied to all surfaces separately, that is, walls on the windward side, leeward side, roof, projecting elements, and so on. The sum of the effects determines the overall effect on the structure such as overturning moment and base shear. The advantages and disadvantages of this method are as follows:

Advantages: It is versatile and can be applied to structures of complex geometry, architectural projections, and irregularities. The 1997 UBC establishes no height or other limitations for its use and states that it "may be used for any structure."

Disadvantages: It is more laborious; the information is collected element by element.

Method 2: Projected-Area Method (1621.3)   Wind pressures are assumed to act on the silhouette or image projected on a vertical surface, almost like designing a two-dimensional object; however, one must not forget the vertical load—often suction on the roof as a whole. The advantages and disadvantages of this method are as follows:

Advantages: It is fast and easy to use for a regular building such as a box type and it is not necessary to collect wind pressures element by element or projection by projection.

Disadvantages: It can be used only up to 200 ft in height; usage for gabled rigid frames is not allowed.

Once you have determined which method to use, the pressure coefficient C_q can be read from Table 16-H. The coefficient C_e reflects to what extent the structure is exposed to wind in relation to its environment:

Exposure D represents the most severe exposure, for instance, a structure close to the shoreline facing large masses of water, even a lake extending over a mile in front of the exposed building.

Exposure B means the structure is protected by its surroundings in the form of a forest or other structure for at least 1 mile.

Caution. Beware of wind tunnel effect in large towns where other buildings, rather than sheltering your project, channel the wind toward it with great amplification. Several cooling towers in England collapsed under increased wind velocities and pressure caused by surrounding cooling towers that created a veritable wind tunnel.

Exposure C represents an intermediate, average exposure condition such as flat and open terrain extending one-half mile or more from the building site.

Basic wind pressure q_s is measured at a standard 33-ft height and a function of wind velocities measured in the specific geographical region where the project is located. Figure 16-1 provides minimum basic wind speeds in miles per hour on page 2-36 of the 1997 UBC.

Note. Check with the building officials about local conditions; some areas are notorious for catching high winds and, because of the microclimate effect, you might have to adjust your figures.

Consider the occupancy or function-dependent importance factor I_w. From the public point of view other projects may be more important than yours; for instance, fire and police stations, hospitals, and emergency units that need a higher factor of safety have a wind importance factor I_w of 1.15. Hazardous facilities housing toxic or explosive materials also have a wind importance factor I_w of 1.15. Obviously we do not want more damage added to the one caused by wind or earthquake.

The example that follows illustrates how the 1997 UBC regulations such as wind pressure and force computations are applied to the project.

4.7 EXAMPLE: WIND ANALYSIS OF FOUR-STORY BUILDING

Evaluate the wind pressure coefficients and horizontal forces acting in the north–south direction for the first and top stories of the project. Use 1997 UBC method 2.

Figure 4.2 shows the dimensions of the building.

Basic wind speed 70 mph, exposure C
Maximum building height $h = 54$ ft
From Table 16-F, $q_s = 12.6$ psf

First story: height range 0–13.5 ft.

From Table 16-G for a maximum height ≤ 15 ft $\qquad C_e = 1.06$
From Table 16-H and noting that structural height $h > 40$ ft $\qquad C_q = 1.4$
From Table 16-K, occupancy category 4 $\qquad I_w = 1.0$

Hint. To find I_w, use the process of elimination. If your project does not seem to fit any of the above categories go down the lines; it might be under "All structures . . . not listed in Category 1, 2 or 3." The wind pressure is determined as

$$p = C_e C_q q_s I_w$$
$$= (1.06)(1.4)(12.6)(1.0) = 18.7 \text{ psf}$$

Uppermost story: height range 40.5–54.0 ft.
From Table 16-G and interpolating for the average height of 47.25 ft:

$$C_e = 1.310 + 0.043 = 1.353$$

$$C_q = 1.4 \quad \text{(unchanged)}$$

$$I_w = 1.0 \quad \text{(unchanged)}$$

$$p = (1.353)(1.4)(12.6)(1.0) = 23.9 \text{ psf}$$

Now we complete our wind pressure evaluation for the rest of the building. The lateral wind pressure at the second story is given as

$$p = (1.13)(1.4)(12.6)(1.0) = 20.0 \text{ psf}$$

and the lateral wind pressure for the third story as

$$p = (1.26)(1.4)(12.6)(1.0) = 22.2 \text{ psf}$$

Figure 4.3 shows the lateral wind pressure distribution per unit width of wall. Figure 4.4 shows the north–south lateral wind pressure distribution for the entire structure.

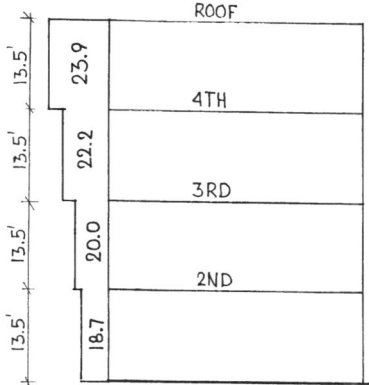

Figure 4.3   Wind design pressures (psf).

The total base shear in the north–south direction is

$$V = 13.5(2318 + 2154 + 1940 + 1814) = 111 \text{ kips} < 280 \text{ kips}$$

Here $E_h/1.4 = 280$ kips is the base shear for the earthquake.

Earthquake design forces govern.

4.8   SEISMIC ZONES 3 AND 4

Before we do the seismic force analysis it is appropriate to point out some of the basic differences and similarities between *seismic zones 3 and 4,* in

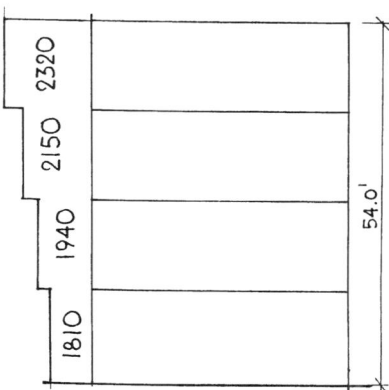

Figure 4.4   North–south wind pressure forces (lb/ft) acting on the building.

general and in particular, related to moment-resisting frames. Earthquake design forces are determined from the 1997 UBC Tables 16-I through 16-U, including Tables 16-Q and 16-R, which provide the seismic coefficients C_a and C_v, respectively, a function of soil type and seismic zone.

Differences

The major difference is that, while for seismic zone 4 near-source factors N_a and N_v must be used in conjunction with the values in the tables, N_a and N_v do not apply to zone 3. Therefore the values of C_a and C_v can be obtained from the tables and used for evaluation of the base shear in formulas (30-4)–(30-7).

Similarities

In steel design, UBC 1997 allows you the option of using either the Special Moment Resisting Frame (SMRF) or the ordinary moment-resisting frame (OMRF) for both zones 3 and 4, but with different R values. The SMRF design will enjoy a *more favorable R* value of 8.5 while the OMRF will be penalized by a *less favorable R* = 4.5, which no doubt will result in a more costly design.

When the SMRF is opted for the special seismic provisions, the design must follow the requirements of both CHAP. 16 and CHAP. 22, "Detailed Systems Design Requirements." Among the seismic provisions, operators such as ρ and Ω_0 for both seismic zones are meant to improve performance and ensure a sound structural design.

4.9   EARTHQUAKE ANALYSIS OF FOUR-STORY OFFICE BUILDING

The 1997 Design Project Data

Total building height	54.0 ft
Total building weight	3025 kips
Seismic zone	4
Seismic source type	A (8 km from source)
Soil type	Very dense, per geotechnical report

The weight of the structure was based on the following:

- Floor/roof area: 11,350 ft^2
- Roof deck: 56 psf
  $4\frac{1}{2}$-in. lightweight concrete on metal sheeting, including roofing, suspended ceiling, ducts and pipes, curtain walls, and steel framing.

- Weight of roof: 640 kips
- Floor deck: 70 psf

    $4\frac{1}{2}$-in. lightweight concrete on metal sheeting, including flooring, ducts and pipes, suspended ceiling, curtain walls, built-in partitions, and steel framing.

- Weight of floor: 795 kips
- Other parameters are:

$Z = 0.4$	Table 16-I
Soil type S_C	Table 16-J
$N_v = 1.4$ (interpolated)	Table 16-T
$C_v = 0.56N_v = 0.56(1.4) = 0.784$	Table 16-R
$N_a = 1.1$ (interpolated)	Table 16-S
$C_a = 0.4N_a = 0.4(1.1) = 0.44$	Table 16-Q
$R = 8.5$	Table 16-N
$I = 1.0$	Table 16-K
$C_t = 0.035$	Section 1630.2.2
$T = C_t\,(h_n)^{3/4} = 0.035(54)^{3/4} = 0.697\ \text{s} \approx 0.70\ \text{s}$	Formula (30-8)
$V = \left(\dfrac{C_v I}{RT}\right)W = \left[\dfrac{0.784 \times 1.0}{(8.5 \times 0.7)}\right] 3025 = 400\ \text{kips}$	Formula (30-4)
$R = $ EQ response modification coefficient	DIV. IV 2-11

Note that the total design base shear need not exceed

$$V = \left(\frac{2.5C_a I}{R}\right)W$$

$$= \left(\frac{2.5 \times 0.44 \times 1.0}{8.5}\right) 3025 \times 392 \text{ kips*} \qquad \textbf{GOVERNS}$$

or be less than

$$V = 0.11C_a IW \qquad\qquad\qquad (30\text{-}6)$$

$$= 0.11 \times 0.44 \times 1.0 \times 3025 = 146.4 \text{ kips}$$

$$= \left(\frac{0.8ZN_v I}{R}\right)W \qquad\qquad\qquad (30\text{-}7)$$

$$= \left(\frac{0.8 \times 0.4 \times 1.4 \times 1.0}{8.5}\right)3025 = 160 \text{ kips}$$

*$V = 392$ kips is at "strength level."

Lateral Force Distribution

The north–south lateral force distribution at strength level is determined as

$$V = F_t + \sum_{i=1}^{n} F_i \tag{30-13}$$

$$F_t = 0 \qquad \text{for } T \leq 0.7 \tag{1630.5}$$

$$F_x = \frac{(V - F_t)w_x h_x}{\sum_{i=1}^{n} w_i h_i} \tag{30-15}$$

$$F_5 = \left(\frac{(392 - 0)640 \times 54}{640 \times 54 + 795(40.5 + 27.0 + 13.5)} \right) = 136.9 \text{ kips}$$

$$F_4 = \frac{392 \times 795 \times 40.5}{640 \times 54 + 795(40.5 + 27.0 + 13.5)} = 127.5 \text{ kips}$$

$$F_3 = \frac{392 \times 795 \times 27}{640 \times 54 + 795(40.5 + 27.0 + 13.5)} = 85 \text{ kips}$$

$$F_2 = \frac{392 \times 795 \times 13.5}{640 \times 54 + 795(40.5 + 27.0 + 13.5)} = 42.5 \text{ kips}$$

Figures 4.5 and 4.6 show the north–south lateral force distribution for the building and one of the frames.

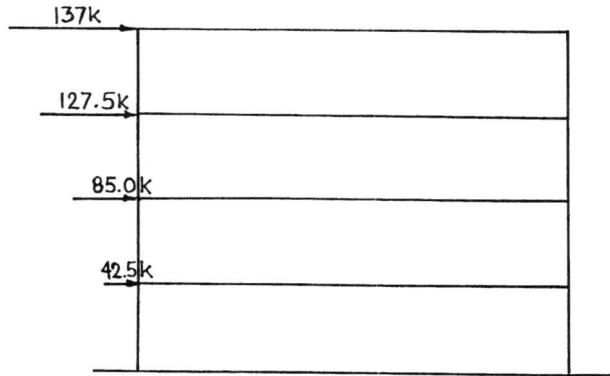

Figure 4.5   North–south strength-level earthquake forces acting on building, by 1997 UBC.

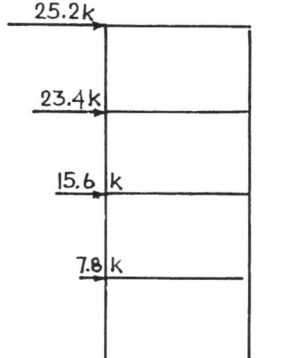

Figure 4.6   North–south strength-level forces acting on SMRF 1, by 1997 UBC.

The preceding results will now be used for the final design.

Wind will *not* be considered.

4.10   DESIGN FOR EARTHQUAKE

Table 4.1 summarizes the north–south lateral earthquake forces acting on the building and frame 1. The values represent the largest horizontal load acting on an individual frame SMRF 1 based on:

(a) Load sharing of frames at floor level
(b) Torsional effect

Before going any further we will apply the reliability/redundancy factor ρ test as required by the 1997 UBC (1630.1) to ensure that there are sufficient

TABLE 4.1   Lateral Earthquake Force F_i Acting on Building and Frame 1[a]

Level	Acting on Building	Acting on Frame 1	
		4 frames/floor	6 frames/floor
5	137.0	37.8	25.2
4	127.5	35.2	23.4
3	85.0	23.4	15.6
2	42.5	11.7	7.8

[a] Values include torsion as well as redundancy factor $\rho = 1.0$.

elements to resist the lateral component of the earthquake and that none of the elements carries an unjustifiably large portion of the total load:

$$\text{Check } \rho = 2 - \frac{20}{r_{max} \sqrt{A_B}} \qquad (30\text{-}3)$$

The design story shear at the first level is

$$\Sigma F_i = (137 + 127.5 + 85 + 42.5) = 392 \text{ kips}$$

There is about 10.0 kips of additional shear due to earthquake-caused torsion acting on frame 1* that makes this frame the most heavily loaded *single element* in the four-SMRF structural system.

The maximum element-story shear carried by frame 1 evaluated at the first level (Table 4.1) including torsional shear is

$$S = \left(\frac{392}{4}\right) + 10.0^* = 108 \text{ kips}$$

The maximum element-story shear ratio is

$$r_{max} = \frac{108}{392} = 0.2757$$

$$\rho = 2 - \frac{20}{0.1837\sqrt{11,350}} = 2 - 0.68 = 1.32 > 1.25 \quad \text{Allowed (1630.1.1)}$$

Because 1997 UBC does not allow ρ larger than 1.25 for SMRFs, we add one more frame to each side, increasing the total number of frames to *six per floor* in the north–south direction. Figure 4.7 shows the new 1997 layout.

Clearly the 1997 UBC provision is to ensure better load sharing among the earthquake (E_h) resisting elements by providing more redundancies and more elements against the seismic load. Indirectly, but effectively, the provision makes you reflect on the soundness of the initial structural scheme to avoid that certain structural elements carry an unduly large share of the total horizontal earthquake load.

---

* 1997 UBC, Section 1630.7, "Horizontal Torsional Moments": "Provisions shall be made for increased shears resulting from horizontal torsion where diaphragms are not flexible. The most severe load combinations for each element shall be considered for design."

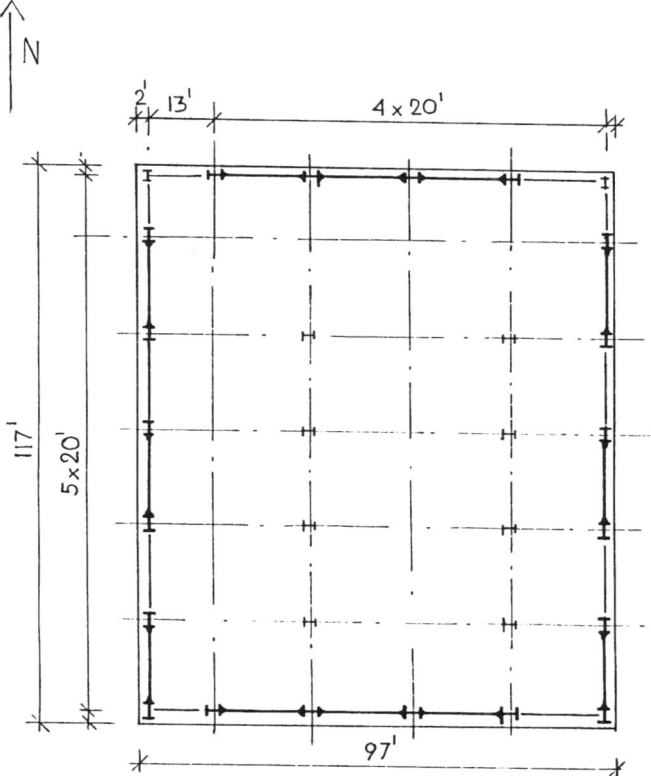

Figure 4.7 Revised structural layout. Two SMRFs added to meet reliability criteria (30-3), 1997 UBC.

The test is naturally carried out floor by floor in computing r_i. The largest value of r_i, r_{max}, represents the worst scenario and will be the controlling factor.

Unlike previous editions, the 1997 UBC urges you to review your initial disposition of structural elements and, if proved inadequate, to go back to the drawing board and come up with a new scheme before going any further with detailed design.

Note. Keep track of your strategy in countering EQ forces at the very early stage of your design.

Once the new structural scheme has been worked out, in our case by increasing the number of frames, we need to compute new values for r_{max} and ρ. The maximum element-story shear defined by 1630.1.1 is the seismic design shear at the most heavily loaded single frame subjected to maximum horizontal and torsional seismic forces at the first level (Table 4.1 in this text):

$$S = 25.2 + 23.4 + 15.6 + 7.8 = 72.0 \text{ kips}$$

$$r_{\max} = \frac{72.0}{392} = 0.1837$$

$$\rho = 2 - \frac{20}{0.1837\sqrt{11,350}} = 0.977 < 1.0$$

Because 1997 UBC does not allow $\rho < 1.0$, we will use $\rho = 1.0$ for our computation.

You will notice a major change in design philosophy. The 1997 UBC calls for a more robust design to resist earthquakes, which is reflected in the configuration and disposition of the major lateral-force-resisting system. *Four SMRFs in the north–south direction would have satisfied the requirements of the 1994 edition.*

Figure 4.7 shows the new north–south lateral force distribution on frame 1 for the six-SMRF-per-floor arrangement. In addition, the following gravity loads are superimposed on the frame beams:

Dead Load[*,†]

Uniformly distributed roof load acting on a tributary width of 8.0 ft
Uniformly distributed floor load acting on a tributary width of 8.0 ft
Roof: $w_R = 550 \text{ lb/ft} \approx 0.55 \text{ k/ft}$
Floor: $w_F = 594 \text{ lb/ft} \approx 0.60 \text{ k/ft}$

Live Load

20-psf roof (Table 16-C)
50-psf office use (Table 16-A)
Roof: $w_{Lr} = L_r \times$ tributary width
$\qquad w_{Lr} = 20 \times 8.0 = 160 \text{ lb/ft} = 0.16 \text{ k/ft}$
Floor: $w_L = L \times$ tributary width
$\qquad w_L = 50 \times 8.0 = 400 \text{ lb/ft} = 0.40 \text{ k/ft}$

Two sets of load combinations for the computer analysis were prepared with the above data:

---

[*] The 0.5-ft strip on the roof and floor will not receive a live load because it is occupied by parapet and curtain walls, respectively.
[†] Note that w_R and w_F include the weight of the heating, ventilation, and air-conditioning system (HVAC).

1. For the LRFD design (1612.2) using strength-level factored loads
2. For the detailed systems design requirements (1633) and the maximum inelastic response displacements (1630.9) applying strength-level factored loads (1612.2) amplified by factors such as $0.7R$ (Section 1630.9.2) and $0.4R$ (Section 6.1, CHAP. 22, DIV. IV), respectively

The next major step before preparing the detailed component design is to comply with the 1997 UBC story-drift limitation (1630.9 and 1630.10) of the displacement of the structure caused by the horizontal vector components of the earthquake (E_h) in an effort to prevent buildup of large internal stresses and strain caused by excessive deformation.

What is involved in the analysis of the maximum story-drift control, called interstory drift in the 1999 edition of the SEAOC *Blue Book?* First, the story drift, termed maximum inelastic response displacement (1630.9.2), is evaluated at strength level; that is, the deformations Δ_S are calculated under limit-state or strength-level earthquake forces. The drift so obtained is further magnified by a factor of $0.7R$, thus further increasing it to more than eight times larger than the elastic deformation obtained by a theoretical allowable stress design analysis using ASD load combinations (12-7)–(12-11), CHAP. 16, Section 1612.3.1. A very strong restriction indeed which calls for a comparatively more robust design. (See Figures 4.7 and 4.8.)

The story-drift check (1630.9 and 1630.10) is an overall stiffness review for the *entire* structure; it involves all structural elements affecting the deformation characteristics of the lateral-force-resisting system. It would be pointless to refine the design of components if the entire system fails the test.

For the initial drift analysis—computer analysis part II—we first applied sectional properties of the 1994 steel frame prototype (Figure 4.2) to the six-frame system. *The test failed.* The computer analysis part II, "Drift Evaluation," shows excessive first-floor story-drift of

$$\Delta_M = 0.7R\Delta_S \tag{30-17}$$

$$\Delta_M = 0.380 \text{ ft} = 4.56 \text{ in.} \quad \text{larger than allowable 4.05 in.}$$

$$\Delta_{\text{allowable}} = 0.025 \times \text{story height} = 4.05 \text{ in.}$$

for a fundamental period of less than 0.7 s (1997 UBC, Section 1630.10.2). The fundamental period of the structure was found to be 0.697 s.

Based on the foregoing, new input data were generated. The new input data contained two types of major changes: *first-floor beam and column size increase and reduction of member size at upper levels.* The most significant modifications to the inherited 1994 prototype frame are discussed below.

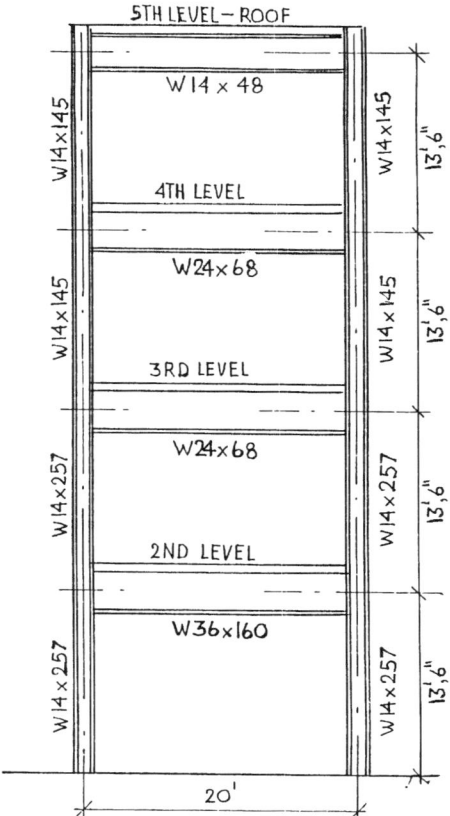

Figure 4.8 Revised frame to meet reliability criteria, 1997 UBC.

4.11 SIGNIFICANT CHANGES IN 1997 DESIGN

Beam and Column Size Increase

First- and second-story columns increased from W14 × 193 to W14 × 257 and second-floor beam increased from W36 × 135 to W36 × 160 to fit the 1997 UBC stringent drift control requirements. By now it is obvious that the 1997 UBC story-drift control will not let you go any further with a detailed design—*even* if it meets the strength requirements—unless you have complied with the story-drift limitations *and* the reliability/redundancy factor (ρ) test.

Reduction of Member Size at Upper Levels

It became imperative to economize on member sizes inherited from the 1994 design. In carrying out this second task, we had to meet two objectives:

(a) *Comply with the 1997 UBC story-drift control at all stories.* This meant to curb reduction in member sizes so as not to violate 1630.9 and 1630.10. The values in Table 4.2 (in Section 4.27 on column design) verify that the 1997 UBC provisions for maximum allowable story drift (1630.9 and 1630.10) were satisfied.

(b) *Ensure that the revised member sizes meet the AISC and UBC requirements.* AISC *Specification Load and Resistance Factor Design,* (second edition) and UBC "Detailed Systems Design Requirements" (1633 and CHAP. 22, DIV. IV).

With the second upgrading the structure is ready for final analysis and design. The first upgrading, as described earlier, occurred when a pair of frames were added to each floor level.

Figures 4.9(*a*) and (*b*) show the new mathematical model and applied loads (1612.2). A partial fixity representing about 15% of flexural member stiffness was applied to column joints 1 and 2 to represent *real-life* condition.

A reduced modulus $E = 26,000$ ksi was used taking into account results of cyclic tests that showed pinched hysteresis loops and implied reduction of E values. The modulus of elasticity for steel was reduced from 29,000 to 26,000 ksi to reflect softening of the metal under the effect of hysteresis caused by earthquake cyclic loading. A more accurate value of the reduced

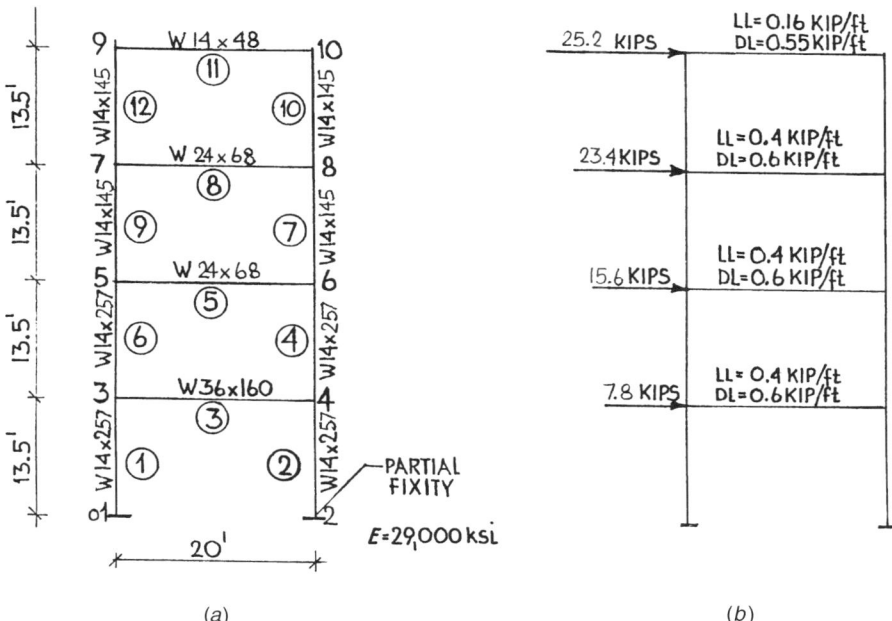

(a)   (b)

Figure 4.9 (*a*) Mathematical model showing final member sizes. Member numbers are circled. (*b*) Seismic design loads.

E could be the subject of research. Obviously a revised E is a complex issue; for instance, it is a function of the softening of the metal as well as a function of bolt slippage and damage to beam-to-column connection.

All of these factors had an overall effect in reducing the lateral stiffness of the moment frame. By reducing the E value, we attempted to demonstrate that a simplified corrected E value can be utilized in the elastic time-history analysis performed (UBC 1631.6.2). A number of nonlinear time-history analysis computer programs are available (UBC 1631.6.3); however, it is doubtful that all of the complexities involved are taken into account, specifically the deterioration of the moment connection itself due to earthquake-imposed cyclic loading.

The computer analysis for LRFD strength-level design is by 1612.2, CHAP. 22, DIV. IV, and the story-drift analysis per 1630.9.2.

4.12 1997 VERSUS 1994 DESIGN

A comparison between the two designs is enlightening. Remember that in the 1997 UBC standards the 1994 four-frame scheme was deemed inadequate because it did not have enough structural elements to counter earthquake forces (the ρ, r_{max} control).

Responding to the 1997 UBC warning about insufficient system redundancy, we were forced to provide more resisting structural elements and upgrade the structural system to a *six-frame* scheme, keeping the original structural elements of the 1994 design in place. In doing so we expected that the structure would pass the second test—the 1997 UBC maximum story-drift limitation test (1630.9 and 1630.10).

However, despite a major change in structural layout complying with the restriction imposed by the reliability/redundancy factor ρ, the first-story drift was still too large for the 1997 UBC. Thus we were left with no alternative but to increase both beam and column sizes for the first story.

The ρ, r_{max} restriction—intended to prevent member overstress—and the maximum story-drift provision make the 1997 UBC far more conservative than earlier editions, a marked departure from previous design philosophy.

Before going into the actual design it is appropriate to review the 1997 UBC load combination provisions applicable to LRFD design, that is, a set of load combinations (12-1)–(12-6) applied in conjunction with CHAP. 22, DIV. II, "Design Standard for Load and Resistance Factor Design Specification for Structural Steel Buildings," per the AISC *Manual of Steel Construction, Load and Resistance Factor Design,* Volumes I and II (second edition, 1998). The specific set entitled "Load Combinations Using Strength Design or Load and Resistance Factor Design" (1612.2) is

$$1.4D \tag{12-1}$$

$$1.2D + 1.6L + 0.5 \ (L_r \ \text{or} \ S) \tag{12-2}$$

$$1.2D + 1.6(L_r \ \text{or} \ S) + (f_1L \ \text{or} \ 0.8W) \tag{12-3}$$

$$1.2D + 1.3W + f_1L + 0.5(L_r \ \text{or} \ S) \tag{12-4}$$

$$1.2D + 1.0E + (f_1L + f_2S) \tag{12-5}$$

$$0.9D \pm (1.0E \ \text{or} \ 1.3W) \tag{12-6}$$

where

$$f_1 = \begin{cases} 0.5 & \text{for floor live loads} \\ 1.0 & \text{for floor live loads in excess of 100 psf} \end{cases}$$

$$f_2 = \begin{cases} 0.2 & \text{for roofs} \\ 0.7 & \text{for roofs that do not shed off the snow} \end{cases}$$

and, in addition, a set of load combinations (3-1)–(3-8) and (6-1) and (6-2) for the "Detailed Systems Design Requirements." As mandated by CHAP. 16, 1633, this set operates in conjunction with the "Detailed Systems Design Requirements" found in CHAP. 22, DIV. IV, "Seismic Provisions for Structural Steel Buildings."

It must be emphasized that the specific detailing instructions of CHAP. 22, DIV. IV, address only seismic design and require that modified strength-level forces be applied to specific earthquake-resisting elements, such as columns, beam-to-column joint, and panel zone. In addition, CHAP. 16 contains specific seismic restrictions such as maximum inelastic response displacement Δ_M (Section 1630.9 and 1630.10).

4.13 SUMMARY OF PROCEDURE

We will use load combinations (12-1)–(12-6), Section 1612.2.1, for "Load and Resistance Factor Design." These combinations include earthquake and wind forces in addition to gravity forces and are applicable to this specific project, subjected to winds up to 70 mph, exposure C, seismic zone 4, source type A. The design procedure must be adjusted to local conditions if higher winds prevail on a specific site.

We will use a *second* set of load combinations (3-1)–(3-6) for the "Detailed Systems Design Requirements" of CHAP. 22, DIV. IV (1633).

Yet a *third* set of load combinations is needed —although not specifically termed as such in the 1997 UBC—for column strength,

$$P_u \geq 1.0P_D + 0.5P_L + 0.4RP_E \qquad (6\text{-}1)$$

$$P_u \geq 0.9P_{DL} - (0.4R)_o P_E \qquad (6\text{-}2)$$

as mandated by 2211.4.6.1, CHAP. 22, DIV. IV, and if you consider that the 1997 UBC calls for augmented earthquake forces for the design of the panel zone (2211.4.8.2a) or for a magnified inelastic displacement $(0.7R)\Delta_S$ (1630.9.2), each of these specific requirements corresponds to yet another set of load combination not listed under 1612.2.1 and expressed, for instance, as

$$0.0D + (0.7 \times 8.5)E = 5.95E \qquad (30\text{-}17)$$

We entered these relevant special load combinations in the computer analysis part II.

CHAPTER 22, DIV. IV, "Seismic Provisions for Structural Steel Buildings," contains special recommendations for the design strength of members and connections developed in conjunction with strength-level factored external loads and resisted by the nominal strength of the member multiplied by $\phi \leq 1.0$ resistance factors. Of particular concern is the design of:

(a) Columns
(b) Beam-to-column connections
(c) Panel zone

We will address each issue when we design the particular element or connection.

In general, factored strength-level forces are called for in the overall design of the structure. In addition, the 1997 UBC requires that for "specific elements of the structure, as specifically identified in this Code," the minimum design strength should be the product of seismic amplification factors and design forces set forth in Section 1630. In accordance with this rule, a $0.4R$ seismic amplification is applied to CHAP. 22, Section 6.1, column design formulas as

$$1.2P_D + 0.5P_L + 0.2P_s + 0.4R \times P_E \qquad \phi_c P_n \qquad (6\text{-}1)$$

$$0.9P_D - 0.4R \times P_E \qquad \phi_t P_n \qquad (6\text{-}2)$$

or to formulas (3-7) and (3-8) of CHAP. 22, Section 3.1, as

$$1.2D + 0.5L + 0.2S \pm 0.4R \times E \tag{3-7}$$

$$0.9D \pm 0.4R \times E \tag{3-8}$$

The reader will note that we applied the preceding load combination for the column design. It will also be noted that, by using load combinations (3-7) and (3-8), a number of other specific requirements of CHAP. 22 are waived.

An example of such a specific provision is the beam-to-column joint design requirement, CHAP. 22, Section 8.2.b, that involves superposition of load cases. The requirement calls for a shear strength based on the application of the beam end moment of either M_p or the moment derived from the factored shearing resistance of the panel zone, UBC 1997, CHAP. 22, formula (8-1). To the seismic shear thus obtained we must add the reaction of factored gravity loads $1.2D + 0.5L + 0.2S$.

As indicated earlier, the alternative to the individual load superposition would be to use load combinations (3-7) and (3-8) permitted by CHAP. 22, Section 8.2.b, probably resulting in a more organized bookkeeping of engineering calculations that would lead to a direct design.

These specific 1997 UBC requirements, under CHAP. 22, DIV. IV, "Seismic Provisions for Structural Steel Buildings," are to ensure that vital elements of the structure such as columns, beam-to-column connections, and panel zone will perform to guard the overall safety of the structure.

4.14 DESIGN STRATEGIES

(a) Apply loads as required by the 1997 UBC load combinations (12-1)–(12-6), (3-1)–(3-6), (6-1), and (6-2).

(b) Comply with CHAP. 22, DIV. IV, "Seismic Provisions for Structural Steel Buildings," as mandated by "Detailed Systems Design Requirements" (1633).

(c) Ensure adequacy of structural components and connections under the provisions of the AISC *Manual of Steel Construction, Load and Resistance Factor Design,* second edition, 1998 (the *Specification*), as mandated by CHAP. 22, DIV. II, of UBC 1997.

4.15 DESIGN OF BEAMS: CODE REQUIREMENTS

The 1997 UBC detailed systems design requirements, CHAP. 22, DIV. IV, Section 2211.4.8, states, under "Requirements for Special Moment Frames (SMF)":

8.1. Scope. Special moment frames (SMF) shall have a design strength as provided in the *Specification* to resist the Load Combinations 3-1 through 3-6 as modified by the following added provisions:

8.2. Beam-to-Column Joints

8.2.a. The required flexural strength, M_u, of each beam-to-column joint shall be the lesser of the following quantities:

1. The plastic bending moment, M_p, of the beam.
2. The moment resulting from the panel zone nominal shear strength, V_n, as determined using Equation 8-1.

The joint is not required to develop either of the strengths defined above if it is shown that under an amplified frame deformation produced by Load Combinations 3-7 and 3-8, the design strength of the members at the connection is adequate to support the vertical loads, and the required lateral force resistance is provided by other means.

8.2.b. The required shear strength, V_u of a beam-to-column joint shall be determined using the Load Combination $1.2D + 0.5L + 0.2S$ plus the shear resulting from M_u, as defined in Section 8.2.a, on each end of the beam. Alternatively, V_u shall be justified by a rational analysis. The required shear strength is not required to exceed the shear resulting from Load Combination 3-7.

8.2.c. The design strength, ϕR_n of a beam-to-column joint shall be considered adequate to develop the required flexural strength, M_u, of the beam if it conforms to the following:

1. The beam flanges are welded to the column using complete penetration welded joints.
2. The beam web joint has a design shear strength ϕV_n greater than the required shear, V_u, and conforms to either:
   a. Where the nominal flexural strength of the beam, M_n, considering only the flanges is greater than 70% of the nominal flexural strength of the entire beam section [i.e., $b_f t_f (d - t_f) F_{yf} \geq 0.7 M_p$]; the web joint shall be made by means of welding or slip-critical high strength bolting, or;
   b. Where $b_f t_f (d - t_f) F_{yf} < 0.7 M_p$, the web joint shall be made by means of welding the web to the column directly or through shear tabs. That welding shall have a design strength of at least 20 percent of the nominal flexural strength of the beam web. The required beam shear, V_u, shall be resisted by further welding or by slip-critical high-strength bolting or both.

8.2.d. Alternate Joint Configurations: For joint configurations utilizing welds or high-strength bolts, but not conforming to Section 8.2.c, the design strength shall be determined by test or calculations to meet the criteria of Section 8.2.a. Where conformance is shown by calculation, the design strength of the joint shall be 125% of the design strengths of the connected elements.

4.16   SECOND-FLOOR BEAM

The maximum moment corresponding to the 1997 UBC load combination (3-5) is $M_u = 646$ kip-ft, derived from computer load combination [10].

Caution. The 1997 UBC formulas (3-5) and (3-6) contain both vertical and horizontal earthquake design components E_v and E_h (30-1)

The E_v component either magnifies or reduces (uplift) the effect of gravity load and is expressed as

$$0.5C_a ID = 0.5(0.44)(1.0)D = 0.22D$$

Therefore

$$1.2D + 1.0(E) + 0.5L = 1.2D + 1.0(E_v + E_h) + 0.5L$$

$$= (1.2D + 0.22D) + 0.5L + E_h$$

$$= 1.44D + 0.5L + E_h \qquad (3\text{-}5)$$

Formula (3-5) was entered into the computer analysis as load combination [10].

The LRFD moment capacity of the W36x160 with

$$L_b = 9.0 \text{ ft} \leq L_p = 10.4 \text{ ft}$$

$$\phi_b M_p = 1680 \text{ kip-ft} > 646 \text{ kip-ft applied} \qquad \text{(ASTM A36 steel)}$$

was obtained from the load factor design selection AISC Table 4-16. The section satisfies the 1997 UBC strength requirements. The 8-in. dimension deducted from the clear span represents the width of the lateral bracing provided at midspan of the beam.

Because the beam deformation is subjected to reversals during the entire duration of the earthquake, design and detailing of the bracing must be such that full lateral support to top and bottom flanges is ensured (1997 UBC, 2211.4.8.8).

According to the AISC *Specification,* because $L_b < L_p$, full plastic moment response could be expected along the entire length of the beam. However, a cautionary comment is relevant at this point: *The $L_b < L_p$ mathematical statement alone is no guarantee of yielding; it only assures that no torsional instability occurs before the plastic response can be achieved, provided other* *conditions of mechanics of materials are satisfied. Analysis by the author of the effect of complex boundary conditions at the beam-to-column interface indicates that joint behavior is complex and beyond simple design manual rules.*[1,2]

4.17 BEAM-TO-COLUMN JOINT

Three basic issues need to be addressed:

(a) The moment of resistance of the connection at the beam–column interface
(b) The joint shearing resistance to vertical reactions
(c) The horizontal shearing resistance of the panel zone, defined between column flanges and continuity plates

4.18 FLEXURAL RESISTANCE OF BEAM-TO-COLUMN JOINT

As described earlier, UBC 1997 (2211.4) gives the option of designing the joint for the lesser of the following values:

1. Plastic moment of resistance M_p of the beam:

As $\phi_p M_p = 1680$ kip-ft   (AISC LRFD selection table 4-16)

$$M_p = \frac{\phi_p M_p}{\phi_p} = \frac{1680}{0.9} = 1867 \text{ kip-ft} = 22{,}404 \text{ kip-in.}$$

2. Moment resulting from the panel zone nominal shear strength V_n as determined using Equation (8-1) of 2211.4

Using formula (8-1) and dividing both sides by $\phi_v = 0.75$,

$$V_n = 0.6 F_y d_c t_p \left[1 + \frac{3 b_{cf} t_{cf}^2}{d_b d_c t_p} \right]$$

where t_p = total thickness of panel zone = 1.175 in. column web thickness
d_c = overall column depth = 16.38 in.
d_b = overall beam depth = 36.0 in.
b_{cf} = width of column flange = 16.0 in.
t_{cf} = thickness of column flange = 1.89 in.
F_y = specified yield strength of panel zone steel column (A36)

$$V_n = 0.6 \times 36 \times 16.38 \times 1.175 \left[1 + \frac{3 \times 16.0 \times 1.89^2}{36 \times 16.38 \times 1.175} \right]$$

$$= 518.6 \text{ kips}$$

$$M_u = V_n(d - t_f) = 518.6 (36.01 - 1.02) = 18,145 \text{ kip-in.}$$

We will attempt to demonstrate that it is virtually impossible—unless the 1997 UBC seismic provisions are violated—to design a fully restrained moment connection relying on a beam-to-column flange complete joint penetration weld only, without improving the moment connection joint.

The statement in 2211a and b implicitly requires a M_u joint flexural resistance larger than can be provided by the complete joint penetration (CP) weld of the beam flanges. This is quite evident from the inequality equation, which can be expressed as

$$M_u = \phi M_f < \phi M_p < M_p$$

$$\phi M_f \neq M_p$$

Although perhaps less evident at first glance, 2211.4.a.2 involves an inequality that urges the designer to come up with an improved connection. It appears that the quoted 1997 UBC provisions prevent the designer from reverting to the pre-Northridge moment connection that caused much concern in the engineering community.

The first step to joint design is to determine the moment of resistance of the CP groove weld joint connecting the beam flange to the column flange interface. Noting that the tensile resistance of the CP joint weld is identical to the base metal and with resistance factor, $\phi = 0.9$,

$$\phi M_f = \phi \, b_f t_f (d - t_f) F_Y = 0.9 \times 12.0 \times 1.02(36.01 - 1.02) \times 36$$

$$= 13,880 \text{ kip-in.} < M_p = 18,145 \text{ kip-in.}$$

where ϕM_f is the design resistance of the CP weld of beam flanges to the column. Since the moment of resistance of the CP weld connection is insufficient, top and bottom plates will be attached by welding to the flanges of the beam using fillet welds between flange and plate and complete joint penetration weld between the plate and the column flange.

Attach a 10-in.-wide, $\frac{3}{8}$-in. top plate tapered along its length to improve stress transfer between plate and flange. The design tensile strength of the plate is given as

$$\phi H_p = \phi b_p t_p F_y = 0.9 \times 10.0 \times 0.375 \times 36$$

$$= 121.5 \text{ kips} \quad \textit{for yield}$$

$$\phi H_p = \phi b_f t_p U F_u = 0.9 \times 10 \times 0.375 \times 0.75 \times 58$$

$$= 146.8 \text{ kips} \quad \textit{for fracture}$$

The 121.5 kips tensile strength by yielding   GOVERNS.

Assuming the same size for the bottom plate, the lever arm of the attached plates will be approximately

$$L_a = d + t_p = 36.01 + 0.375 = 36.38 \text{ in.}$$

The additional design moment of resistance offered by the plate reinforcement is given as

$$\phi M_p = \phi H_p(d + t_p) = 121.5 \times 36.38 = 4420 \text{ kip-in.}$$

The total design strength of the moment connection is

$$\phi_b M_n = \phi_b M_p = \phi M_F + \phi_b M_p = 13{,}880 + 4420$$

$$= 18{,}300 \text{ kip-in.} > 18{,}145 \text{ kip-in.}$$

The minimum fillet weld size permitted by the AISC *Manual* Table J2.4 for 1.0 in. flange thickness is $\frac{5}{16}$ in. The design resistance of a $\frac{5}{16}$-in. fillet weld (Table J2.5) is

$$\phi_w \, 0.707(\tfrac{5}{16})(0.60 \times F_{EXX}) = 0.75 \times 0.707 \times (\tfrac{5}{16})0.60 \times 70 = 6.96 \text{ kip/in.}$$

The required weld length (two lengths, one on each side of the plate) is

$$L = \frac{121.5}{2 \times 6.96} = 8.73 \text{ in.}$$

Provide 9-in.-long fillet welds on each side with ($\frac{5}{8}$-in. end returns to comply with the AISC LRFD *Manual of Steel Construction* J2b fillet weld termination requirement: "fillet welds shall be returned around the side or end for a distance not less than two times the weld size."

The reader will note that the end of the plate pointing toward the midspan of the beam will not be welded. This is because researchers in general believe that in this manner a smoother stress transfer can be achieved.

The width of the bottom cover plate will be wider than the lower beam flange to allow for a flat horizontal fillet welding position. Dictated by practical reasons, the plate will be rectangular.

The design resistance of a 13.0-in.-wide, $\frac{5}{16}$-in. A36 plate is given as

$$\phi H_{5/16} = 0.9 \times \tfrac{5}{16} \times 13 \times 36 = 131.6 \text{ kips} > 121.5 \text{ kips required}$$

$$\phi H_{5/16} = 0.9 \times \tfrac{5}{16} \times 13 \times 0.75 \times 58 = 159.0 \text{ kips}$$

Specifying a $\tfrac{1}{4}$-in. fillet weld, the required weld length (on two sides) is

$$\frac{121.5}{5.57} = 21.9 \text{ in.} \qquad (11.0 \text{ in. each side})$$

Both top and bottom A36 cover plates will be attached to the column flange by a complete-joint-penetration-groove weld. It is recommended that this operation be carried out in a separate phase, after welding of beam flanges to the column is complete. Figure 4.10 shows cover plate dimensions and welds.

The reader will note that the back-up bar that facilitates erection and welding of the upper beam flange has been removed. Although adding to construction costs, it is believed that the removal of back-up bars will improve performance of the joint, because presence of the temporary erection piece might evoke stress concentrations. Finite-element models also attest to it.

Although logical to scholars of engineering, the foregoing has never been fully proven. Brittle fracture of the joint during cyclic loading has been equally observed even in large-scale specimens, despite thorough preparation such as carefully removed back-up bars, gouged-out weld root pass, filled with new weld material, and smooth ground at these locations. Finite-element models are seldom capable of reproducing the complexities of boundary conditions and related stress disturbances in the beam column joint.

A complex load path and stress distribution may also exist in built-up joints. During the author's visit to the large-scale testing laboratory at the University of California at San Diego in the fall of 1999, Dr. Chia-Ming Uang indicated that one of the characteristics of the cover-plated joint is that, while the top cover plate undergoes tension due to downward loading of the frame beam, a considerable portion of the beam top flange, normally considered to be in full tension, will be subjected to compression.

Complex boundary conditions that evoke three-dimensional stresses exist in a uniaxially loaded specimen when fixed into a machine platen that prohibits lateral expansion or contraction.[3,4] The beam flange solidly welded to a robust column flange will react somewhat similar to a specimen attached by interlocking to an almost infinitely stiff machine platen.

The reader will note that the 1997 UBC provisions 2211.4, Section 8.2.a., are strength related. They aim at making the welded connection stronger than the simple beam flange-to-column full-penetration weld that failed in the Northridge earthquake.

The SEAOC and Applied Technology Council (ATC) procedures, channeled through the 1997 UBC provisions, are far from setting forward an energy-dissipating system. Like a car without shock absorbers, any energy

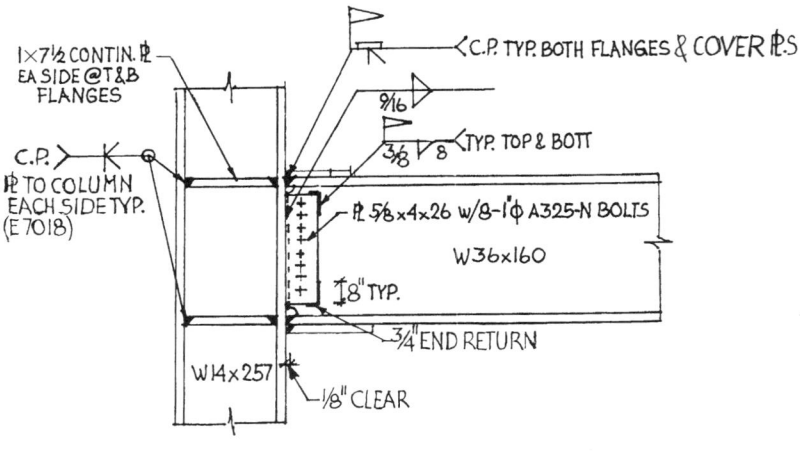

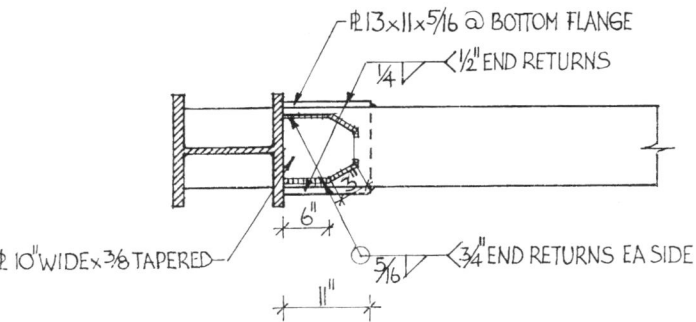

Figure 4.10 *Top:* second-floor beam-to-column connection. *Bottom:* plan view of cover plates.

dissipation is achieved at the high cost of permanent damage to beam–column joints including the base material. A better approach would be calibrated dampers to release the enormous stresses on the connection by the swaying mass of the structure during an earthquake.

4.19 SHEAR TAB DESIGN

Once adequate flexural resistance has been provided by a flange-to-flange connection capable of resisting moment and horizontal forces only, the engineer has to provide sufficient strength to resist the vertical forces of the beam-to-column joint. These forces are:

1. Gravity forces—the reaction caused by dead weight of floor and structure and a portion of live load that has to be supported even during the most severe earthquake
2. Reactions of the frame beam subjected to sway and bending due to lateral seismic or wind forces

The procedure allows treating force components—moment and vertical shear—separately. After all, the three basic equations

$$\sum H = 0 \qquad \sum M = 0 \qquad \sum V = 0$$

that pertain to the free-body diagram must be obeyed. Using these equations, however, could lead in a pitfall *unless* the law of compatibility of deformations is also observed. This issue will be analyzed later.

The quoted CHAP. 22, DIV. IV, Section 8.2.b, requires designing the joint for a gravity load of

$$1.2D + 0.5L = 1.2(0.6) + 0.5(0.4) = 0.92 \text{ k/ft}$$

uniformly distributed load and, in addition, a shear resulting from M_u defined in 8.2.a applied "on each end of the beam."

By 8.2.a, M_u was defined earlier as the lesser of the following quantities:

1. The plastic bending moment M_p of the beam
2. The moment resulting from the panel zone nominal shear strength V_n as determined using Equation (8-1)

The value of M_u was found to be

$$M_u = 18{,}145 \text{ kip-in.} = 1512 \text{ kip-ft}$$

The shear resulting from the uniform load (no snow load present) is

$$V_W = \frac{(20.0 - 1.37) \times 0.92}{2} = 8.6 \text{ kips}$$

The shear resulting from M_u end moments is

$$V_{mu} = \frac{2 \times 1512}{20.0 - 1.37} = 162.4 \text{ kips}$$

The total shear to be resisted by the joint is

$$V_u = 8.6 + 162.4 = 171.0 \text{ kips}$$

From AISC *Manual of Steel Construction,* Volume II, "Connections," Table 9-158, eight 1-in. ϕ 325N bolts with standard holes, $\frac{7}{16} \times 24$ single plate, and $\frac{3}{8}$-in. weld size have a shear capacity of 171 kips. However, the shear tab must also resist a moment of

$$0.2(M_u) = 0.2(1512 \times 12) = 3629 \text{ kip-in.}$$

As the nominal flexural resistance of the beam flanges is less than 70% of the nominal flexural strength of the entire beam section,

$$b_f t_f(d - t_f)F_y = \frac{12.0 \times 1.0 \times (36.0 - 1.0)36}{12} = 1260 < 0.7M_p$$

$$= 0.7(1867) = 1307 \text{ kip-ft}$$

The 1997 UBC, Section 2211.4, 8.2.c.2.b, states: "the web joint shall be made by means of welding the web to the column directly or through shear tabs. The welding shall have a design strength of at least 20% of the nominal flexural strength of the beam web." Stated otherwise, the single plate and its weld, designed to function as a simple shear connection as tabulated in AISC *Manual of Steel Construction,* Volume II, must be modified to resist both shear and the 20% flexural resistance of the beam web to counteract the secondary effects caused by frame action.

The 20% nominal flexural strength of the beam web having a cross section of $t_w T = 0.65 \times 32.12$ is given as

$$0.2M_{n,\text{web}} = 0.2 \times 0.65 \times 32.12^2 \times \tfrac{36}{4} = 1207 \text{ kip-in.}$$

By increasing the length of the shear tab to 26 in., the additional shear tab thickness to resist the 20% $M_{n,\text{web}}$ is given as

$$t = \frac{1207}{26.0^2 \times (36/4)} = \frac{4 \times 1207}{26.0^2 \times 36} = 0.198 \text{ in.}$$

However an extra thickness of $\frac{3}{16} = 0.188$ in. will be adequate because, by increasing the shear tab length, we have gained a reserve shearing resistance for the connection. Thus the following total shear web thickness is deemed adequate for the combined bending and shear:

For shear only $\frac{7}{16}$ in.   From AISC, "Connections", Table 9-158
For bending only $\frac{3}{16}$ in.

$$t_{\text{total}} = \tfrac{10}{16} \text{ in.} = \tfrac{5}{8} \text{ in.}$$

4.20 SHEAR TAB-TO-BEAM WELDED CONNECTION

The strength of a $\frac{1}{16}$-in. weld is given as

$$\phi(0.707)(\tfrac{1}{16})(0.6)F_{EXX} = 0.75(0.707)(\tfrac{1}{16})(0.6)(70) = 1.392 \text{ kips/in.}$$

The fillet weld to connect the shear tab to a $\frac{5}{8}$-in.-thick beam web must be at least $\frac{1}{4}$ in. thick following the minimum weld sizes of AISC Table J2.4. The limit-state equation will be derived assuming a $\frac{3}{8}$-in. fillet weld with a design strength of 8.352 kips/in. resistance and noting that the internal arm is

$$d - X = 26.0 - X$$

for a fully plasticized tension/compression zone of uniform thickness. The equation of the required length X of the $\frac{3}{8}$-in. fillet weld applied to top and bottom of the shear tab-to-beam web connection,

$$X(8.352)(26.0 - X) = 0.2M_u = 1207 \text{ kip-in.}$$

reduces to

$$X^2 - 26X + 144.5 = 0$$

giving

$$X = 8.0 \text{ in.}$$

Thus provide an 8-in.-long, $\frac{6}{16}$-in. weld top and bottom of the shear tab. See Figure 4.10.

The regulation that requires welding in addition to bolts stems from test results at the University of California at Berkeley and the University of Texas at Austin. During cyclic loading the tests exhibited considerable slippage of bolts in shear tab connections not welded to the beam web.

4.21 SECOND-FLOOR PANEL ZONE

The 1997 UBC, CHAP. 22, DIV. IV, Section 8.3.a, mandates load combinations (3-5) and (3-6) for the panel zone design. Of these, (3-5) defines the bending moment acting on structural element 3, represented by [10], which yields the largest value for the panel zone design shear.

The moment computed by [10] is

$$M_u = 645.4 \text{ kip-ft}$$

coupled with a floor shear component $S = 35.3$ kips. The load combination formulas

$$1.2D + 1.0E + 0.5L + 0.2S \tag{3-5}$$

$$0.9D - (1.0E \text{ or } 1.3W) \tag{3-6}$$

contain E, which is well defined in the 1997 UBC, CHAP. 16, DIV. IV, Section 1630.1.1, under "Earthquake Loads" as

$$E = \rho E_h + E_v \tag{30-1}$$

as the sum of *all* earthquake effects, horizontal and vertical.

Regrettably, Section 2211, rather than keeping the well-organized system of CHAP. 16, adopts AISC "Seismic Provisions for Structural Steel Buildings," June 15, 1992, without incorporating it into the body of the 1997 UBC. In the adoption, E is defined as the "earthquake load (where the horizontal component is derived from base shear Formula $V = C_s W_g$)," taking away the clear-cut mathematical definition of CHAP. 16 and leaving the engineer in the uncertainty as to whether to take the vertical component into consideration or not.

Formulas (3-5) and (3-6) are just a variation of the 1997 UBC, Section 1612.2.2.1, formulas (12-5) and (12-6), where E is clearly defined under the section "Notations" as the sum of all earthquake effects in the quoted Section 1630.

The C_s does not appear in the UBC notation or its formulas. It comes from the transcribed notations of the AISC "Seismic Provisions" without being adjusted and properly incorporated into the framework of the Code.

In general, the local building code [NBC by BOCA, SBC by Southern Building Code Congress International (SBCCI), UBC by ICBO] takes precedence over a national standard (ACI, AISC, NDS) and often modifies its recommendations. In this text the 1997 UBC well-defined E will be used, including E_v, the vertical effect of earthquake.

It would be unwise to ignore the physics of earthquakes and the strong message sent by the San Fernando earthquake of nearly 100% g vertical component and the Northridge earthquake with nearly 200% g measured at specific locations. Still vivid in the author's memory are the news of the nearly 100% g vertical acceleration readings by strong-motion seismograph at the Pacoima Dam during the 1971 San Fernando earthquake.

The panel zone shear is expressed by the difference of two vector components:

1. the reaction of the beam flanges H that can be conservatively approximated by $M_u/(d - t_f)$ ignoring any contribution of the beam web,

$$H = \frac{M_u}{d - t_f} = \frac{645.4 \times 12}{(36.0 - 1.0)} = 221.3 \text{ kips}$$

less

2. the floor shear of 35.3 kips transferred by the column above the joint.

The resultant shear that intends to shear across the column web and flanges, represented by the second-order term of UBC formula (8-1),

$$V_u = 221.3 - 35.3 = 186.0 \text{ kips}$$

must be less than or equal to the panel zone shear strength given by UBC (8-1),

$$\phi_v V_n = 0.6\phi_v f_y d_c t_p \left[1 + \frac{3 b_{cf} t_{cf}^2}{d_b d_c t_p} \right] = 0.6(0.75)(36)(16.4)(1.175)$$

$$\times \left[1 + \frac{3 \times 16 \times 1.89^2}{36 \times 16.4 \times 1.175} \right] = 389 \text{ kips}$$

As the computed panel zone shear is less than the shear strength, 186 k < 389 k, the panel zone meets the 1997 UBC strength requirements. The UBC symbols for formula (8-1) are

t_p = total thickness of panel zone including doubler plate(s)

d_c = overall column depth

d_b = overall beam depth

t_{cf} = column flange thickness

F_y = specified yield strength of panel zone steel

In addition, the 1997 UBC 8.3.b, (8-2), requires a minimum panel zone thickness

$$t_z \geq \frac{d_z + w_z}{90} = \frac{34 + 12.6}{90} = 0.518 \text{ in} \tag{8-2}$$

As the column web thickness t_{wc} constitutes the panel zone thickness (no doubler plates provided) and

$$t_{wc} > t_z$$
$$= 1.175 \text{ in.} > 0.518 \text{ in.}$$

this provision is also met.

For a better load transfer of the rather concentrated beam flange reactions impacting the column flanges during cyclic sway of the frame, we will provide horizontal continuity plates between the column flanges, top and bottom of the panel zone, in line with the centerline of beam flanges. It is good practice to match the size of the continuity plates with the size of beam flanges rather than relying on column flange resistance to compensate for undersized continuity plates. Earthquake damage survey recorded by the author involved badly damaged column flanges torn from the column web. It was a clear manifestation that the column flange-to-web joint, in general, is not strong enough to resist the large impacting forces of a wildly swaying joint. This action delivers an impact reaction of the beam flange during dynamic, earthquake-generated cyclic loading, not quasi-static laboratory cyclic loading.

To disperse the concentrated effect of horizontal beam-end reactions, we will provide 5-in.-wide, 1-in.-thick continuity plates closely matching the beam flange sizes. Figure 4.10 shows the horizontal continuity plates snugly fitted and welded to both column flanges and column web.

Observation of the impact of the Northridge earthquake on SMRF structures revealed surprisingly extensive panel zone damage. In a significant number of cases tearing of the column flange occurred, especially where continuity plates had not been provided.

It is common sense to spread the rather large and concentrated beam flange reaction load among the individual column components: the inner and outer column flanges and the column web. It is also a fact that the dynamic impact of the beam flange reaction caused by the cyclic sway of the frame at story levels can be significantly larger than predicted by the static force procedure (Section 1630.2). A critical view and analysis of this problem are presented in Chapters 1 and 12 of this book.

4.22 THIRD-FLOOR BEAM

The flexural strength requirements are dictated by 1997 UBC, CHAP. 22, load combinations (3-5) and (3-6), reflected in our computer load combinations [10] [11]. Member 5, which is the largest design moment, derives from load combination (3-5) [10],

$$M_u = 313 \text{ kip-ft}$$

As

$$L_p = 7.8 \text{ ft} < L_b = 9.0 \text{ ft} < L_r = 22.4 \text{ ft}$$

for the W24 × 68 braced at midspan, the full plastic moment capacity can be obtained providing the following is satisfied: The moment capacity comes from AISC LRFD equations (F1-2) and (F1-3),

$$C_b = \frac{12.5M_{max}}{2.5M_{max} + 3M_A + 4M_B + 3M_C}$$

$$= \frac{12.5 \times 313}{2.5 \times 318 + 3 \times 173 + 4 \times 28 + 3 \times 117} = 2.23 \quad \text{(F1-3)}$$

$$\phi_b M_n = \phi_b C_b[M_p - BF(L_b - L_p)]$$

$$= 0.9 \times 2.23[478 - 12.1(9.0 - 7.5)] = 923 \text{ kip-ft}$$

but must be smaller than or equal to

$$\phi_b M_p = 478 \text{ kip-ft}$$

The flexural capacity of the beam is

$$\phi_b M_p = 478 \text{ kip-ft} > 313 \text{ kip-ft applied}$$

Note. Beam size has been increased from W21 × 68 to W24 × 68 to meet the 1997 UBC story-drift criteria (Sections 1630.9 and 1630.10). When the *drift* criteria overrule the *strength* criteria, the moment capacity of the member normally exceeds the demand. This is clearly illustrated in the preceding example.

4.23   THIRD-FLOOR SHEAR TAB CONNECTION

Design for vertical shear must satisfy requirements 8.2.a and 8.2.b of UBC 1997, CHAP. 22, DIV. IV.

Design Shear Tab for Vertical-Shear V_n Only

The 1997 UBC, CHAP. 22, DIV. IV, Section 8.2.a, gives three options for determining the vertical shear at the beam–column joint:

(a) Design for the plastic bending moment capacity M_p applied at both ends.

(b) Design for "the moment resulting from the panel zone nominal shear strength V_n as determined using Equation 8-1."

The result of these choices must be superimposed to the shear caused by 1.2D *+ 0.5L + 0.2S load combination.*

(c) Provide design strength to balance the reactions resulting from 1997 UBC load combinations (3-7) and (3-8), which represent the amplified frame deformation per UBC Section 1630.9 and 1630.10.

The design of the joint will be determined by the lesser of the above values. Of these choices we have

(a) $M_u = \dfrac{\phi_b M_p}{\phi_b} = \dfrac{478}{0.9} = 531$ k-ft   (LRFD Selection Table 4-18)

(b) $V_n = 0.6F_y d_c t_p \left[1 + \dfrac{3b_{cf}t_{cf}^2}{d_b d_c t_p} \right]$

$\quad = 0.6 \times 36 \times 16.38 \times 1.175 \left[1 + \dfrac{3 \times 16.0 \times 1.89^2}{23.73 \times 16.38 \times 1.175} \right]$

$\quad = 572$ kips

By using internal lever arm $= 0.95d_b$ following the AISC recommendation (Commentary, p. 6-235),

$$M_u = \frac{572 \times 0.95(23.73)}{12} = 1075 \text{ k-ft}$$

(c) $M_{u5} = 932$ kip-ft and $M_{u6} = 1004$ k-ft, where subscripts refer to the joint number in member 5.

Item (a) represents the lesser of the choices allowed by 1997 UBC, Section 8.2.a, and will be used for the determination of the design shear. The $2M_u/L$ component is given as

$$\frac{2M_u}{L} = \frac{2 \times 531}{18.7} = 56.8 \text{ kips}$$

The gravity load component of the shear is given as

$$V = (1.2D + 0.5L + 0.0S)\frac{L}{2} = [1.2(0.6) + 0.5(0.4)] \times \frac{18.7}{2} = 8.6 \text{ kips}$$

$$V_u = 8.6 + 56.8 = 65 \text{ kips}$$

From the AISC LRFD, Volume II, *Connections,* Table 9-153, we select 5 − $\frac{3}{4}\phi$ 325N standard bolts with $4 \times 15 \times \frac{1}{4}$-in. nominal single plate with $\frac{3}{16}$-in. fillet weld (standard holes, "Rigid Support Condition"). However, we must add a $\frac{1}{16}$-in. weld to each side of the plate to counteract the 20% unbalanced moment of the web flexural resistance. The total theoretical weld size is

$$t = \tfrac{3}{16} + \tfrac{1}{16} = \tfrac{1}{4} \text{ in.}$$

Provide a $\frac{1}{4}$-in. fillet weld on each side of the shear tab. The reader will note that the $\frac{1}{4}$ in. *nominal* shear tab thickness was also increased to $\frac{1}{2}$ in. The formal proof of the adequacy of the shear tab to resist both the vertical and $0.2M_{p,\text{web}}$ forces will be left to the designer.

Experience from recent California earthquakes has verified that earthquakes can cause larger than predicted forces on connections and that shear tabs are quite vulnerable in an earthquake. In the course of inspections of moment frame connections, the author observed numerous shear tabs that cracked and split in the Northridge earthquake. Figure 4.11 shows the complete connection detail.

Design Shear Tab Connection for Bending

$$0.7M_p = \frac{0.7(\phi_b M_p)}{\phi_b} = 372 \text{ kip-ft} \qquad \text{for W24} \times 68 \text{ beam}$$

$$b_f t_f (d - t_f) F_y = 8.965 \times 0.585(23.73 - 0.585) \times \tfrac{36}{12} = 364 \text{ kip-ft}$$

Therefore the beam web must be welded to the upper and lower ends of the shear tabs to provide at least 20% of the web nominal flexural strength. The 20% of M_p of the beam web is given as

$$M_{p,\text{web}} = 0.415 \times (21.0)^2 \left(\tfrac{36}{4}\right) = 1647 \text{ kip-in.}$$

$$0.2M_p = 0.2(1647) = 329 \text{ kip-in.}$$

Assuming a shear tab length of 18 in. and a $\frac{1}{4}$-in. fillet weld for the moment connection, the required fillet weld length X will be found by using a basic limit-state plastic moment of resistance equation for identical rectangular tension–compression zones having a 5.7276-kip/in. weld capacity:

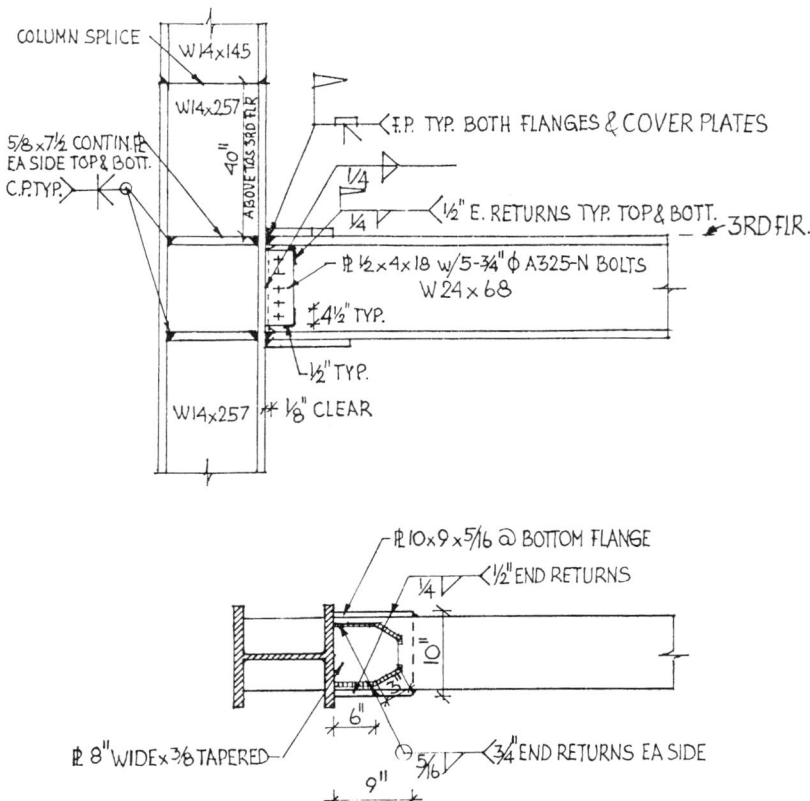

Figure 4.11   *Top:* Third-floor beam-to-column connection. *Bottom:* Plan view of cover plates.

$$X(5.7276)(18 - X) = 329 \text{ kip-in.}$$

$$5.7276X^2 - 103.10X + 329 = 0$$

This reduces to

$$X^2 - 18X + 57.5 = 0$$

resulting in

$$X = 4.15\text{-in. weld length} \qquad \text{say } 4.5 \text{ in.}$$

The fillet weld provided for pure vertical shear for the shear tab-to-column connection would also be increased by $\frac{1}{16}$ in. on each side.

Note. The $\frac{1}{16}$-in. weld must be added to the minimum fillet weld size derived previously under "Design Shear Tab for Vertical Shear V_n Only."

4.24   THIRD-FLOOR BEAM-TO-COLUMN MOMENT CONNECTION

The reader is reminded that designing the beam and designing the beam-to-column moment connections are two distinctively different issues that need a different design methodology. For moment connections, only the beam flanges, *not the entire section,* are welded to the column.

CHAPTER 22, DIV. IV, Section 8.2.a, of the 1997 UBC requires that the moment connection—*the beam flange–column connection*—must have a flexural strength that is either

(a) M_p, the unreduced nominal strength of the beam, or

(b) the moment resulting from the panel zone nominal shear strength V_n.

That is:

(a) $M_p = \dfrac{\phi_b M_p}{\phi_b} = \dfrac{478.0}{0.9} = 531.0 \text{ kip-ft} = 6374 \text{ kip-in.}$

(b) $\phi_v V_n = (0.6)(0.75)(36)(16.38)(1.175)\left(1 + \dfrac{3 \times 16.0 \times 1.89^2}{23.73 \times 16.38 \times 1.175}\right)$

$= 429 \text{ kips}$

$M_u = V_n(d - t_f) = \left(\dfrac{429}{0.75}\right)(23.73 - 0.585)$

$= 13{,}239 \text{ kip-in.} > 6374 \text{ kip-in.}$

Therefore the flexural resistance of the beam-to-column connection must be at least 6374 kip-in.

The flexural resistance of the full-penetration butt welds of the beam flanges to the column can be expressed as

$M_f = 0.9 \times 2 \times 36(8.96 \times 0.585)\dfrac{(23.73 - 0.585)}{2} = 3931 \text{ kip-in.}$

Consequently the flange-to-column connection must be reinforced. One solution is to add top and bottom cover plate to the flanges.

The moment to be resisted by the added cover plates is given as

$$M_u = M_p - M_f = 6374 - 3931 = 2443 \text{ kip-in.}$$

Assuming a lever arm of

$$L_a = 24.0 \text{ in.} \qquad \text{for added cover plates}$$

and

$$b = 8.0 \text{ in. for width of top plate}$$

the required top-cover plate thickness is

$$\phi t b U F_y L_a = M_p - M_f$$

$$t = \frac{M_p - M_f}{\phi F_u b U L_a} = \frac{2443}{0.9 \times 0.75 \times 58 \times 8 \times 24} = 0.325 \text{ in.}$$

or

$$t = \frac{M_p - M_f}{\phi F_y b L_a} = \frac{2443}{0.9 \times 36 \times 8 \times 24} = 0.393 \text{ in.} \qquad \textbf{GOVERNS}$$

say a $\frac{3}{8}$-in.-thick plate.

Use an 8-in.-wide, $\frac{3}{8}$-in. plate of length and taper similar to the cover plate used for the second-floor connection. The required length of a $\frac{5}{16}$-in. fillet weld to attach the cover plate to the beam flange on both sides of the cover plate is

$$L = 0.9 \times 36 \times 8 \times \frac{0.393}{2 \times 6.9} = 7.38 \text{ in.}$$

However, use a weld length and configuration similar to the one used for the cover plate of the second-floor connection. Use two $\frac{5}{16}$-in. fillet welds with $\frac{3}{4}$-in.-long end returns on both sides of the $8.0 \times \frac{3}{8}$ tapered plate. Similarly a 10.0-in.-wide, $\times$ 9.0-in.-long, $\frac{5}{16}$-in. bottom plate is welded to the bottom flange with a 9.0-in.-long, $\frac{1}{4}$-in. fillet weld applied to both sides of the plate. Figure 4.11 shows the connection detail.

4.25 THIRD-FLOOR PANEL ZONE

The design shear is derived from load combinations (3-5) and (3-6) per the 1997 UBC, CHAP. 22, DIV. IV, Section 8.3.a. Of these load combinations, (3-5), represented by [10], gives higher design load at joint 6, member 5:

$$M_u = 313 \text{ kip-ft} = 3756 \text{ kip-in.}$$

The horizontal flange reaction gives

$$\frac{M_u}{0.95d} = \frac{3756}{0.95 \times 23.73} = 167 \text{ kips}$$

less the floor shear component carried by the column above,

$$S = 27 \text{ kips}$$

The design panel shear acting on the panel zone is given as

$$H - S = 167 - 27 = 140 \text{ kips} < \phi_v V_n$$
$$= 0.75(572) = 429 \text{ kips}$$

In addition, the panel zone thickness t_p must be equal to or larger than

$$t_z = \frac{d_z + w_z}{90} = \frac{[(23.73) - (2 \times 0.625)] + [(16.38) - (2 \times 1.89)]}{90}$$
$$= 0.39 \text{ in.} < 0.415 \text{ in.}$$

Therefore 1997 UBC, CHAP. 22, DIV. IV, Section 8.3.b, is also met and the panel zone meets the 1997 UBC requirements.

4.26 DESIGN OF COLUMNS

We now analyze the steps involved in our design of columns—and beam–columns connections—subjected to bending and axial compression. The design typically involves:

(a) Designing the column for moments obtained by the moment-resisting frame analysis
(b) Designing the column for the axial force component of the frame analysis

We will use the basic equations (H1-1a) and (H1-1b) of the AISC LRFD Specification (the *Specification*), Volume I, Chap. H, for the analysis:

$$\frac{P_u}{\phi_c P_n} + \frac{8}{9}\left(\frac{M_{ux}}{\phi_b M_{nx}} + \frac{M_{uy}}{\phi_b M_{ny}}\right) \leq 1.0 \qquad \text{(H1-1a)}$$

$$\frac{P_u}{2\phi_c P_n} + \left(\frac{M_{ux}}{\phi_b M_{nx}} + \frac{M_{uy}}{\phi_b M_{ny}}\right) \leq 1.0 \qquad \text{(H1-1b)}$$

The entire procedure is summarized in the LRFD column design flowchart (see Figure 4.13). The formulas involve complex issues that will be discussed next.

Issue of Bending

The external moments M_{nt} and M_{lt} have to be magnified due to the presence of axial force according to Chap. C of the *Specification*. Moreover, following Chap. F of the *Specification,* the moment of resistance of the column has to be treated as the flexural resistance of a beam subject to flexural buckling.

Issue of Axial Force

The axial resistance of the column has to be evaluated for the worst slenderness ratio, that is, taking into account the effective length and radius of gyration with respect to major and minor axes. The 1997 UBC, CHAP. 22, DIV. IV, "Detailed Systems Design Requirements," must be followed for the internal moments and forces.

Dr. Gabor Kasinczy,[5] a Hungarian engineer, introduced the basic principle of the plastic behavior of structural steel as early as 1914. The Kasinczy plastic theory was based on the concept that, if a steel element is bent under extreme loading—today named *limit load*—it will undergo plastic deformation and yield. In yielding, the steel element would maintain a constant plastic moment of resistance until point B was reached on the stress–strain diagram (Figure 4.12). Beyond this point an increased moment of resistance would occur as the metal enters the strain-hardening stage.

Dr. Kasinczy had thus found an ideal material that would remain friendly when stretched to its yield limit and would give enough warning by large

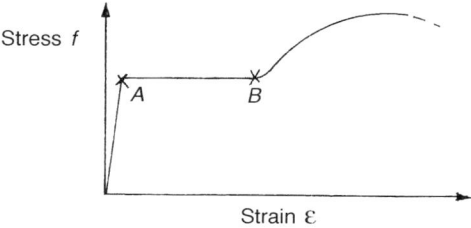

Figure 4.12   Idealized stress–strain diagram.

permanent deformations rather than undergo sudden catastrophic failure. The years that followed proved that the entire issue is far more complex and that a beam stressed to its limit may evoke other less friendly mechanisms. What are these mechanisms? We will look at the behavioral pattern of a *compact* steel beam (*Specification* definition), a W or box section, when the laterally unsupported span of the beam is increased.

Assume a simply supported beam as an example. Hold the ends firmly against torsional rotation but leave the entire span L laterally unsupported. Apply a large vertical external load that will cause the extreme fibers of the metal to reach the extreme of its elastic limit (point A, Figure 4.12) at a specific critical section. If the unbraced (laterally unsupported) length L is short, the overstressed section will develop a full plastic moment of resistance to counter the overload and restore equilibrium. In fulfilling its *fail-safe* function, the beam will not collapse; however, it will display a permanent, unsightly deformation telling us that it has been subjected to an abnormal load not meant for the normal use of the structure.

The AISC *Specification* fully recognizes the importance of this well-defined unbraced length, calling it L_p (the subscript stands for plasticity). Stated otherwise, if the unbraced length—the distance of lateral supports to the compression flange—stays within its limit L_p, the *Specification* allows use of a full plastic moment of resistance M_p.

Now gradually increase the unbraced length. As we exceed L_p, we cannot expect a full plastic moment of resistance at the most critically stressed and strained crosssection but rather the in-between response of a partially plasticized section. The moment of resistance M_r belongs to a region that corresponds to an unbraced—and laterally unsupported—length L_r (Specification) that can only guarantee a partially plasticized moment of resistance.

If we increase the unbraced length of the beam beyond L_r, the mechanism of structural response will alter dramatically. The compression flange will act like a column with a tendency to buckle when its slenderness ratio reaches a critical value. Naturally, the phenomenon is complex, involving a number of factors that include the interaction of the web and tensile chord of the beam. When the compression flange reaches such a critical state, it will buckle rather than promoting development of a plastic moment of resistance at the critically bent section. Of course, because the tensile flange is not quite ready to buckle with the compression flange, buckling will appear as a torsional type of failure.

We then consider zones that represent distinctive response patterns corresponding to short, medium, and long unbraced lengths. As stated before, the boundaries of these zones are defined by L_p, L_r brace lengths, which are well defined in the LRFD beam design tables, Part 3 of the *Specification*. Should the actual unbraced length L_b fall between L_p and L_r, a transition stage is reached and the resistance of the section needs to be modified between the values M_r (corresponding to the L_r unsupported length) and M_p (full plastic response). The LRFD equation (F1-2) of the *Specification* allows this transition for an intermediate value for $\phi_b M_n$ (see also p. 4-11, Part 4):

$$\phi_b M_n = C_b [\phi M_p - BF(L_b - L_p)] \qquad \text{(F1-2)}$$

where L_b is the actual unbraced length

$$BF = \frac{\phi_b(M_p - M_r)}{L_r - L_p} \qquad \textit{Specification } 4\text{-}11$$

$$C_b = \frac{12.5 M_{max}}{2.5 M_{max} + 3 M_A + 4 M_B + 3 M_C} \qquad \text{LRFD equation (F1-3)}$$

with M_A, M_B, M_C the absolute values of moments at the $\frac{1}{4}$, $\frac{1}{2}$, and $\frac{3}{4}$ points of the unbraced beam segment, respectively.

In the LRFD equation (F1-3), equation C_b is a modification factor or modifier that corresponds to the effective, rather than nominal, unbraced length, and its value is inversely proportional to the effective unbraced length. That is, the shorter the effective unbraced length, the larger the C_b moment of resistance multiplier. For the function of C_b, assume a simply supported beam with end moments that cause a single curvature deformation. It is understood that the ends of the beam are firmly supported in the lateral direction. Obviously the unsupported length is L. On the other hand, a beam with end moments that will cause a double curvature corresponds to an effective unbraced length smaller than L.

While C_b is unity for the first case, it is larger than 1 for the second case, reflecting allowance for a moment of resistance larger than allowed for case 1. Naturally, the limit is $\phi_b M \leq \phi_b F_y Z$. In other words, C_b is capable of adjusting design parameters to the effective, rather than nominal, unbraced length.

Having evaluated the above bending moment terms, the engineer is almost ready for the LRFD interaction equations (H1-1a) and (H1-1b):

$$\frac{P_u}{\phi_c P_n} + \frac{8}{9}\left(\frac{M_{ux}}{\phi_b M_{nx}} + \frac{M_{uy}}{\phi_b M_{ny}}\right) \leq 1.0 \qquad \text{(H1-1a)}$$

$$\frac{P_u}{2\phi_c P_n} + \left(\frac{M_{ux}}{\phi_b M_{nx}} + \frac{M_{uy}}{\phi_b M_{ny}}\right) \leq 1.0 \qquad \text{(H1-1b)}$$

As stated earlier, to account for the $P\Delta$ effect, the applied moments M_{nt}, M_{lt} must be augmented to

$$M_u = B_1 M_{nt} + B_2 M_{lt} \qquad \text{(C1-1)}$$

where $B_1 = C_m/(1 - P_u/P_{e1})$
$\quad M_{nt}$ = required flexural strength in member, assuming no lateral translation of frame

B_2 = moment magnification factor per (C1-4) or (C1-5) of AISC LRFD *Manual*

M_{lt} = required flexural strength in member as result of lateral translation of frame only

P_{e1} = Euler buckling strength evaluated in plane of bending

P_u = required axial compressive strength for member

C_m = $0.6 - 0.4M_1/M_2$

M_1/M_2 = ratio of smaller to larger end moment of member unbraced in plane of bending, positive when member bends in reverse curvature

Alternatively, a nonlinear computer program that provides $P\Delta$ analysis can be used.

Axial Force Component

The second major term in interaction equation (H1-1a) of the *Specification* is the axial force component $P_u/\phi P_n$. We enter into the numerator the axial force augmented by the addition of E_v, the vertical design earthquake component per CHAP. 16, formula (30-1), and CHAP. 22, DIV. IV, formula (6-1) of the 1997 UBC.

Remember that an LRFD seismic design must contain the axial force components caused by both E_h and E_v superimposed to the axial loads corresponding to $1.2D + 0.5L$. In the denominator of (H1-1a) or (H1-1b) equations, $\phi_c P_n$ can be obtained directly from the column design tables, Part 3, of the *Specification* using the smaller of the values corresponding to the most critical slenderness ratio.

When the terms of the axial force components are defined, we are ready to apply the LRFD interaction equation (H1-1a) or (H1-1b):

$$\frac{P_u}{\phi_c P_n} + \frac{8}{9}\left(\frac{M_{ux}}{\phi_b M_{nx}} + \frac{M_{uy}}{\phi_b M_{ny}}\right) \leq 1.0 \qquad \text{(H1-1a)}$$

Code Requirements

To carry out the column design, we must adhere to 1997 UBC, CHAP. 22, DIV. IV, Sections 2211.4.8.6 and 2211.4.8.7.a, as follows:

1. The column–beam moment ratio by formulas (8-3) and (8-4) to ensure that the SEAOC fundamental *strong column–weak beam* principle is observed
2. Whether lateral support to column flanges require lateral support only at the level of the top flanges of the frame beam or at the level of both top and bottom beam flanges [formula (8-3) or (8-4) must yield a ratio ≥ 1.25]

Column–Beam Moment Ratio (UBC 2211.4.8.6)

The following will illustrate the importance of this 1997 UBC provision and the consequences if the structure was designed based on compliance with the strength and drift provisions only, ignoring the *strong column–weak beam* issue.

Assume that both strength and drift criteria can be met by substituting a weaker column instead of the W14 × 257 indicated for the second story. Assume that instead of splicing the column above the third floor, as indicated in the finalized design, we will splice above the second floor and have a W14 × 125 column for the second story. The following data were obtained from the AISC *Manual* and previous work:

$$
\begin{aligned}
Z_c &= 260 \text{ in.}^3 &&\text{Plastic modulus of W14} \times 145 \text{ column above}\\
Z_c &= 487 \text{ in.}^3 &&\text{Plastic modulus of W14} \times 257 \text{ column below}\\
Z_{b3} &= 177 \text{ in.}^3 &&\text{Plastic modulus of W24} \times 68 \text{ beam}\\
Z_{b2} &= 624 \text{ in.}^3 &&\text{Plastic modulus of W36} \times 160 \text{ beam}\\
A_g &= 42.7 \text{ in.}^2 &&\text{Gross area of W14} \times 145 \text{ column}\\
A_g &= 75.6 \text{ in.}^2 &&\text{Gross area of W14} \times 257 \text{ column}\\
F_{yc} &= 36.0 \text{ ksi} &&\text{Specified minimum yield strength of column}\\
F_{yb} &= 36.0 \text{ ksi} &&\text{Specified minimum yield strength of beam}\\
P_{uc} &= 255 \text{ kips} &&\text{Required axial column strength, second story}\\
P_{uc} &= 485 \text{ kips} &&\text{Required axial column strength, first story}\\
H &= 162 \text{ in.} &&\text{Average of floor heights above and below joint}
\end{aligned}
$$

$$
\frac{\sum Z_c(F_{yc} - P_{uc}/A_g)}{\sum Z_b F_{yb}} = 0.768 \le 1.0 \qquad \text{does not comply with (8.3)}
$$

as

$$
\sum Z_c\left(F_{yc} - \frac{P_{uc}}{A_g}\right) = 22{,}157 \text{ kip-in.}
$$

and

$$
\sum Z_b F_{yb} = 28{,}836 \text{ kip-in.}
$$

An increased column size is needed to comply with 2211.4, Section 8.6.

4.27   COLUMN FINAL DESIGN DATA

$$Z_c = 487 \text{ in.}^3 \qquad \text{Plastic modulus of W14} \times 257 \text{ column above}$$
$$Z_c = 487 \text{ in.}^3 \qquad \text{Plastic modulus of W14} \times 257 \text{ column below}$$
$$Z_{b3} = 177 \text{ in.}^3 \qquad \text{Plastic modulus of W24} \times 68 \text{ beam}$$
$$Z_b = 624 \text{ in.}^3 \qquad \text{Plastic modulus of W36} \times 160 \text{ beam}$$
$$A_g = 75.6 \text{ in.}^2 \qquad \text{Gross area of W14} \times 257 \text{ column}$$
$$F_{yc} = 36.0 \text{ ksi} \qquad \text{Specified minimum yield strength of column}$$
$$F_{yb} = 36.0 \text{ ksi} \qquad \text{Specified minimum yield strength of beam}$$
$$P_{uc} = 255 \text{ kips} \qquad \text{Required axial column strength, second-story column}$$
$$P_{uc} = 494 \text{ kips} \qquad \text{Required axial column strength, first-story column}$$
$$H = 162 \text{ in.} \qquad \text{Average of floor height above and below joint}$$

$$\frac{\Sigma \, Z_c(F_{yc} - P_{uc}/A_g)}{\Sigma \, Z_b F_{yb}} = 487 \, \frac{(36 - 494/75.6) + (36 - 255/75.6)}{36 \times 801}$$

$$= 1.05 \geq 1.0$$

We have satisfied the 1997 UBC formula (8-3) requirements.

With regard to Section 2211.4.8.7.a, "Beam-to-Column Connection Restraint," indicating that both column flanges "shall be laterally supported at the levels of both top and bottom beam flanges," because the derived ratio of (8.3) was not greater than 1.25 (8.7.a.1.a), it cannot be proved that the "column remains elastic when loaded with Load Combination 3-7."

The reader will note, without formal proof, that, for the upper floor connections, the "ratios calculated using Equations 8-3 or 8-4 are greater than 1.25"; therefore the requirement of "column flanges at a beam-to-column connection require lateral support only at the level of top flanges of the beams" is also satisfied.

The gravity load-carrying system of beams welded to the beam–column joint orthogonal to the plane of the moment frame will be used as lateral support at TOS (top of steel). Providing such practical lateral column support might be an advantage over providing top and bottom lateral flange support, which could be more costly and time consuming.

Having complied with 1997 UBC, Sections 2211.4.8.6 and 2211.4.8.7, beam–column proportioning criteria and previously with formula (30-3) of Section 1630.1.1 reliability/redundancy—the ρ criteria—a final interstory-drift analysis is made by computer load combination [5] to limit the "Maximum Inelastic Response Displacement" of the floor, UBC Sections 1630.9 and 1630.10. Table 4.2 shows the results. All story drifts are less than the 0.025 times the story height requirement of UBC 1997 for "structures having a fundamental period less than 0.7 seconds," which is the case (1630.10.2).

TABLE 4.2 Maximum Inelastic Response Displacement of Frame and Allowable Story Drift: Six-Frame Structural System, UBC 1997 Design

Story	Maximum Inelastic Response Displacement, $\Delta_M = 0.7R\,\Delta_s$ in. (30-17), Computer Analysis Load Combination [7]	Allowable Story Drift (1630.10.2), $0.025 \times$ Story Height (in.), $T < 0.7$ s
1	$\Delta_M = 3.72$	$\Delta_{ALLOWED} = 0.025 \times 162 = 4.05$
2	$\Delta_M = 2.94$	$\Delta_{ALLOWED} = 4.05$
3	$\Delta_M = 3.97$	$\Delta_{ALLOWED} = 4.05$
4	$\Delta_M = 3.29$	$\Delta_{ALLOWED} = 4.05$

A summary of the process is as follows:

1. Comply with the 1997 UBC formula (30-3) reliability/redundancy criteria.
2. Comply with the 1997 UBC Sections 2211.4.8.6 and 2211.4.8.7 beam–column proportioning criteria.
3. Comply with the 1997 UBC Sections 1630.9 and 1630.10 story-drift limitation criteria.

Having satisfied the above requirements, we were ready for the final computer run. Figure 4.8 has shown us the member sizes established by the process and used for the final analysis.

4.28 FIRST-STORY COLUMN DESIGN FOR COMPRESSION: MAJOR AXIS

We will use the 1997 UBC load combination (3-7), CHAP. 22, DIV. IV:

$$1.2D + 0.5L + 0.2S + 0.4R \times E = 1.2D + 0.5L + 0 + 3.4(E_v + \rho E_h) \tag{3-7}$$

as

$$E_v = \pm\, 0.5C_aID = (0.5)(0.44)(1.0)D = \pm\, 0.22D$$

After rearranging terms, the expression becomes

$$1.2D + 3.4(0.22D) + 0.5L + 3.4E_h = 1.948D + 0.5L + 3.4E_h$$

and likewise (3-8) can be expressed as

$$0.9D - 3.4(0.22D) + 3.4E_h = 0.152D + 3.4E_h \qquad \text{for uplift}$$

The above expressions are represented by computer load combinations [12] and [13] in Computer Analysis, Part I.

As mentioned earlier, usage of (3-7) has certain advantages in simplifying engineering design procedures; additional specific requirements need not be applied and may be waived (Sections 8.2.a and 8.2.b) if this conservative-enough load combination is used. The maximum axial design force is

$$P_u = 494 \text{ kips}$$

coupled with $M_u = 1292$ kip-ft bending moment for column 2 from computer load combination [12].

Note. The UBC column strength formula (6-1) gives identical results because (6-1) may be considered as a derivative of (3-7) for maximum axial force design.

The moment component that pertains to no lateral translation is

$$M_{nt} = 12 \text{ kip-ft}$$

caused by dead load plus live load and

$$M_{lt} = 1280 \text{ kip-ft}$$

the required flexural strength component that corresponds to the lateral translation of the frame. However, instead of the theoretical values, only 85% of these moments are effective on the column, following the moment gradient drop at the bottom of the beam flange where the clear column span begins,

$$M = M_{nt} + M_{lt} = 10 + 1097 = 1107 \text{ kip-ft moment} \qquad [12]$$

that must be modified for the $P\Delta$ effect as required by Equation (C1-1) of the *Specification:*

$$M_u = B_1 M_{nt} + B_2 M_{lt} \qquad \text{(C1-1)}$$

Evaluation of the B_1 and B_2 coefficients yields

$$B_1 = \frac{C_m}{1 - P_u/P_{e1}} \geq 1.0 \qquad \text{(C1-2)}$$

where

$$C_m = 0.6 - 0.4\left(\frac{M_1}{M_2}\right) = 0.6 - 0.4\left(\frac{1.0}{10}\right) = 0.56 \qquad \text{(C1-3)}$$

and

$$P_{e1} \text{ with } \frac{Kl}{r_x} = \frac{1.0(144)}{6.71} = 21.5$$

$$= 620 \times 75.6 = 46{,}872 \text{ kips} \qquad \text{From Table 8, } \textit{Specification}$$

$$B_1 = \frac{0.56}{1 - 494/46{,}872} = 0.567$$

$$B_1 M_{nt} = 0.567 \times 12 = 6.80 \qquad \text{say } 7.00 \text{ kip-ft}$$

and

$$B_2 = \frac{1}{1 - \Sigma P_u \left(\dfrac{\Delta_{oh}}{\Sigma HL}\right)} \qquad \text{(C1-4)}$$

$$B_2 = \frac{1}{1 - 3245\left(\dfrac{3.72}{(0.7R)(392)(162)}\right)}$$

$$B_2 = \frac{1}{1 - 3245\left(\dfrac{3.72}{(5.95)(392)(162)}\right)} = 1.033$$

Note. The story drift and base shear are expressed as $0.7R \times$ actual values. Alternatively, use

$$B_2 = \frac{1}{1 - \Sigma P_u/\Sigma P_{e2}} \qquad \text{(C1-5)}$$

As the *Specification* allows the option of either B_2 value,

$$B_2 M_{lt} = 1.033 \times 1097 = 1133 \text{ kip-ft}$$

$$M_{ux} = B_1 M_{nt} + B_2 M_{lt} = 7 + 1133 = 1140 \text{ kip-ft}$$

that is, the required flexural capacity augmented by the $P\Delta$ effect.

Effective length calculation in the plane of bending, as specifically stated in "C1. Second Order Effects" of the AISC *Specification* (6-41), yields

$$I_{c1} = I_{c2} = 3400 \text{ in.}^4$$

$$I_{B2\text{floor}} = 9700 \text{ in.}^4$$

$$\Sigma \left(\frac{I_c}{L_c} \right) = 2 \times \frac{3400}{144} = 47.2$$

Note that L_c is taken from the base to the beam centerline,

$$\Sigma \left(\frac{I_g}{L_g} \right) = \frac{9700}{226} = 43.2$$

$$G_{2\text{floor}} = \frac{47.2}{43.2} = 1.10$$

From the alignment chart of the AISC *Specification* (6-186),

$$\text{With } G_{\text{base}} = 1.0 \text{ k} = 1.33$$

and the effective column length is

$$\frac{K_x L_x}{r_x / r_y} = \frac{1.33(13.5 - 1.5)}{1.62} = 9.85 \text{ ft}$$

The effective length of the column about its weak axis assuming full lateral support at the centerline of the beam and K conservatively taken as unity disregarding partial fixity at the base is

$$K_y L_y = 1.0(12.0) = 12 \qquad \textbf{GOVERNS}$$

The value of ϕP_n needs to be based upon the lesser capacity in either direction,

$$\phi P_n = 2170 \text{ kips}$$

from the design axial strength table (3-18) of the AISC *Specification*,

$$\frac{P_u}{\phi P_n} = \frac{494}{2170} = 0.231$$

as

$$\frac{P_u}{\phi P_n} > 0.20$$

Equation (H1-1a) is valid and must be used for the interaction formula as

$$L_p = 17.5 \text{ ft} > L_b = 12 \text{ ft}$$

the full plastic moment of resistance $\phi_b M_p = \phi_b M_{nx}$ can be utilized and entered in the denominator of the second term of the expression for the W14 × 257 column:

$$\phi_b M_{nx} = 0.9(Z_x)(F_y) = \frac{0.9(487)(36)}{12} = 1315 \text{ kip-ft}$$

$$\frac{P_u}{\phi P_n} + \frac{8}{9}\left(\frac{M_{ux}}{\phi_b M_{nx}}\right) = \left(\frac{494}{2170}\right) + \frac{8}{9}\left(\frac{1140}{1315}\right) = 0.998 < 1.0$$

Hence the column meets the most severe strength requirements of CHAP. 22, DIV. IV, (6.1).

4.29   COLUMN DESIGN FLOWCHART

The simplified flowchart in Figure 4.13 illustrates the column design procedure. Only the major steps are numbered in sequential order.

4.30   DESIGN OF THIRD-STORY COLUMN FOR COMPRESSION

The principles and methodology demonstrated for the design of the first-story column were applied to the third-story column design. Formal proof has been omitted for brevity.

4.31   DESIGN OF THIRD-STORY COLUMN SPLICE

Note the change in column dimension above the third floor to compensate for reduced forces at the upper levels. Effective connection will be achieved by a pair of splice plates to counter uplift and by partial penetration weld provided to butting column webs to resist moments.

   The tensile resistance of the column does not have to be verified since the column was designed for higher axial force and moment in compression. The column splice, however, must be designed in compliance with CHAP. 22, DIV. IV, 6.1.b, of UBC 1997; the splice will be designed for uplift resulting from load combination (6-2) of CHAP. 22, DIV. IV:

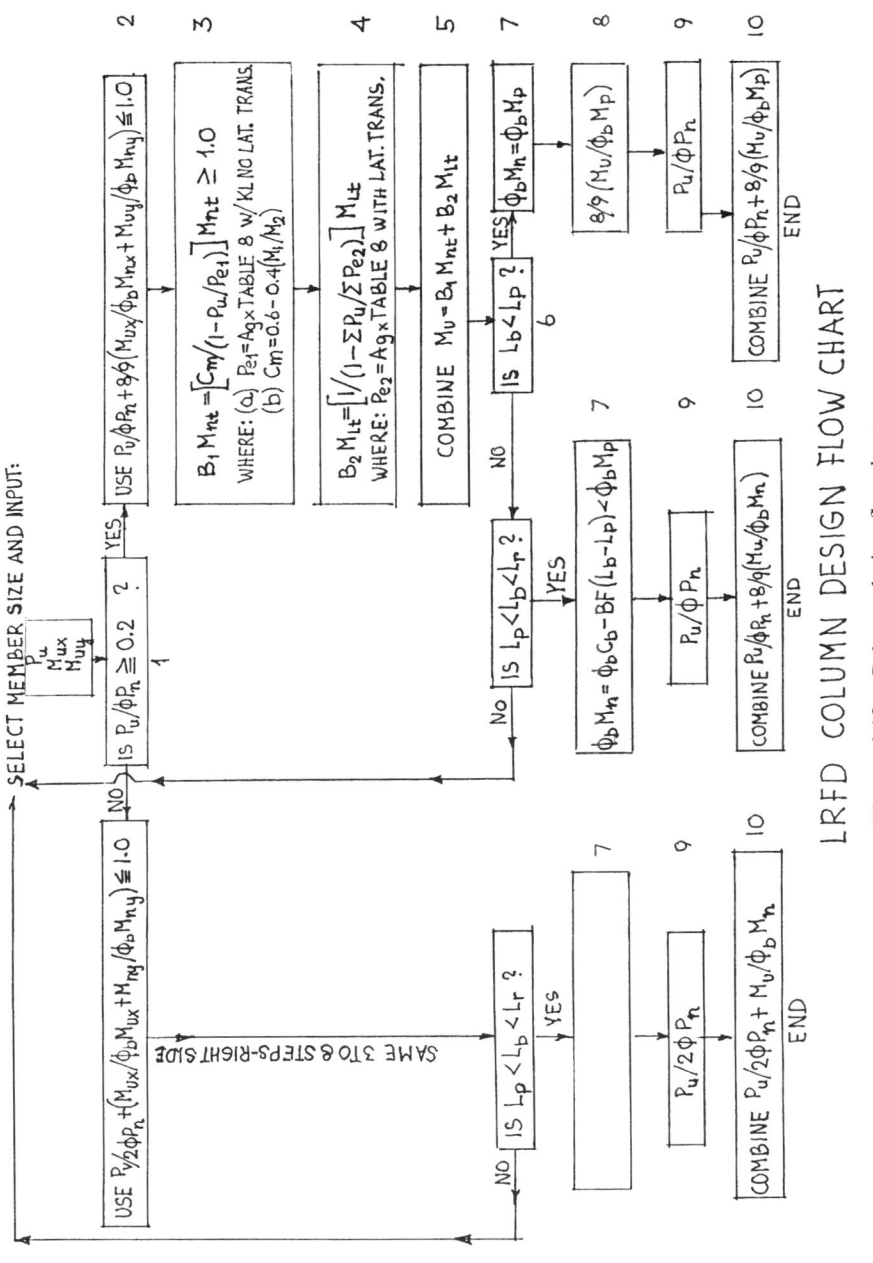

Figure 4.13 Column design flowchart.

$$P_{min} = 0.90P_{DL} - 0.4R \times P_E = \phi_t P_n$$

where the subscript min refers to forces associated with maximum uplift. The provisions call for a design that follows the applicable specific "Detailed Systems Design Requirements" of CHAP. 22, DIV. IV, 6.1 and 6.2, of UBC 1997. The maximum tensile force in the column is determined by UBC formula (6-2) represented by computer load combination [13]; that is,

$$0.9D + (0.4 \times R \times E) = 0.9D + 3.4(E_v + E_h)$$

Because E_v represents a negative acceleration

$$-0.5C_a ID = -0.22D$$

the expression becomes

$$0.9D - 3.4(0.22D) + 3.4E_h = 0.9D - 0.748D + 3.4E_h = 0.152D + 3.4E_h$$

$$P_{min} = 115 \text{ kips}$$

from load combination [13] of the computer analysis.

We will attach splice plates to each side of the column web and design for the uplift axial force. Partial penetration welds applied to the column flanges will be provided for the moment component acting on the splice, 40 in. above the third-floor TOS level. Design of the welded flange connection will be left as an exercise for the reader.

The partial penetration welds of the column flanges must be able to resist 150% of the applied moment component with

$$\phi_w = 0.8$$

and

$$F_w = 0.6F_{EXX} \qquad \text{From CHAP. 22, DIV. IV, 6.2.b.1}$$

We start our splice design by determining dimensions of the splice plate assuming

$$t = \tfrac{3}{16} \text{ in. thick}$$

and using the basic strength relationship

$$\frac{P_u}{2} = \phi_t F_y A_g = \phi_t F_y t\, w = (0.9)(36)(0.1875)w = 6.075w$$

$$w = \frac{115}{2(6.075)} = 9.465 \text{ in.}$$

say 10-in.-wide plates that fit within the column flanges, $T = 11\tfrac{1}{4}$.

The AISC *Specification* requires that the strength of welded connection plates be the lesser of the expressions (Section B.3)

$$P_u = \phi_t F_y A_g$$

where A_g is the gross section area and $\phi_t = 0.9$ and

$$P_u = \phi_t F_u A_e = \phi_t F_u U A_g$$

where A_e = effective area
   U = reduction factor with respect to shear lag, stress concentration, etc., and a function of L/W such as
   When $1.5w > L > w$ $U = 0.75$
   When $2w > L \geq 1.5 \ w$ $U = 0.87$
   When $L > 2w$ $U = 1.0$
   L = weld length in the direction of stress

Assume $U = 0.75$ will be selected for the reduction factor:

$$P_u = 2\phi_t F_u U A_g = 2(0.75)(58)(0.75)(0.1875)(10.0) = 122 \text{ kips} > 115 \text{ kips}$$

The plates have sufficient strength.

The two $\frac{3}{16} \times 10$-in. plates will be welded, one to each side of the column web. To obtain the required plate length, we need to design the weld. By choosing a $\frac{1}{8}$-in. weld size, we will determine the required weld length L and associated plate length. The design strength (across the throat) of a $\frac{1}{8}$-in. fillet weld in shear per 1-in. length is given by

$$0.707 t \phi_w F_w = 0.707 t (\phi_w)(0.6 F_{EXX}) = 0.707(1/8)(0.75)(0.6 \times 70)$$

$$= 2.7838 \text{ kip-in.}$$

where t is weld size. Thus the required weld length (four lengths to match two splice plates) is

$$L = \frac{115}{4 \times 2.7838} = 10.32 \text{ in.} \qquad \text{say 11 in.}$$

Figure 4.14 shows details of the column splice.

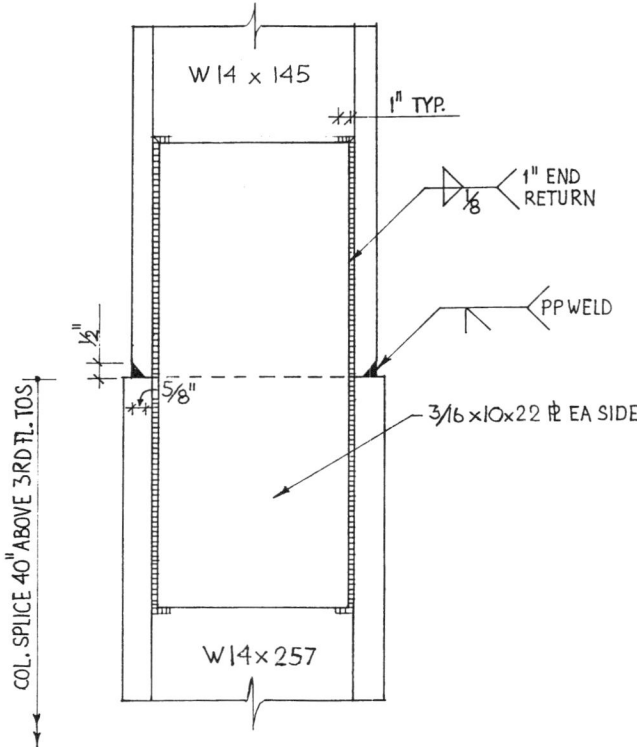

Figure 4.14 Column splice.

4.32 REEXAMINATION OF PRE- AND POST-NORTHRIDGE MOMENT CONNECTION RESEARCH AND LITERATURE

In the "Recommended Lateral Force Requirements and Commentary," pp. 307–308, 1996 Edition, SEAOC states: "During the Northridge earthquake, more than 100 buildings with welded steel moment-resisting frames suffered beam-to-column connection failures. Observation of these buildings indicated that, in many cases, fractures initiated within the connections at very limited levels of inelastic behavior."

Pre-Northridge Research

How did research begin on these bearm-to-column connections? Perhaps the first report of experiments on welded moment connections was issued by Popov and Stephen,[6] followed by Bertero et al.[7] in 1973.

Those reports described full-size joint tests up to W24 × 76 beams welded to wide-flange columns under *quasi-static* cyclic loading. The results of the test looked promising: inelastic joint behavior associated with a plastic joint rotation close to 2 radians, and reassurance of plastic yielding of the panel zone. As discussed in Chapter 12, this zone was considered rigid and unyielding in European literature, which referred to it as a *rigid joint.*

For a time research focused mostly on the beam-column interface, assuming that full plastic yielding of the beam can be achieved at the interface.

Additional reports based on finite-element analysis reassured us that even the panel zone would be capable of full yielding under the shearing effect of beam-flange reactions and would further increase inelastic joint rotation.

Based on such favorable recommendations from the academia the scenario was set for widespread commercial application of the structural system without proof of specific experimental tests or further testing.

Numerous reports followed through the 1980s such as Popov and Tsai[8] reporting results similar to the initial Popov et al. investigations. Occasional brittle fractures mentioned in these reports received cursory treatment and were not considered true representatives of joint behavior but rather freak occurrences due to workmanship flaws, among others. Had these fractures received more attention they could have served as early warning of troubles to come.

Additional research was conducted just before the Northridge earthquake by Engelhardt and Husain[9] on W18 and W24 beams with flanges groove-welded to W12 × 136 column-beam webs bolted to shear tabs. The experiments revealed problems.

All specimens had been produced by competent steel fabricators using certified welders and welds ultrasonically tested by certified inspectors. Surprisingly, some joint assemblies exhibited almost no ductile hysteretic behavior and fractured at the very beginning of cyclic loading. This was reported on Specimen 4 by Engelhardt and Husain in 1993.

Similar disappointing joint failures were reported by Uang[10] in 1995 and Uang and Bondad[11] in 1996 in their post-Northridge experimental tests of the joint.

Post-Northridge Research

A number of solutions to the moment frame connection problem have been proposed, likely to evolve further in coming years.[12–19]

Two key strategies were developed to circumvent problems associated with the *pre-Northridge moment frame connection.*

1. Strengthening the connection by means of cover plates, ribs, haunches, and side plates.
2. Weakening the beam that frames into the connection, that is, the *dog-bone* connection. The aim of this modification was that, by reducing—weakening—the beam flanges, the location of the plastic hinge could

be moved away from the critical beam-column interface.
Two reduced flange shapes were developed under this proposal: linear profile and circular profile.

Among the strengthening alternatives:

(a) Plate rib reinforcement welded to beam flanges.
(b) Haunch reinforcement welded to beam flanges.
(c) Cover plate reinforcement welded to beam flanges.
(d) Side plate reinforcement welded to beam.

Test data on some of the described alternatives revealed a percentage of brittle fractures, such as those reported by Engelhardt and Sabol[19] in 1996.

In July of 2000 FEMA issued the FEMA-353 *Recommended Specifications and Quality Assurance Guidelines for Steel Moment-Frame Construction for Seismic Applications*"[18] that superseded their 1995 FEMA 267.[13] "Many of the specification provisions contained in Part I—explains FEMA—are already included in industry standard specification and building code requirements."

FEMA 2000 *Recommendations* pay particular attention to materials—the quality of steel, as well as welding methods and bolting procedures for seismic applications of steel moment-frame construction.

Welding

In the design examples of our project we used special E7018 or E70TG-K2 flux core electrodes for the beam-to-column moment connections. The AISC "Modification of Existing Welded Steel Moment Frame Connections for Seismic Resistance"[12] mentions in reference to Flange Weld Modifications, that "Past tests on RBS (Reduced Beam Section) connections both for new construction and for modification of existing connections, have generally employed the self shielded flux cored arc welding process (FCAW), using either the E70TG-K2 or E71T-8 electrodes." . . . and "In addition, successful tests on other types of connections have employed the shielded metal arc welding (SMAW) process using an E7018 electrode . . ."

FEMA-353[18] recommends that "All electrodes, fluxes and shielding gases should conform to the latest American Welding Society (AWS) A5-series specifications."

Chapters 12 and 13 in this book offer further discussion on the subject and a view on new trends in seismic engineering developments.

REFERENCES

1. Erdey, C. K. 1999. "Performance of Steel Moment Frames in an Earthquake." Paper presented at Eurosteel '99, Second European Conference on Steel Structures, Prague.

2. Erdey, C. K. 1999. "Performance of SMRF and SCBF Structures Subjected to Earthquakes," Paper read at the University of Bratislava.

3. Erdey, C. K. 1979. "A New Load-Transmitting Medium to Measure Strength of Brittle Materials." *Journal of Testing and Evaluation.* Vol. 7, No. 6, pp. 317–325.

4. Erdey, C. K. 1980. "Finite Element Analysis and Tests with a New Load-Transmitting Medium to Measure Compressive Strength of Brittle Materials." *Materials and Structures,* Vol. 13, No 74. pp. 83–90.

5. McCormac, J. C. 2003. *Structural Steel Design.* Upper Saddle River, NJ: Prentice-Hall.

6. Popov, E. P., and Stephen, R. M. 1972. "Cyclic Loading on Full-Size Steel Connections." American Iron and Steel Institute (AISI), *Bulletin No.* 21, Washington, D.C.

7. Bertero, V. V., Krawinkler, H., and Popov, E. P. 1973. "Further Studies on Seismic Behavior of Steel Beam-to-Column Subassemblages." Report No. UCB/EERC-73/27. Earthquake Engineering Research Center, University of California, Berkeley.

8. Popov, E. P., and Tsai, K. C. 1989. "Performance of Large Seismic Steel Moment Connections under Cyclic Loads." *Engineering Journal,* American Institute of Steel Construction (AISC), Vol. 26, No. 2, pp. 51–60.

9. Engelhardt, M. D., and Husain, A. S. 1993. "Cyclic-Loading Performance of Welded Flange-Bolted Connections." *Journal of Structural Engineering,* ASCE, Vol. 119, No. 12, pp. 3537–3550.

10. Uang, C. M. 1995. "Dynamic Testing of Large-Size Steel Moment Connections," VHS video of tests. University of California, San Diego.

11. Uang, C. M., and Bondad, D. M. 1996. "Dynamic Testing of Full-Scale Steel Moment Connections." Paper presented in the *Eleventh World Conference on Earthquake Engineering,* Acapulco, Mexico. Division of Structural Engineering, University of California, San Diego.

12. AISC, 1999, "Modification of Existing Welded Steel Moment Frame Connections for Seismic Resistance," *Steel Design Series,* No. 12.

13. FEMA, 1995, *Interim Guidelines: Evaluation, Repair, Modification and Design of Steel Moment Frames,* FEMA-267, SAC 96-03, Washington, D.C., August. These *Interim Guidelines* were later superseded by FEMA 353 of July 2000.

14. FEMA, 1997, NEHRP *Provisions,* Publication 302, Washington, D.C.

15. FEMA, 1998, *Update on the Seismic Safety of Steel Buildings, A Guide for Policy Makers,* Washington, D.C.

16. FEMA, 1999, "*Seismic Design Criteria for New Moment-Resisting Steel Frame Construction,*" January, 50% Draft.

17. FEMA, 2000 *Recommended Seismic Evaluation and Upgrade Criteria for Existing Welded Steel Moment-Frame Buildings,*" FEMA 351, July.

18. FEMA, 2000 *Recommended Specifications and Quality Assurance Guidelines for Steel Moment-Frame Construction for Seismic Applications,* FEMA 353, July.

19. Engelhardt, M. D., and Sabol, T. A. 1996. "Reinforcing of Steel Moment Connections with Cover Plates: Benefits and Limitations," *Proceedings, US-Japan Seminar on Innovation in Stability Concepts and Methods for Seismic Design in Structural Steel,* Honolulu, Hawaii.

CHAPTER 5

SEISMIC STEEL DESIGN: BRACED FRAMES

5.1 INTRODUCTION

When conventional steel structures were adopted in seismic areas such as California, the impact that strong ground movements would have on these structures was practically unknown. Some earthquakes of relatively large magnitude had occurred in scarcely populated areas where no modern steel structures had been built. With the increase of population new areas were developed along with the demand for large engineered steel structures—moment-resisting frames (MRF) and concentrically braced frames (CBF)—to meet the compelling needs of developing communities for office buildings, hospitals, and other high-rise structures. At the onset of these developments SEAOC led the efforts of the engineering community to adjust conventional steel structures to earthquake impact. Of particular significance is the 15 years of analytical and experimental research conducted by SEAOC in collaboration with the University of Michigan Department of Civil Engineering. Other institutions such as the University of California at Berkeley conducted parallel research. Concentrically braced frames were subjected to cycling loading to simulate earthquake-generated ground motions; structural response and areas in need of improvement were noted and efforts made to enhance structural resistance to earthquakes.

Coincidental with the research-and-development (R&D) work, a number of strong ground motion earthquakes hit populated areas in California where modern engineered steel structures, including CBFs, had been built. Valuable information was gathered and analyzed by the FEMA-sponsored SAC joint venture—Structural Engineers Association of California, Applied Technology Council, and California Universities for Research in Earthquake Engineering—as well as by other researchers.

Today, significant improvement in earthquake-related design has been achieved. It is chiefly reflected in the SEAOC *Recommended Lateral Force Requirements and Commentary,* and in the 1997 UBC, which, as mentioned earlier, can be considered the model Code in seismic design that opened the road for its IBC successors.

5.2 PROJECT DESCRIPTION: FOUR-STORY LIBRARY ANNEX

The design presented here will use the LRFD method in accordance with the provisions of the Load and Resistance Factor Design 1997 UBC and AISC *Manual of Steel Construction, Load and Resistance Factor Design,* second edition, 1998. The building will serve as annex to a main library and will be used exclusively for reading rooms. The specifications are as follows:

Concrete floors on metal deck
Ground and typical floor area: $104 \times 104 = 10,820$ ft^2
Structural system:
    Special concentrically braced frame: UBC 1997, Table 16-N
    Steel: ASTM A36 and A500B
Total building height: 54 ft
Seismic zone: 4
Seismic source: type A (5 km from source)
Soil type: S_C (very dense)
Basic wind speed: 85 mph, exposure D

Figures 5.1 and 5.2 show the structural layout of a typical floor and dimensions of the braced frame, respectively.

5.3 WIND ANALYSIS

We have already analyzed in detail the design wind pressures for buildings and structures (1997 UBC, CHAP. 16) in Chapter 4 of this text when we performed the design of a SMRF structure. Therefore we will go ahead with the seismic design for the braced-frame project as we have demonstrated that seismic forces govern.

Table 5.1 summarizes the wind pressure evaluation for the entire building. Figure 5.3 shows the lateral wind pressure (psf) distribution per unit width of the wall and Figure 5.4 the lateral wind force distribution (lb/ft) for the entire braced-framed structure.

The total wind-generated base shear at strength level is given as

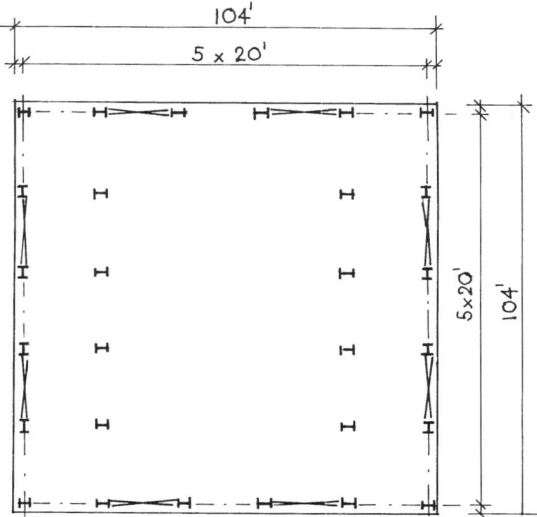

Figure 5.1   Structural layout of a typical floor.

$$V = 199 \times 1.3 = 259 \text{ kips} < 670 \text{ kips}$$

Note that 670 kips is the base shear for the earthquake.

Earthquake design forces GOVERN.

5.4   EARTHQUAKE ANALYSIS

The plan and methodology recommended in Chapter 4 of this text for the design of a SMRF structure, which consisted of six steps, apply to the seismic design of a braced-frame structure. Hence we will omit it here for simplicity.

The braced frame in Figure 5.2 was designed per the LRFD *Manual of Steel Construction* (2nd ed.), in accordance with the 1997 UBC CHAP. 16 seismic load distribution. However, the designer did not take into account the special seismic "Detailed Systems Design Requirements" of CHAP. 22, DIV. IV. Figure 5.7 in Section 5.12 of this text gives member sizes pertaining to this incomplete design.

Our task will be to incorporate the special "Detailed Systems Design Requirements" of CHAP. 22, DIV. IV, of the 1997 UBC to produce a complete code-complying seismic design.

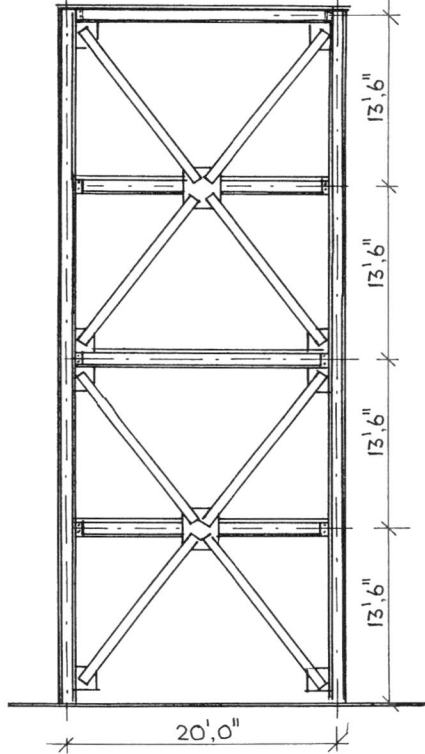

Figure 5.2 Typical project braced frame.

TABLE 5.1   Wind Pressure Evaluation for Building

		Horizontal Wind Pressures	
Story	Height (ft)	$p = C_e C_q q_s I_w$ (psf)	Wind Load Acting on Building, $p \times$ width (lb/ft)
1	0–13.5	$p = (1.39)(1.4)(18.6)(1.0) = 36.2$	3765
2	13.5–27.0	$p = (1.45)(1.4)(18.6)(1.0) = 37.8$	3932
3	27.0–40.5	$p = (1.58)(1.4)(18.6)(1.0) = 41.0$	4243
4	40.5–54.5	$p = (1.69)(1.4)(18.6)(1.0) = 44.0$	4570

Wind Design Forces Acting at Floor Level

Roof (Fifth level)	$(13.5/2 + 0.5)(4.5) = 32.6$ kips
Fourth level	$13.5(4500 + 4243)/2 = 59$ kips
Third level	$13.5(4243 + 3932)/2 = 55.4$ kips
Second level	$13.5(3932 + 3765)/2 = 52.0$ kips
Total base shear	$V = 199$ kips

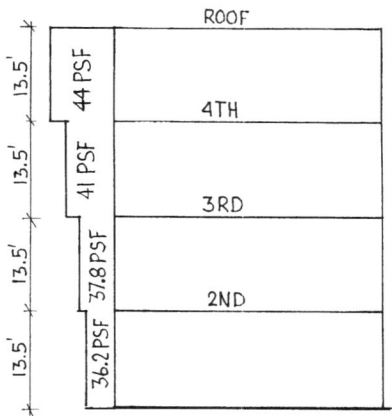

Figure 5.3 Wind design pressures (psf).

Project Data

Unit weight	Units expressed in service load level: load factor 1.0
Roof deck	60 psf; includes weight of suspended ceiling, roof insulation, ducts, etc.
Floor deck	90 psf; includes weight of suspended ceiling, ducts, curtain walls, etc.
Weigh of roof	650 kips
Weight of floor	973 kips
Weight of building	3570 kips

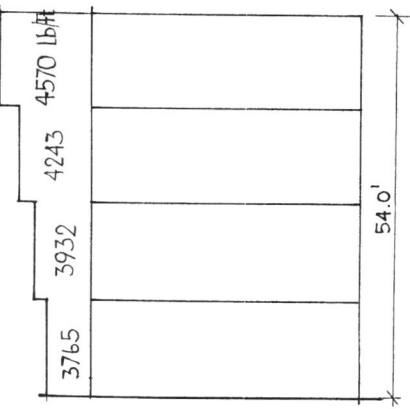

Figure 5.4 Wind forces (lb/ft).

Base-Shear Calculations

$$Z = 0.4$$

$$\text{Soil} = S_C \qquad\qquad\qquad\qquad \text{Table 16-J}$$

$$N_v = 1.6 \qquad\qquad\qquad\qquad \text{Table 16-T}$$

$$C_v = 0.56N_v = 0.56(1.6) = 0.896 \qquad\qquad \text{Table 16-R}$$

$$N_a = 1.2 \qquad\qquad\qquad\qquad \text{Table 16-S}$$

$$C_a = 0.4\ N_a = 0.4(1.2) = 0.48 \qquad\qquad \text{Table 16-Q}$$

$$R = 6.4 \qquad\qquad\qquad\qquad \text{Table 16-N}$$

$$I = 1.0 \qquad\qquad\qquad\qquad \text{Table 16-K}$$

$$C_t = 0.020$$

$$T_s = \frac{C_v}{2.5C_a} = \frac{0.896}{2.5 \times 0.48} = 0.747 \text{ s}$$

$$T = C_t(h_n)^{3/4} = 0.02(54)^{3/4} = 0.4 \text{ s} < T_s$$

$$V = \left(\frac{C_v I}{RT}\right)W$$

$$= \left(\frac{0.896 \times 1.0}{6.4 \times 0.4}\right)3570 = 1250 \text{ kips}$$

The total need not exceed deisgn base shear

$$V = \left(\frac{2.5C_a I}{R}\right)W$$

$$= \left(\frac{2.5 \times 0.48 \times 1.0}{6.4}\right)3570 = 670 \text{ kips} \leftarrow \textbf{GOVERNS}$$

but should not be less than

$$V = 0.11\ C_a\ I\ W$$

$$= 0.11 \times 0.48 \times 1.0 \times 3570 = 188 \text{ kips}$$

Also, for seismic zone 4, it should not be less than

$$V = \left(\frac{(0.8Z\,N_v I)}{R}\right)W$$

$$= \left(\frac{0.8 \times 0.4 \times 1.6 \times 1.0}{6.4}\right)3570 = 286 \text{ kips}$$

Note that $V = 670$ kips is at strength level. When used in combination with dead load and live load, these loads have to be brought up to strength level as well, multiplying their respective values by the appropriate load factors given in the 1997 UBC, 1612.2.1.

Lateral Force Distribution

$$V = F_t + \sum F_i$$

$$F_t = 0 \qquad \text{for } T \le 0.7$$

$$F_x = \frac{(V - F_t)\,w_x h_x}{\sum h_i}$$

$$F_5 = \frac{670 \times 650 \times 54}{650 \times 54 + 974(40.5 + 27.0 + 13.5)} = 206 \text{ kips}$$

$$F_4 = \frac{670 \times 974 \times 40.5}{650 \times 54 + 974(40.5 + 27.0 + 13.5)} = 232 \text{ kips}$$

$$F_3 = \frac{670 \times 974 \times 27}{650 \times 54 + 974(40.5 + 27.0 + 13.5)} = 155 \text{ kips}$$

$$F_2 = \frac{670 \times 973 \times 13.5}{650 \times 54 + 973(40.5 + 27.0 + 13.5)} = 77.3 \text{ kips}$$

Table 5.2 shows horizontal wind and earthquake forces acting on the typical braced frame. Table 5.3 shows maximum inelastic response displacement (1630.9.1 and 1630.9.2) and allowable story-drift values (1630.10.2) applicable to our project.

TABLE 5.2   Horizontal Wind and Earthquake Forces (F_i) Acting on Typical Braced Frame (kips)

Level	Wind at Service Level[a]	Earthquake Force (E_h) at Strength Level[a]
5	9.8	62.0
4	17.7	70.0
3	16.6	46.6
2	15.6	23.3

[a] Values include torsion and redundancy factor $\rho = 1.0$

TABLE 5.3 Maximum Inelastic Response Displacement and Allowable Story Drift

Story	Maximum Inelastic Response Displacement,[a] $\Delta_M = 0.7R\Delta_s$ in. (30-17)	Allowable Story Drift (1630.10.2), $0.025 \times$ Story Height (in.), $T < 0.7$ s
1	$\Delta_M = 0.064 \times 12 = 0.77$	$\Delta_{ALLOWED} = 0.025 \times 162 = 4.05$
2	$\Delta_M = (0.165 - 0.064) \times 12 = 1.21$	$\Delta_{ALLOWED} = 4.05$
3	$\Delta_M = (0.278 - 0.165) \times 12 = 1.36$	$\Delta_{ALLOWED} = 4.05$
4	$\Delta_M = (0.372 - 0.278) \times 12 = 1.13$	$\Delta_{ALLOWED} = 4.05$

Note: Displacement values are within UBC 1997 required values.
[a] Results of computer analysis.

5.5 WIND AND EARTHQUAKE LOADS

As is normally the case for a great number of projects in UBC seismic zone 4, or even seismic zone 3, wind might not be an issue. However, evaluation of both wind and earthquake loads is essential, particularly in high-wind areas, to determine whether wind or seismic governs the design. Should wind govern, the engineer is still bound to comply with the seismic design and detailing rules of the appropriate materials sections of the code. The actual design will consist of two parts:

1. Establish loads and general rules per the 1997 UBC, CHAP. 16 (analysis and preliminary design stage).
2. Carry out the actual design in compliance with the AISC *Specification* and the "Detailed Systems Design Requirements" pertaining to the specific materials sections, 1997 UBC, CHAPS. 18–23. The "Design Standards for Load and Resistance Factor Design Specification for Structural Steel Buildings" are contained in CHAP. 22, DIVS. II and IV of the code.

5.6 RESPONSE OF BRACED FRAMES TO CYCLIC LATERAL LOADS

When the externally applied loading becomes complex, for example, time dependent, resulting in stress reversals, recent research work[1-4] indicates that the rules of simple geometry no longer work and are replaced by more powerful laws such as compatibility of deformations and other factors.

The braced frame of this project is a split-braced frame; in a graphical sense, the diagonal brace elements run between the first and third floors and then from the third to the fifth level (roof). In other words, the brace system connects the odd-numbered floors and crisscrosses at every even-numbered floor (in our building, second and fourth floors). If we were to not connect the crisscrossing joint to the even-numbered floors, these floors would be free to displace laterally and cause the frame columns to bend considerably between the brace points of first and third and the third and fifth levels. It would be a hybrid structure, partly braced frame, partly MRF, and the values of a braced frame that provide economy and stiffness against displacement would be lost.

What makes this a *split-braced frame* is that the crisscrossing points of the diagonal braces are also tied into the second and fourth floors, which provides sufficient stiffness to the system and lateral support to the columns at each floor level. If the frame were subjected only to lateral static loads, the vertical and horizontal force components of the braces would balance out between each other and the externally applied load at each nodal point, including the crisscrossing point.

If asked whether the brace reactions cause bending of the second-floor beam connected to the diagonal bracings, a reader with structural background would answer, without hesitation, that there will be no bending if the CGs of the members intersect at the *work point*—our basic assumption for this project. However, the issue is complex in an earthquake because we are not dealing with statically applied lateral loads. The earthquake causes serious reversals of structural deformations, inertial loads, and internal forces.

As the ground movement at the base reverses rapidly, as much as two to four cycles per second, the structure sways left and right with high-frequency response. A picture of, say, the second floor taken with a sufficiently sensitive high-speed camera would demonstrate that the floor has displaced to the right while the base of the structure has moved to the left and the structure is leaning heavily on the right diagonal brace of the inverted V, putting the brace in compression. Due to the tendency of a compressed member to buckle, the brace will bow out under compression (Figure 5.5).

Tests on structures subject to similar cyclic loading showed a marked drop in element stiffness in the compression brace as compared to the stiffness of the corresponding diagonal tension brace of the inverted-V system. Because of the marked difference in element stiffness, the diagonal tension brace picks up a considerably larger share of the lateral inertial loads acting on the structure as compared to the compression brace.

Research has also proved that, because of the imbalance of axial loads in the diagonal braces, a considerable unbalanced vertical-load component exists at the floor-beam/diagonal brace connection. The unbalanced vertical reaction causes bending of the connecting floor beam, a phenomenon that was not envisaged under the purely theoretical concepts of braced frames subjected to static loading. The bending of the beam intensifies as the stiffness of the compression brace deteriorates under cyclic loading.

If the floor beam is not properly engineered it might undergo local or torsional buckling. Therefore it is imperative that

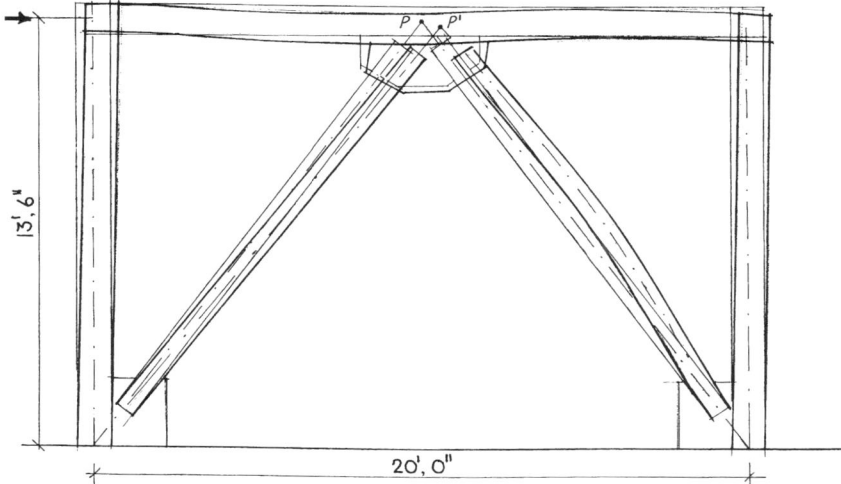

Figure 5.5 Deformation of a one-story brace frame module subjected to lateral load.

(a) the brace-to-beam joint be laterally supported against buckling and
(b) the choice of beam be such that the section is compact and the laterally unsupported length is equal to or smaller than L_p to allow a fully plasticized moment response.

The special provisions of the 1997 UBC in CHAP. 22, DIV. IV, reflect research results when calling for compact sections and a minimum strength for lateral bracing to beam flanges equal to 1.5% of the nominal yield strength times the beam flange area to resist lateral buckling of the beam.

The above-described phenomena, demonstrated on one story of the structure, apply to all floor beams and diagonal braces. The bowing out of the diagonal brace has other serious consequences such as stress reversals; these are generated mostly in the diagonal bracings when the structure undergoes repetitive cyclic loading. When the cross-bracing bows out under compression, the structure, subjected to reversals, displaces in the opposite direction, inducing *high tensile force* into the diagonal member that was, an instant before, *in compression*. Because of its relatively sluggish response to stress reversals, the bowed-out member will not straighten out fast enough to respond to the high-tension demand of the force reversal. The result is a whiplash of dynamic impact prone to cause joint damage. This phenomenon was observed in experimental tests on steel specimens as well as in structures damaged by the Northridge earthquake.

The slenderer the diagonal brace element, the stronger the dynamic impact and potential damage to the joint. Recent research work has verified this fact. Consequently, the 1997 UBC, CHAP. 22, DIV. IV, special detailing provision 9.2.a, limits the slenderness ratio of the brace member to

$$\frac{L}{r} \le \frac{720}{\sqrt{F_y}}$$

In addition, test results and observation of earthquake-damaged buildings have shown that *slender* components of brace members could undergo local buckling, resulting in brace damage. Thus the 1997 UBC does not allow slender components that do not fall into the category of *compact* or *noncompact* sections.

To avoid local buckling and subsequent brittle fracture, 1997 UBC, CHAP. 22, DIV. IV, 9.2.d, restricts the diameter-to-wall thickness and flat-width-to-wall thickness ratios of structural pipes and tubing. In addition, the code encourages the designer to increase the "design strength of the brace member . . . at least 1.5 times the required strength using Load Combinations 3-5 and 3-6 . . . [for] V and Inverted V type bracing" (CHAP. 22, DIV. IV, 9.4.a.4). These load combinations are similar in nature to formulas (12-5) and (12-6) of CHAP. 16, 1612.2.1. When increasing the brace member design strength by 50%, UBC 1997 clearly asks the designer to compensate for the dynamic impact on deflected shape and for the stress reversals normally not taken into account in structural calculations.

5.7 1997 UBC PROVISIONS

The code recognizes the limitations of analytical methods (static force method, response spectrum analysis) used to predict the actual forces that will act on the structure during an earthquake as well as numerous unknown factors related to a specific earthquake. To compensate for limitations and unknown factors the 1997 UBC provides added strength requirements in CHAPS. 16 and 22.

When using the static force method remember that you are substituting static for dynamic forces. Dynamic forces produce time-dependent structural oscillations influenced by a number of factors such as relative frequency of ground motion versus natural frequencies and inherent damping of the structure. These factors are capable of causing significant amplification to vibration-related structural deformations and inertial forces.

There are other unknown causes that will impact the structure, such as the nature of the earthquake; its magnitude, frequency, and superimposed vibrations carried by the spectrum of *p* and *s* waves that determine the intensity of ground shaking; and duration of the earthquake: The longer it works on some inherent structural weaknesses, the more it is likely to hurt the structure. These considerations should forewarn the engineer to adhere to the 1997 UBC requirements, which are based on the experience and knowledge gained from actual earthquakes.

Some specific provisions of the "Detailed Systems Design Requirements" of CHAP. 22, DIV. IV, of the 1997 UBC are reproduced on the following pages as a reference.

5.8   RULES APPLICABLE TO BRACING MEMBERS

CHAPTER 22, DIV. IV, of the 1997 UBC, "Seismic Provisions for Structural Steel Buildings" states:

9.2 Bracing Members

9.2.a. Slenderness: Bracing members shall have an

$$\frac{L}{r} \le \frac{720}{\sqrt{F_y}} \text{ except as permitted in Sect. 9.5.}$$

9.2.b. Compressive Design Strength: The design strength of a bracing member in axial compression shall not exceed $0.8\phi_c P_n$.

9.2.c. Lateral Force Distribution: Along any line of bracing, braces shall be deployed in alternate directions such that, for either direction of force parallel to the bracing, at least 30% but no more than 70% of the total horizontal force shall be resisted by tension braces, unless the nominal strength, P_n, of each brace in compression is larger than the required strength, P_u, resulting from the application of the Load Combinations 3-7 or 3-8. A line of bracing, for the purpose of this provision, is defined as a single line or parallel lines whose plan offset is 10% or less of the building dimensions perpendicular to the line of bracing.

9.2.d. Width-Thickness Ratios: Width-thickness ratios of stiffened and unstiffened compression elements in braces shall comply with Sect. B5 in the Specification. Braces shall be compact or non-compact, but not slender (i.e., $\lambda < \lambda_p$). Circular sections shall have an outside diameter to wall thickness ratio not exceeding $1{,}300/F_y$; rectangular tubes shall have a flat-width to wall thickness not exceeding $110/\sqrt{F_y}$ unless the circular section or tube walls are stiffened.

In addition, the 1997 UBC requires that bracing members be designed to at least 1.5 times the required strength. A summary of the provisions in Section 9.4.a, "V and Inverted V Type Bracing," is as follows:

1. The design strength of the brace members should be at least 1.5 times the required strength using load combinations 3-5 and 3-6.

2. The beam intersected by braces should be continuous between columns.
3. A beam intersected by V braces should be capable of supporting all tributary dead and live loads assuming the bracing is not present.
4. The top and bottom flanges of the beam at the point of intersection of V braces should be designed to support a lateral force equal to 1.5% of the nominal beam flange strength ($F_y b_f t_f$).

For the strength of bracing:

9.3.a. Forces: The required strength of bracing joints (including beam-to-column joints if part of the bracing system) shall be the least of the following:

1. The design axial tension strength of the bracing member.
2. The force in the brace resulting from the Load Combinations 3-7 or 3-8.
3. The maximum force, indicated by an analysis, that is transferred to the brace by the system.

For net area:

9.3.b. Net Area: In bolted brace joints, the minimum ratio of effective net section area to gross section area shall be limited by:

$$\frac{A_e}{A_g} \geq \frac{1.2\alpha P_u^*}{\phi_t P_n} \tag{9-1}$$

where:
A_e = Effective net area as defined in Equation B3-1 of the *Specification*.
P_u^* = Required strength on the brace as determined in Sect. 9.3.a.
P_n = Nominal tension strength as specified in Chapter D of the *Specification*.
ϕ_t = Special resistance factor for tension = 0.75
α = Fraction of the member force from Sect. 9.3.a. that is transferred across a particular net section.

For Gusset plates:

9.3.c. Gusset Plates:

1. Where analysis indicates that braces buckle in the plane of the gusset plates, the gusset and other parts of the connection shall have a design strength equal to or greater than the in-plane nominal bending strength of the brace.
2. Where the critical buckling strength is out-of-plane of the gusset plate, the brace shall terminate on the gusset a minimum of two times the gusset thickness from the theoretical line of bending which is unrestrained by the column or beam joints. The gusset plate shall have a required compressive strength to resist the compressive design strength of the brace member without local buckling of the gusset plate. For braces designed for axial load only, the bolts

or welds shall be designed to transmit the brace forces along the centroids of the brace elements.

5.9   COLUMN STRENGTH REQUIREMENTS

In addition to load combinations (12-1)–(12-6) of CHAP. 16 and the above quoted special load combinations of CHAP. 22, the 1997 UBC requires:

6.1. Column Strength:
When $P_u/\phi P_n > 0.5$, columns in seismic resisting frames, in addition to complying with the *Specification,* shall be limited by the following requirements:

6.1.a. Axial compression loads:

$$1.2P_D + 0.5P_L + 0.2P_S + 0.4R \times P_E \leq \phi_c P_n \qquad (6\text{-}1)$$

where the term $0.4R$ is greater or equal to 1.0.
   Exception: The load factor on P_L in Load Combination 6-1 shall equal 1.0 for garages, areas occupied as places of public assembly, and all areas where the live load is greater than 100 psf.

6.1.b. Axial tension loads:

$$0.9P_D - 0.4R \times P_E \leq \phi_t P_n \qquad (6\text{-}2)$$

where the term $0.4R$ is greater or equal to 1.0.

However, the 1997 UBC says:

6.1.c. The axial Load Combinations 6-1 and 6-2 are not required to exceed either of the following:

1. The maximum loads transferred to the column, considering 1.25 times the design strengths of the connecting beam or brace elements of the structure.
2. The limit as determined by the foundation capacity to resist overturning uplift.

5.10   DESIGN FOR EARTHQUAKE

The first step is to evaluate ρ, the reliability/redundancy factor, which controls formula (30-1). As stated before, ρ is a function of r_{max}.
   In CHAP. 16, DIV. IV, 1630.1.1, the 1997 UBC defines:

r_{max} = the maximum element-story shear ratio. For a given direction of loading, the element-story shear ratio is the ratio of the design story shear in the most heavily loaded single element divided by the total design story shear. For any given Story Level i, the element-story shear ratio is denoted as r_i. The maximum element-story shear ratio r_{max} is defined as the largest of the element story shear ratios, r_i, which occurs in any of the story levels at or below the two-thirds height level of the building.

For braced frames, the value of r_i is equal to the maximum horizontal force component in a single brace element divided by the total story shear.

The maximum horizontal force component for the braced frame is 101 kips. This is the vector component of the force induced by the design earthquake in brace 3, represented by computer analysis load combination [3]. The largest story shear is the base shear for the building:

$$V = 670 \text{ kips}$$

$$r_{max} = \frac{101}{670} = 0.15075$$

$$\rho = 2 - \frac{20}{(0.15075)(104)} = 0.724 < 1.0 \qquad (30\text{-}3)$$

Because the value of ρ is less than 1, following the 1997 UBC restriction "ρ shall not be taken less than 1.0 and need not be greater than 1.5" (CHAP. 16, DIV. IV, 1630.1.1), $\rho = 1.0$ will be used in our computations for the code basic formula

$$E = \rho E_h + E_v \qquad (30\text{-}1)$$

As $\rho = 1.0$ for the project, it will have no augmenting effect on the lateral earthquake forces applied to the structure. Table 5.2 shows values of lateral wind and earthquake forces with $\rho = $ unity.

5.11 STRATEGIES FOR BRACE MEMBER DESIGN

The first step is to select member sizes for the diagonal brace system (Figure 5.6). Notice that there is an ordered pair of internal forces for each pair of diagonal braces.

When one brace is in compression, the other is in tension.

Earthquake forces are reversible. As the ground motion reverses in rapid succession, each diagonal brace is subjected to an internal axial compression force followed by a tensile force of similar magnitude. This cyclic loading lasts until the vibration of the structure stops. The bottom line is that each diagonal brace must be designed for both tensile and compressive axial forces and associated secondary effects, moments, and shear. However, in a braced

CONVENTION: WIDTH OF STRUCTR.
TUBING IO TIMES THE ACTUAL
SIZE, WALL THICKNESS EX-
PRESSED IN 1/16 IN. TO FIT THE
REQUIRED INPUT OF THE C. PROGR.

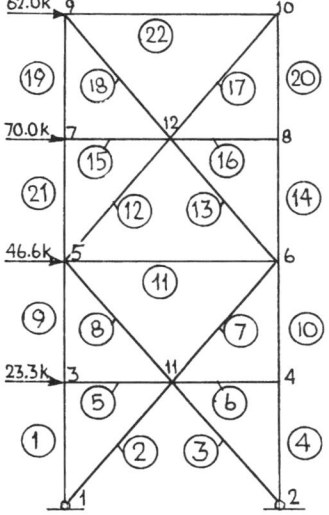

Figure 5.6 Braced-frame computer model with seismic loads.

structure, for practical reasons the value of the secondary effects might not be significant. For instance, for the first-story braces we obtain a maximum design force of 245 kips tension for component 2 and 275 kips axial compression for component 3. We will design each brace for a set of complementary compressive and tensile forces obtained by analysis assuming single-direction oriented lateral forces. We are using this approach with the knowledge that earthquake forces are reversible and the structure is symmetrical.

This approach as well as some forthcoming restrictive brace design regulations may appear unfamiliar to readers well acquainted with the design of structures subjected to static forces; however, the 1997 UBC static force procedure in 1630.2 is only an alternative to the more complex dynamic analysis procedures in (1631); therefore, when using the static force procedure, we need to bear in mind that:

1. We are not dealing with static forces but vibration induced by dynamic forces and hence we need to make necessary design adjustments.
2. We should adhere to all code provisions for analysis, design, and detailing.

Convention: The width of structural tubing is 10 times the actual size. Wall thickness is expressed in $\frac{1}{16}$ in. to fit the required input of the computer program.

5.12 BRACE MEMBERS 2 AND 3

The governing load combinations for the design are

$$(1.5)(1.2D + 0.5L \pm 1.0 \times E) \qquad \text{From 9.4.a.1, CHAP. 22, DIV. IV}$$

$$(1.5)(0.9D \pm 1.0E) \qquad \text{From 9.4.a.1, CHAP. 22, DIV. IV}$$

The internal member forces thus obtained should be

$$P_u \leq 0.8\phi_c P_n \qquad \text{for compression*}$$

$$\leq \phi_t P_n \qquad \text{for tension}$$

From CHAP. 22, DIV. IV ("Bracing Members"), "**9.2.b. Compressive Design Strength:** The design strength of a bracing member in axial compression shall not exceed $0.8\phi_c P_n$." For the design of connections:

Compression:

$$1.2P_D + 0.5P_L + 0.4R\ P_E < \phi_c P_n \qquad \text{Implied by (3-7)}$$

Tension:

$$0.9P_D - 0.4R\ P_E \geq \phi_t\ P_n \qquad \text{Implied by (3-8), CHAP. 22, DIV. IV, 3.1}$$

Figures 5.7 and 5.8 show the resulting member sizes.

5.13 BRACE MEMBERS 3 AND 2: FIRST STORY

Brace 3: Design for Compression

Load combination [9] of the computer analysis produces the 1.5 times magnified axial compression in the member per CHAP. 22, DIV. IV, 9.4.a.1:

$$P_{\max} = 275 \text{ kips} \qquad KL = 13.5 \text{ ft}$$

The member compressive strength per (9.2.b) is

---

*The 0.8 multiplier of $\phi_c P_n$ is from 9.2.b, "Bracing Members," CHAP. 22, DIV. IV.

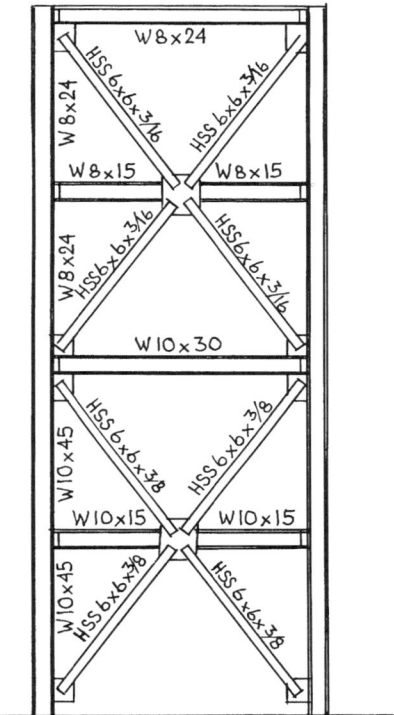

Figure 5.7 Member sizes not complying with 1997 UBC, CHAP. 22, DIV. IV.

$$0.8\phi_c P_n = P_{\max}$$

To use the column table of the LRFD *Specification,* we must divide the above equation by 0.8 to obtain

$$P_u = \phi_c P_n = \frac{P_{\max}}{0.8}$$

$$= \frac{275}{0.8} = 344 \text{ kips}$$

From Part 3, column design tables of the *Specification,* we obtain the sizes of the square structural tubing with $KL = 13.5$ ft that provide an axial strength equal to or larger than P_u:

$$8 \times 8 \times \tfrac{3}{8}(37.69) \text{ and } 10 \times 10 \times \tfrac{5}{16}(40.35) \ 7 \times 7 \times \tfrac{1}{2}(42.05)$$

Although the design convention would utilize the lightest sections, that is, $8 \times 8 \times \tfrac{3}{8}$ (37.69) and $10 \times 10 \times \tfrac{5}{16}$, such choices would be at fault.

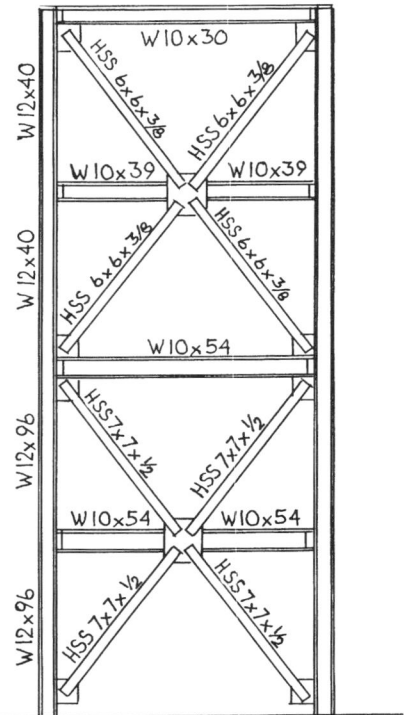

Figure 5.8 Final member sizes in compliance with 1997 UBC seismic regulations, including CHAP. 22, DIV. IV.

Clearly CHAP. 22, DIV. IV, 9.2.d, calls for a flat width-to-wall thickness ratio not exceeding

$$\frac{110}{\sqrt{F_y}} = \frac{110}{\sqrt{46}} = 16.22$$

The $8 \times 8 \times \frac{3}{8}$ structural tubing has a ratio of

$$\frac{8 - 3(\frac{3}{8})}{\frac{3}{8}} = 18.3 > 16.22$$

and the $10 \times 10 \times \frac{5}{16}$ structural tubing a ratio of

$$\frac{9.06}{\frac{5}{16}} = 32 > 16.22$$

Thus none of these options is acceptable per CHAP. 22, DIV. IV, 9.2.d.

The next lightest structural tubing to meet the above criteria is the $7 \times 7 \times \frac{1}{2}$ section with a compressive strength $P_u = 375$ kips which, in addition, has a thicker wall that allows a stronger weld. This will in turn reduce the size of connection and the gusset plate. The $7 \times 7 \times \frac{1}{2}$ section has a flat width-to-wall-thickness ratio of

$$\frac{7 - 3(\frac{1}{2})}{\frac{1}{2}} = 11 < 16.2$$

The 1997 UBC requires that bracing members have

$$\frac{L}{r} \leq \frac{720}{\sqrt{F_y}} = \frac{720}{\sqrt{46}} = 106$$

$$\frac{L}{r} = 13.5 \times \frac{12}{2.62} = 62 < 106$$

Brace 2: Design for Tension

The maximum axial tensile force in brace 2 is

$$P_u = 245 \text{ kips}$$

given by $1.5(0.9D + 1.0E)$ load combination [10] of the computer analysis, that is, basic load combination (3-6) augmented 1.5 times, per CHAP. 22, DIV. IV, 9.4.a.1.

Two steps are involved in proving that the tensile resistance of the brace is larger than P_u:

1. $P_u = \phi_t F_y A_g$ (with $\phi_t = 0.9$)
2. $P_u = \phi_t F_u A_e = \phi_t F_u (UA_n)$ (with $\phi_t = 0.75$)

where A_g = gross sectional area
$\quad\quad A_n$ = sectional net area
$\quad\quad U = 1 - \overline{X}/L$, Equation (B3-2), Section B3 of the *Specification*

For complex and buildup sections see also Section B3 of the Commentary of the *Specification*. To establish U we must know the length of the weld and hence the length of the connection, L. This is determined in the next step, discussed in Section 5.14 below.

Using the recommendations in the Commentary of the *Specification,* the tubing section will be considered as made up of four equal angles (right-hand side of Figure 5.9) of $3 \times 3\frac{1}{2}$-in. legs, $\frac{1}{2}$ in. thick. The shorter legs grip the (assumed) 1-in.-thick gusset plate. Final design will actually produce a $\frac{3}{4}$-in. gusset plate. Carrying out the required computations yields

$$\overline{X} = 3.0 - 0.9 = 2.1 \text{ in.}$$

Assuming a connection length $L = 14$ in. verified by the weld design,

$$U = 1 - \frac{2.1}{14} = 0.85$$

Therefore

(a) $P_u = \phi_t F_y A_g = (0.9)(46)(12.4) = 513$ kips
(b) $P_u = \phi_t F_u A_e = \phi_t F_u (U)(A_n) = (0.75)(58)(0.85)(12.4) = 458$ kips $>$
    245 kips

5.14   DESIGN OF FILLET WELD CONNECTION

The fillet weld connection will be designed for the maximum brace reactions transmitted to the connections by components 2 and 3 in this bay, bound by the external columns and by the first- and third-floor levels.

The design of the connection is controlled by Section 9.3.a, CHAP. 22, DIV. IV, of UBC 1997, which calls for the lesser value of the following:

1. Design axial strength of bracing member
2. Force in brace resulting from load combinations (3-7) or (3-8) calling for $0.4R = 2.56$ times the magnified earthquake load:

$$1.2D + 0.5L + 0.4R \times E \qquad (3\text{-}7)$$

or
$$0.9D - 0.4R \times E \qquad (3\text{-}8)$$

3. Maximum force indicated by analysis that is transferred to the brace by the system

Among the three options, we will use the design forces given by (3-7) and (3-8) for the design of the connection.

Note that options 1 and 3 would produce forces *larger than* option 2, which represents the lesser value of the three options. Option 1 produces 458 kips, and option 3, which includes the vertical component of the earthquake, E_v, produces 453 kips axial force [12]. The design axial force is 450 kips given by formula (3-8), computer analysis load combination [14].

Assuming a $\frac{3}{8}$-in. fillet weld, the effective throat thickness is

$$0.707 \times 0.375 = 0.2651 \text{ in.}$$

The capacity of the weld is given as

$$\phi F_w = (0.75)(0.6 \times 70)(0.2651)(1.0) = 8.35 \text{ kips/in.}$$

Having four equal-length welds connecting the gusset plate to the slotted tubing, the required weld length is

$$L = \frac{450}{4 \times 8.35} = 13.5 \text{ in.} \qquad 13.5 \text{ in.} < 14 \text{ in. } \textbf{provided}$$

Use 14-in. fillet welds with end returns per (J2-2b) of the *Specification*. The reader is reminded of fillet weld terminations at the end of the section, which refers also to cyclic loading.

5.15   DESIGN OF GUSSET PLATE: FIRST AND SECOND STORIES

We will design gusset plates for the maximum brace reactions transmitted to the connections by components 2 and 3 in this bay bound by the external columns and first- and third-floor levels. Figure 5.9 shows the geometry of a typical brace-to-column gusset plate.

Design for Compression

The maximum axial compression load is given as

$$P_u = 450 \text{ kips} \qquad \text{Computer analysis load combination [14]}$$

The length of the theoretical (unrestrained) line of bending is 24 in. (CHAP. 22, DIV. IV, 9.3.c.2). The compression strength of the $\frac{3}{4}$-in. plate is

$$P_u = \phi_c F_y A_g = (0.85)(36)(24 \times 0.75) = 550 \text{ kips} > 450 \text{ kips}$$

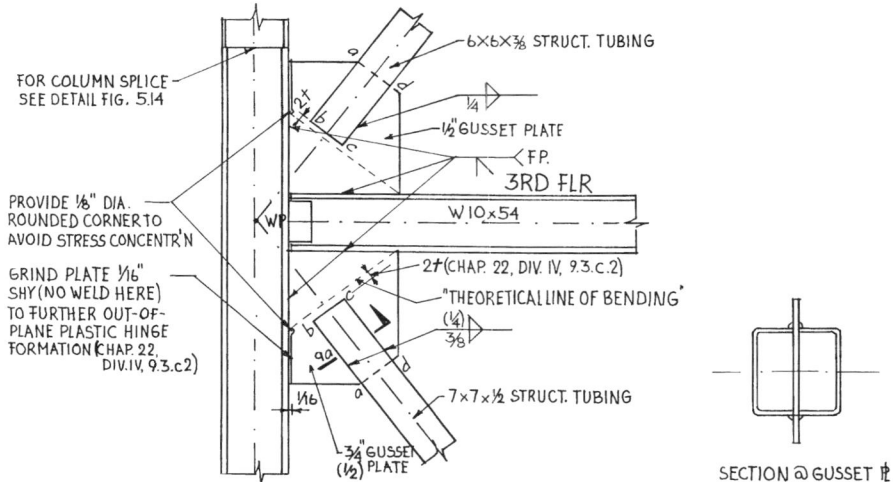

Figure 5.9 Detail of gusset plates. Dimensions at roof level are shown in parentheses.

Design for Block Shear

The design strength of a connection is not always controlled by the strength of the weld connecting the brace to the gusset plate or by the tensile or compressive strength of the plate of a specific boundary. It may instead be controlled by block shear, as described below. Failure may occur by tearing out a portion (*block*) of steel from the gusset plate. This situation occurs when the load in the attached brace member increases and a potential failure surface, marked *a, b, c, d* in Figure 5.9, is created weak enough to promote block shear failure.

Mechanism of Block Shear Failure

There are two modes of block shear failure (J4.3), Part 6 of the *Specification:*

1. When the applied brace force is sufficiently large, it may cause tension yielding along *bc* if this area is relatively small compared to *ab* and *cd* areas subjected to shear. The reader will note that such is the case for the gusset plates of the project. Following tension yielding along *bc* a sudden shear fracture failure will occur along the sheared areas represented by lines *ab* and *cd*.
2. Should the sheared areas *ab* and *cd* be small and the tensile area *bc* be relatively large, shear yielding along *ab* and *cd* will occur first, followed by tension fracture along *bc*.

The LRFD *Specification* provides a *test* equation: If $0.6F_uA_{nv} > F_uA_{nt}$, we have tension yielding and shear fracture (case 1) and the following LRFD equation should be used:

$$\phi R_n = \phi[0.6F_uA_{nv} + F_yA_{gt}] \text{ LRFD equation (J4-3b), } Specification$$

where A_{gv} = gross area subjected to shear
$ A_{gt}$ = gross area subjected to tension
$ A_{nv}$ = net area subjected to shear
$ A_{nt}$ = net area subjected to tension
$ \phi$ = 0.75

If $F_uA_{nt} \geq 0.6F_uA_{nv}$, we will have shear yielding and tension fracture and, in this case, we should use LRFD equation (J4-3a) of the *Specification*:

$$\phi R_n = \phi[0.6F_yA_{gv} + F_uA_{nt}]$$

We will now apply the above outlined procedures to the gusset plate design. The maximum tensile force applied to the connection per formula (3-8) of the 1997 UBC, calling for a $(0.4)R = 2.56 E_h$ earthquake force magnification, is

$$P_u = 0.9D - (0.4)RE = 430 \text{ kips} \qquad \text{Computer load combination [15]}$$

In our case

$$A_{gv} = A_{nv} = 2 (14 \times 0.75) = 21.0 \text{ in.}^2$$

$$A_{gt} = A_{nt} = (7.0)(0.75) = 5.25 \text{ in.}^2$$

The Test

$$0.6 F_uA_{nv} = (0.6)(58)(21.0) = 731 \text{ kips} > (58)(5.25) = 304 \text{ kips}$$

Tension yielding will occur coupled with shear fracture; thus we must use Equation (J4-3b) of the *Specification*:

$$\phi R_n = (0.75)[(0.6)(58)(21.0) + (36)(5.25)] = 690 \text{ kips} > 430 \text{ kips applied}$$

The connection is safe against block shear.

Figure 5.9 shows the dimensions of the gusset plates connecting the brace to the column. Gusset plates at the base are similar but reversed. Figure 5.10 shows details of the brace-to-beam gusset plate connections.

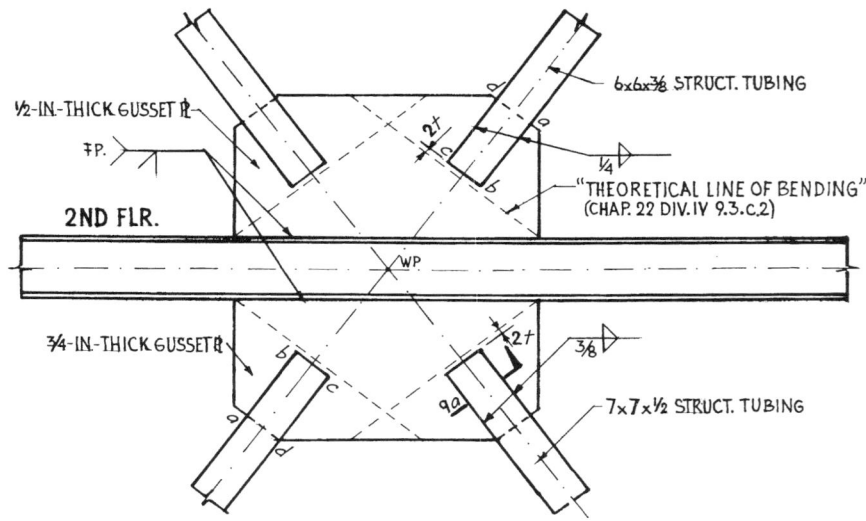

Figure 5.10   Gusset plate detail at second floor.

5.16   BRACE MEMBER 13: THIRD STORY

Design for Compression

The maximum axial load per 9.4.a.1, CHAP. 22, DIV. IV, results from load combination [9] of the computer analysis:

$$1.5(1.2D + 0.5L + 1.0E) = 1.8D + 0.75L + 1.5E$$

that is, from 1.5 times the magnified forces represented by load combination (12-5), 1612.2.1, of UBC 1997:

$$P_{max} = 176 \text{ kips}$$

According to CHAP. 22, DIV. IV, 9.2.b, "the design strength of a bracing member in axial compression shall not exceed $0.8\phi_c P_n = 0.8P_u$." Thus the relationship gives

$$P_u = \frac{P_{max}}{0.8} = \frac{176}{0.8} = 220 \text{ kips}$$

From the column design tables of Part 3 of the *Specification* with $KL = 13.5$ ft, we have the following choices for the lightest square tubing that fits this load:

$6 \times 6 \times \frac{3}{8}$ (27.48) with a capacity of 225 kips

$7 \times 7 \times \frac{5}{16}$ (27.59) with a capacity of 250 kips

$8 \times 8 \times \frac{1}{4}$ (25.82) with a capacity of 231 kips

Of these, only the $6 \times 6 \times \frac{3}{8}$ tubing qualifies for the 1997 UBC flat-width wall thickness ratio of

$$\frac{110}{\sqrt{F_y}} = 16.2 \qquad\qquad (9.2.d)$$

Using the $-3t$ deduction rule of 2.2.1.b of the *Specification* for the design of steel hollow structural sections of the AISC (April 15, 1997), $[6 - 3(\frac{3}{8})]/(\frac{3}{8})$ = 13 < 16.2.
   Finally, we have to comply with the member slenderness provisions of UBC 1997:

$$\frac{L}{r} \leq \frac{720}{\sqrt{F_y}} = 106$$

As $r = 2.27$ for the $6 \times 6 \times \frac{3}{8}$ tubing and $L = 13.5$ ft,

$$\frac{L}{r} = 13.5 \times \frac{12}{2.27} = 72 < 106$$

The $6 \times 6 \times \frac{3}{8}$ brace complies with the 1997 UBC.

Design for Tension

The maximum tensile force per CHAP. 22, DIV. IV, 9.4.a.1, of the 1997 UBC:

$$P_u = 157 \text{ kips} \qquad\qquad [10]$$

As stated, the governing equations of the *Specification* are

$$\text{(a)} \quad P_u = \phi_t F_y A_g$$

$$\text{(b)} \quad P_u = \phi_t F_u (U)(A_e)$$

where

$$U = 1 - \frac{\overline{X}}{L} = 1 - \frac{1.82}{13} = 0.86$$

with

$$\overline{X} = 1.82 \text{ in.} \qquad L = 13.0 \text{ in.}$$

and the area of the $6 \times 6 \times \frac{3}{8}$ structural tubing is

$$A_g = A_e = 8.08 \text{ in.}^2 \qquad \text{From Table 3-41 of } \textit{Specification}$$

Thus

(a) $P_u = (0.9)(46)(8.08) = 334$ kips

(b) $P_u = (0.75)(58)(0.86)(8.08) = 302$ kips $>$ 157 kips applied

The $6 \times 6 \times \frac{3}{8}$ structural tubing has adequate design strength in both tension and compression.

5.17 FILLET WELD DESIGN: THIRD- AND FOURTH-STORY GUSSET PLATES

We will design the fillet weld connection for the maximum brace reactions transmitted to the connections by components 12 and 13 in this bay bound by the external columns, third floor, and roof (CHAP. 22, DIV. IV, 9.3.a.2). The maximum load for the connection design is

$$P_u = 288 \text{ kips} \qquad \text{Computer analysis load combination [14]}$$

corresponding to the 1997 UBC special load combination of

$$1.2D + 0.5L + 0.4RE = 1.2D + 0.5L + 2.56E \qquad (3\text{-}7)$$

The capacity of a $\frac{1}{4}$ fillet weld/in. is 5.567 kips/in. with four sides of the tubing–gusset plate contact welded. The required length of fillet weld is

$$L = \frac{288}{4 \times 5.567} = 12.94 \text{ in.} \qquad \text{say 13 in.}$$

5.18 GUSSET PLATE DESIGN: THIRD AND FOURTH STORIES

We will design the gusset plates for the maximum brace reactions transmitted to the connections by components 12 and 13 in this bay bound by the external columns, third floor, and roof.

Design for Compression

The maximum compressive force is

$$P_u = 288 \text{ kips} \qquad \text{Computer analysis load combination [14]}$$

The length of the theoretical line of bending (5.9) as defined by 9.3.c.2, CHAP. 22, DIV. IV, is

$$L = 24 \text{ in.}$$

Assuming a $\frac{1}{2}$-in. plate, the area of resistance is

$$A = 0.50 \times 24.0 = 12 \text{ in.}^2$$

The compressive strength of the plate is

$$P_u = \phi_c F_e A_g = \phi_c F_y A_g = (0.85)(36)(12) = 367 \text{ kips} > 288 \text{ kips}$$

The plate is safe in compression.

Design for Block Shear

The maximum tensile force acting on the connection following formula (3-8), CHAP. 22, DIV. IV, which calls for a 256% increase in lateral seismic load, is

$$P_u = 275 \text{ kips} \qquad \text{Computer load combination [15]}$$

The potential failure surfaces are shown in Figure 5.9 with

$$ab = 13.0 \text{ in.} \qquad bc = 6.0 \text{ in.} \qquad cd = 13 \text{ in.}$$

The Test

With

$$A_{nv} = 2(13)(0.5) = 13.0 \text{ in.}^2 \qquad A_{nt} = (6.0)(0.5) = 3.0 \text{ in.}^2$$

$$0.6F_u A_{nv} = (0.6)(58)(13.0) = 452 \text{ kips} > F_u A_{nt} = (58)(3.0) = 174 \text{ kips}$$

tension yielding and shear fracture will occur. Therefore we must use Equation (J4-3b):

$$\phi R_n = 0.75[(0.6)(58)(13) + (36)(3.0)] = 420 \text{ kips} > 275 \text{ kips}$$

The plate is safe for block-shear failure.

5.19 VERTICAL COMPONENT

So far we have not addressed the vertical component of the design earthquake, E_v. For ASD design, E_v is not taken into consideration "and may be taken as zero" (1630.1.1). Not so when you are doing LRFD steel design. In that basic formula

$$E = \rho E_h + E_v \qquad \qquad (30\text{-}1)$$

$$E_v = 0.5C_a ID \qquad \text{Per 1630.1.1, item 4}$$

Caution. The component E_v is the tributary dead load acting on the frame. It would be inappropriate to include the weight of some other portions of the building in E_v, which applies to the individual frame because:

1. A large portion of the building mass is carried by the internal gravity load-resisting system.
2. The internal gravity load-resisting system consists of simple beams and girders that are attached with a pivoting mechanism to the lateral-force-resisting system [special concentrically braced frams (SCBFs)] that provides support to them in the lateral direction only, that is, against the horizontal component of the earthquake, E_h.
3. The gravity-resisting structural components on the [special concentrically braced frames (SCBFs)] are free to move up and down in an earthquake without absorbing any appreciable vertical inertial loads coming from the gravity load-resisting system.
4. Only the dead load representing the weight of the braced frame structure and portions of floor attached to it will contribute to the E_v component acting on the structure.

5.20 COLUMN DESIGN

As stated before, the column requirements are given in CHAP. 22, DIV. IV, under 6.1, "Column Strength":

6.1.a. Axial compression loads:

$$1.2P_D + 0.5P_L + 0.2P_S + 0.4R \times P_E \leq \phi_c P_n \qquad (6\text{-}1)$$

where the term $0.4R$ is greater or equal to 1.0.

Exception: The load factor on P_L in Load Combination 6-1 shall equal 1.0 for garages, areas occupied as places of public assembly, and all areas where the live load is greater than 100 psf.

6.1.b. Axial tension loads:

$$0.9P_D - 0.4R \times P_E \leq \phi_t P_n \qquad (6\text{-}2)$$

where the term $0.4R$ is greater or equal to 1.0.

6.1.c. The axial Load Combinations 6-1 and 6-2 are not required to exceed either of the following:

1. The maximum loads transferred to the column, considering 1.25 times the design strengths of the connecting beam or brace elements of the structure.
2. The limit as determined by the foundation capacity to resist overturning up-lift.

Before applying (6-1) or (6-2) let us reflect on the following:

P_E means the axial load induced by both lateral E_h and vertical E_v components of the earthquake.

The effect of E_h itself is magnified by ρ. In our case ρ equals unity.

The E_v is defined as $E_v = (0.5C_a I)D$. For our project:

$\quad C_a = 0.48 \qquad$ From our project earthquake analysis.
$\quad D \qquad\qquad$ Tributary dead load on structure
$\quad I = 1 \qquad\quad$ Seismic importance factor (per Table 16-K)

Thus the coefficient for E_v, in parentheses is

$$(0.5C_a I) = (0.5)(0.48)(1.0) = 0.24$$

According to formulas (6-1) and (6-2), both horizontal and vertical earthquake components of P_e must be magnified by the $0.4 \times R = 0.4 \times 6.4 = 2.56$

magnification factor. The earthquake force magnification has been taken care of in computer analysis load combinations [12] and [13].

Note that $1.2D + (2.56)(0.24)D = 1.8144D$ applied to load combination [12] and formula (6-1), CHAP. 22, DIV. IV, 6.1.a, can be written as

$$1.8144P_D + 0.5P_L + 2.56P_{E_h} \leq \phi_c P_n$$

where P_{E_h} indicates the axial component caused by the lateral (horizontal) earthquake load. Similarly, formula (6-2) of CHAP. 22, DIV. IV, can be re-written as

$$0.9P_D - (0.4 \times 6.4)0.24P_D - 2.56P_{E_h} \leq \phi_t P_n$$

$$0.2856P_D - 2.56\ P_{E_h} \leq \phi_t P_n$$

Design of First-Story Column for Compression (Structural Component 4)

(a) *Design of W12 × 96 for Combined Axial Force and Bending about Major Axis* The maximum axial compression complying with CHAP. 22, DIV. IV, formula (6-1), is

$$P_u = 691.0 \text{ kips} \qquad \text{Computer load combination [12]}$$

coupled with

$$M_u = 16.0 \text{ kip-ft}$$

The column is bent in single curvature so that

$$C_m = 1.0$$

For both major and minor axes, $KL = 13.5$ ft $= 162$ in., with

$$M_{nt} = 16.0 \text{ kip-ft}$$

The design parameters of the W12 × 96 are

$$A_g = 28.2 \text{ in.}^2 \qquad r_x = 5.44 \text{ in.} \qquad Z_x = 147 \text{ in.}^3$$

Using the LRFD equation (C1-1) the magnified moment is

$$M_u = B_1 M_{nt}$$

where B_1 is given by the LRFD equation (C1-2),*

$$B_1 = \frac{C_m}{1 - P_u/P_{e1}}$$

where P_u is the maximum limit axial load and $P_{e1} = \pi^2 EI/(KL)^2$, the Euler buckling strength. Values of P_{e1}/A_g can be determined from Table 8 of the *Specification*, then multiplied by A_{gross} to obtain P_{e1}.

$$\frac{KL_x}{r_x} = \frac{162}{5.44} = 29.78$$

(KL_x must be modified as the tables are built for the minor axis)

$$P_{e1} = \left(\frac{P_{e1}}{A_g}\right)A_g = (323)(28.2) = 9108 \text{ kips}$$

$$B_1 = \frac{1.0}{1 - 691/9108} = 1.082$$

$$M_u = 1.082(16) = 17.3 \text{ kip-ft}$$

Since $L_p = 12.9$ ft $< L_b = 13.5$ ft $< L_r = 61.4$ ft,

$$\phi_b M_p = 397 \text{ kip-ft} \qquad \phi_b M_r = 255 \text{ kip-ft}$$

and $BF = 2.91$ (load factor design selection table 4-18, Part 4 of the *Specification*). Applying the LRFD equations (F1-2) and (H1-1a) yields

$$\phi_b M_{nx} = C_b \left[\phi_b M_p - BF(L_b - L_p)\right]$$
$$= 1.0 \left[397 - 2.91(13.5 - 12.9)\right] = 395 \text{ kip-ft}$$

Having evaluated the moment term for the LRFD equation (H1-1a), the axial force component $\phi_c P_n$ needs to be determined from the column design tables of Part 3 of the *Specification*. To achieve this, the value of KL_x obtained above must be modified as the tables are constructed for the minor axis. The modification factor is

---

*The presence of axial force coupled with bending will give rise to secondary moments that augment the initially applied bending moment M_{nt}. The magnification factor to account for the overall effect is B_1. The reader will recognize that the secondary moment is caused by axial force times the eccentricity ($P\Delta$ effect) represented by the deflected elastic curve caused by the initial bending M_{nt}.

$$\frac{r_x}{r_y} = 1.76$$

Therefore we will enter the column table with an effective KL:

$$KL = \frac{KL_x}{r_x/r_y} = \frac{1.0(13.5)}{1.76} = 7.67 \text{ ft}$$

$$\phi_c P_n = 823 \text{ kips}$$

Using the LRFD equation (H1-1a) from the *Specification*,

$$\frac{P_u}{\phi_c P_n} + \frac{8}{9}\left(\frac{M_{ux}}{\phi_b M_{nx}}\right) = \frac{691}{823} + \frac{8}{9}\left(\frac{17.3}{395}\right) = 0.878 < 1.0$$

(b) *Design of Column about Minor Axis*   From the column design table 3-24, Part 3 of the *Specification*,

$$\phi_c P_{ny} = 747 \text{ kips} > 691 \text{ kips applied}$$

The column has adequate strength to counter the maximum compressive design loads about the principal axes.

Design of Third-Story Column for Compression (Structural Component 14)

(a) *Design of W12 × 40 Column for Combined Axial Force and Bending about Major Axis*   The maximum axial compression complying with CHAP. 22, DIV. IV of the 1997 UBC, formula (6-1), gives

$$P_u = 128 \text{ kips} \qquad \text{Computer analysis load combination [12]}$$

with end moments

$$M_1 = 25 \text{ kip-ft} \qquad M_2 = 32 \text{ kip-ft}$$

The column is bent in double curvature. By the LRFD equation (C1-3),

$$C_m = 0.6 - 0.4\left(\frac{M_1}{M_2}\right) = 0.6 - 0.4\left(\frac{25}{32}\right) = 0.287$$

Note that the ratio of moments is positive if beam and column are bent in double curvature. The design parameters of the W12 × 40 are

$$A_g = 11.8 \text{ in.}^2 \quad r_x = 5.13 \text{ in.} \quad Z_x = 57.5 \text{ in.}^3$$

From Table 8 of the *Specification* with $KL_x/r_x = 162/5.13 = 31.58$,

$$P_{e1} = 287.2 \,(11.8) = 3389.0 \text{ kips}$$

$$B_1 = \frac{C_m}{1 - P_u/P_{e1}} = \frac{0.287}{1 - 128/3389} = 0.298 < 1.0$$

Use $B_1 = 1$, the magnified moment by the LRFD equation (C1-1),

$$M_{ux} = B_1 M_{nt} = (1.0)(32.0) = 32.0 \text{ kip-ft}$$

Since $L_p = 8.0 \text{ ft} < L_b = 13.5 \text{ ft} < L_r = 26.5 \text{ ft}$,

$$\phi_b M_p = 155 \text{ kip-ft} \qquad \phi_b M_r = 101 \text{ kip-ft}$$

and $BF = 2.92$ (load factor design selection table 4-19, Part 4 of the *Specification*). Thus

$$\phi_b M_{nx} = C_b[\phi_b M_p - BF(L_b - L_p)]$$

Because the moments cause a reversed deformation curvature, we will use the LRFD equation (F1-3) to determine C_b:

$$C_b = \frac{12.5 \, M_{\max}}{2.5 M_{\max} + 3M_A + 4M_B + 3M_C}$$

$$= \frac{12.5(32)}{2.5(32) + 3(30) + 4(29) + 3(27)} = 1.09$$

where M_A, M_B, M_C are moment values at the $\frac{1}{4}$, $\frac{1}{2}$, and $\frac{3}{4}$ points of the segment. Thus

$$\phi_b M_{nx} = 1.09[155 - 2.92(13.5 - 8.0)] = 151 \text{ kip-ft}$$

For the LRFD interaction equations we need the value of $\phi_c P_n$. To obtain this value from the column design tables of Part 3 of the *Specification,* we will enter a modified KL value into the table:

$$KL = \frac{KL_x}{r_x/r_y} = \frac{1.0(13.5)}{2.66} = 5.08 \text{ ft}$$

Thus

$$\phi_c P_n = 340 \text{ kips}$$

By the LRFD interaction equation (H1-1a), we obtain

$$\frac{P_u}{\phi_c P_n} + \frac{8}{9}\left(\frac{M_{ux}}{\phi_b M_{nx}}\right) = \frac{128}{340} + \frac{8}{9}\left(\frac{32}{151}\right) = 0.565 < 1.0$$

The column has adequate strength for the combined axial compression and bending about the major axis.

(b) *Design of Column About Minor Axis* In this step no moment is coupled with the axial compression. From column design table 3-25, Part 3 of the *Specification,*

$$\phi_c P_{ny} = 249 \text{ kips} > 128 \text{ kips applied}$$

The column has adequate strength to resist the maximum design loads about the principal axes.

Design of Columns for Tension

At this point the reader is encouraged to review the concepts and steps outlined in Section 4.26, "Design of Columns." Our next step is to design the columns for tension; this will involve structural elements 1 and 21 of the mathematical model of the computer analysis. The same LRFD equations (H1-1a) and (H1-1b) presented for *compression* design will apply for *tension*. No $P\Delta$ effect, and hence no magnification factor such as B_1, will be used because, due to cyclic reversals caused by earthquakes and unlike in the case of axial compression, any initial postbuckling or moment-induced crookedness will be reduced, if not eliminated, by the tensile force, which tends to straighten out the deflected chord under tension (See Figures 5.11, 5.12, and 5.13.)

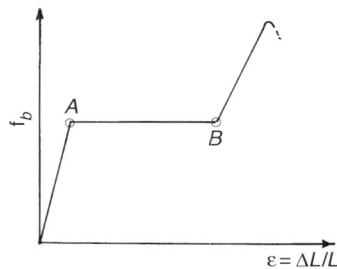

Figure 5.11 Control points of the *idealized* stress–strain diagram.

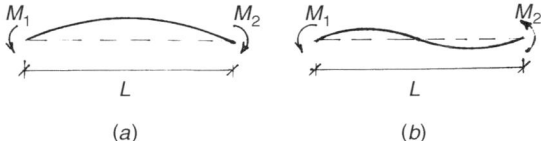

Figure 5.12 Deformed beam shapes determining unsupported length design parameters.

Design of First-Story Column for Tension (Structural Component 1)

(a) *Design of W12 × 96 for Combined Axial Force and Bending about Its Major Axis* ($A_g = 28.2$ in.^{2})   The maximum axial tensile force, mandated by formula (6-2) of CHAP. 22, DIV. IV, 2211.4, UBC 1997, is $P_u = 636$ kips, load combination [13] of the computer analysis, which includes vertical uplift component $E_v = 0.5C_aID$, formula (30-1), CHAP. 16, DIV. IV, of UBC 1997, associated with 16.0 kip-ft bending moment:

1. Bending component term of the interaction equation:

$$L_p = 12.9 \text{ ft} < L_b = 13.5 \text{ ft} < L_r = 61.4 \text{ ft}$$

$\phi_b M_p = 397$ kip-ft   From LRFD selection table 4-18, Part 4

$\phi_b M_r = 255$ kip-ft   From LRFD selection table 4-18, Part 4

$BF = 2.91$ From LRFD selection table 4-18, Part 4

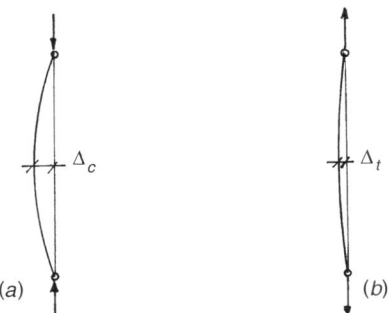

Figure 5.13 *Postbuckled* deformed shapes: (*a*) under compression; (*b*) under tension.

Moment connection corresponding to $L_p < L_b < L_r$:

$$\phi_b M_{nx} = C_b[\phi_b M_p - BF(L_b - L_p)] \leq \phi_b F_y Z$$

$$\phi_b M_{nx} = 1.0[397 - 2.91(13.5 - 12.9)] = 395 \text{ kip-ft}$$

2. Axial force term of the LRFD interaction equation:

$$\phi_t P_n = \phi_t F_y A_g = (0.9)(36)(28.2) = 914 \text{ kips}$$

$$\frac{P_u}{\phi_t P_n} = \frac{636}{914} = 0.70 > 0.2 \qquad \text{Use Equation (H1-1a)}$$

Using the *Specification* interaction equation (H1-1a) yields

$$\frac{P_u}{\phi_t P_n} + \frac{8}{9}\left(\frac{M_{ux}}{\phi_b M_{nx}}\right) = 0.70 + \frac{8}{9}\left(\frac{16.0}{395}\right) = 0.74 < 1.0$$

(b) *Design for Tension about Minor Axis* (*No Moments Present*)   To evaluate $\phi_t P_n$, two conditions must be checked:
1. $\phi_t P_n = \phi_t F_y A_g$ with $\phi_t = 0.9$
   $\phi_t P_n = (0.9)(36)(28.2) = 914 \text{ kips} > 636 \text{ kips} \leftarrow$ **GOVERNS**
2. $\phi_t P_n = \phi_t F_u U A_g$ with $\phi_t = 0.75$
   $\phi_t P_n = (0.75)(58)(0.9)(28.2) = 1104 \text{ kips} > 636 \text{ kips}$

The column is safe against uplift-caused tension.

Design of Third-Story Column for Tension (Structural Component 21)

The maximum design axial tensile force, mandated by CHAP. 22, DIV. IV, 2211.4, formula (6-2), is represented by computer analysis load combination [13]. Two issues will be addressed:

(a) *Combined Axial Force and Bending about Major Axis*   The maximum axial tensile force is

$$P_u = 106 \text{ kips}$$

associated with

$$M_{ux,\text{max}} = 18.0 \text{ kip-ft} \qquad M_{ux,\text{min}} = 15.0 \text{ kip-ft}$$

$$C_b = \frac{12.5 \, M_{max}}{2.5 M_{max} + 3M_A + 4M_B + 3M_C}$$

$$= \frac{12.5(18.0)}{2.5(18.0) + 3(17.3) + 4(16.5) + 3(15.8)} = 1.07$$

The properties of the W12 × 40 column are

$$A_g = 11.8 \text{ in.}^2 \qquad r_x = 5.13 \text{ in.} \qquad r_y = 1.93 \text{ in.}$$

$$L_p = 8.0 \text{ ft} < L_b = 13.5 \text{ ft} < L_r = 26.5 \text{ ft}$$

$$\phi_b M_p = 155 \text{ kip-ft} \qquad \text{From LRFD selection table 4-19, Part 4}$$

$$\phi_b M_r = 101 \text{ kip-ft} \qquad \text{From LRFD selection table 4-19, Part 4}$$

$$BF = 2.92 \qquad\qquad \text{From LRFD selection table 4-19, Part 4}$$

The moment correction corresponding to

$$\phi_b M_n = C_b[\phi_b M_p - BF(L_b - L_p)]$$

$$\phi_b M_n = 1.07[155 - 2.92(13.5 - 8.0)] = 149 \text{ kip-ft}$$

Using the *Specification* interaction equation (H1-1a) yields

$$\frac{P_u}{\phi_t P_n} = \frac{106}{382} = 0.278 > 0.2$$

$$\frac{P_u}{\phi_t P_n} + \frac{8}{9}\left(\frac{M_{ux}}{\phi_b M_{nx}}\right) = 0.278 + \frac{8}{9}\left(\frac{18.0}{149}\right) = 0.385 < 1.0$$

(b) *Design of Column for Tension about Minor Axis*   In the absence of bending moment,

$$P_u = 106 \text{ kips}$$

The following capacity equations apply:

$$\phi_t P_n = \phi_t F_y A_g \qquad \text{with } \phi_t = 0.9$$

$$\phi_t P_n = \phi_t F_u U A_g \qquad \text{with } \phi_t = 0.75$$

$$\phi_t F_y A_g = 0.9(36)(11.8) = 382 \text{ kips} \leftarrow \textbf{GOVERNS}$$

$$\phi_t F_u A_g = (0.75)(58)(0.9)(11.8) = 462 \text{ kips}$$

The balance of this calculation yields

$$\phi_t F_y A_g = 382 \text{ kips} > 106 \text{ kips applied}$$

The column has adequate tensile resistance about the principal axes.

Note: Detailed design presentation of the second- and fourth-story columns (structural elements 9 and 19) is not included since it can be verified that these components are less critically stressed than their counterparts beneath at the corresponding lower story. For instance, while size of the second-story column was kept unchanged—as was the first-story column—the reader will note a dramatic drop in the member force as compared to the column below. Reducing column size at each floor was avoided because frequent splicing would increase cost and slow down construction.

5.21   COLUMN SPLICE DESIGN: THIRD FLOOR

Note the change in column dimension above the third floor to compensate for reduced forces at the upper levels. A pair of splice plates to counter uplift will achieve effective connection. Partial penetration weld will be provided to butting column webs to resist moments. In compliance with CHAP. 22, DIV. IV, 6.1.b, of UBC 1997, we will design the splice for uplift resulting from load combination (6-2) of CHAP. 22, DIV. IV:

$$P_{\min} = 0.90P_{DL} - 0.4R \times P_E \le \phi_t P_n$$

where the subscript min refers to forces associated with maximum uplift. The provisions call for a design to follow the applicable specific "detailed systems design requirements" of CHAP. 22, DIV. IV, 6.1 and 6.2 of UBC 1997.

Assume $P_{\min} = 109$ kips—from load combination [13], computer analysis, and the associated moment at the splice—40 in. above floor level (TOS) (CHAP. 22, DIV. IV, 6.2.b):

$$M_x = 4.75 \text{ kip-ft}$$

However, the partial-penetration welds of the column flanges must be able to resist 150% of the applied force/moment with $\phi_w = 0.8$ and $F_w = 0.6F_{EXX}$ (CHAP. 22, DIV. IV, 6.2.b.1). Therefore the magnified moment is

$$M_{ux} = 1.5 \times 4.75 = 7.20 \text{ kip-ft}$$

We start our design by determining the dimensions of the splice plate.

Assuming a thickness $t = \frac{3}{16}$ in. and using the basic strength relationship,

$$\tfrac{1}{2}P_u = \phi_t F_y A_g = \phi_t F_y tw = (0.9)(36)(0.1875)w = 6.075w$$

$$w = \frac{106}{2(6.075)} = 8.73 \text{ in.} \qquad \text{say 9-in.-wide plate}$$

$$P_u = 2\phi_t F_u U A_g = 2(0.75)(58)(0.75)(0.1875)(9.0) = 110 \text{ kips} > 106 \text{ kips}$$

The plates have sufficient strength.

The two $\tfrac{3}{16} \times 9$ plates will be welded, one to each side of the column web. To obtain the required plate length, the weld needs to be designed. By choosing a $\tfrac{1}{8}$-in. weld size, we will determine the required weld length L and associated plate length. The design strength of the weld across the throat of an $\tfrac{1}{8}$-in. fillet weld in shear per 1-in. length is given by

$$0.707 t \phi_w F_w = 0.707 t (\phi_w)(0.6 F_{EXX}) = 0.707(\tfrac{1}{8})(0.75)(0.6 \times 70)$$

$$= 2.7838 \text{ k/in.}$$

where t = weld size. Therefore the required weld length with four lengths to match two splice plates is

$$L = \frac{106}{4 \times 2.7838} = 9.52 \text{ in.} \qquad \text{say 10 in.}$$

The induced moment, magnified by 150%, will be counteracted by a $\tfrac{3}{8}$-in., 8-in.-long partial-penetration weld (the width of the upper column) with a flexural strength over 10 times larger than applied.

It was felt necessary to provide a weld size of $\tfrac{3}{8}$ in. larger than mandated by strength considerations. The reader is reminded that the column flange thickness of the lower column is 0.9 in. and that specifying a too-small weld might result in a brittle, unsafe weld on account of the cooling-off effect of the mass of the column flange. Table J2.3, Part 6 of the *Specification,* gives a minimum weld size of $\tfrac{5}{16}$ in. for a thickness of $\tfrac{3}{4}$–1 in. Figure 5.14 shows details of the column splice.

5.22 BEAM DESIGN

Design of Second-Floor Beam

Following the provisions of CHAP. 22, DIV. IV, 9.4.a.3: "A beam intersected by V braces shall be capable of supporting all tributary dead and live loads assuming the bracing is not present." The beam will be designed as a simply

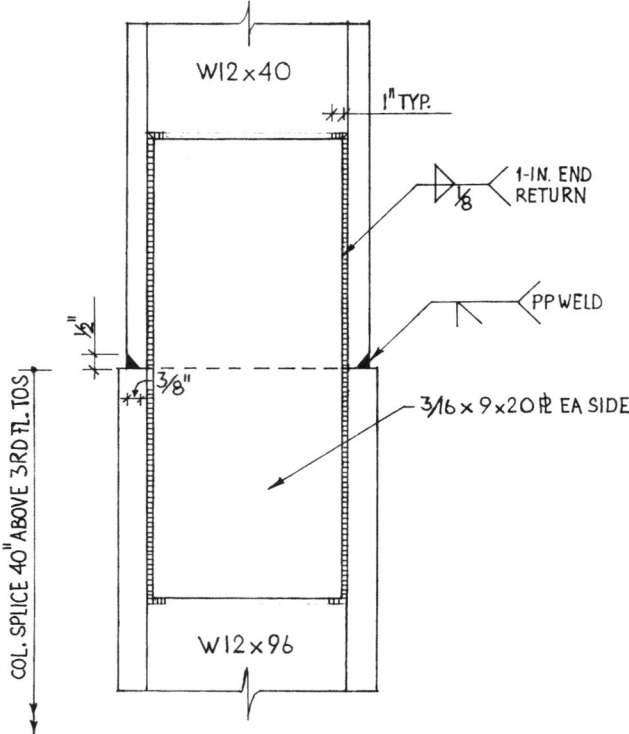

Figure 5.14   Column splice.

supported beam with $L = 19.0$ ft to resist said load. Furthermore and following CHAP. 22, DIV. IV, 9.4.a.4, we will provide fully effective lateral support to the beam at half of its span, that is, the point of intersection with the V braces. With careful design selection,

$$L_b = 9.5 \text{ ft} < L_p = 10.7 \text{ ft}$$

the full plastic moment of resistance M_p can be expected from the beam. Factored gravity loads acting on the beam are

Dead load:	$1.2D = 1.2(0.8) = 0.96$ kip-ft
Live load:	$1.6L = 1.6(0.7) = \underline{1.12}$ kip-ft
	2.08 kip-ft

However, per CHAP. 22, DIV. IV, 9.4.a.1: "The design strength of the brace members shall be at least 1.5 times the required strength using Load Combinations 3-5 and 3-6." We will multiply both moment and axial force in the brace member by 1.5. The maximum simply supported bending moment is

$$M_u = 1.5\left(\frac{wL^2}{8}\right) = 1.5\left(\frac{2.08 \times 19.0^2}{8}\right) = 141 \text{ kip-ft}$$

The maximum associated axial force acting at the centerline of the beam is given as

$$P_x = 15 \text{ kips} \qquad \text{From load case [3], computer analysis}$$

$$P_{ux} = 1.5(15) = 22 \text{ kips}$$

The beam will be designed for combined bending and axial force about the major axis with

$$KL = 9.5 \text{ ft}$$

and for pure axial load about the potential out-of-plane buckling about its minor axis. Going back to the LRFD equations (H1-1a) and (H1-1b) and the LRFD moment magnification equation (C1-5), the properties of the W10 × 54 are

$$A_g = 15.8 \text{ in.}^2 \qquad r_x = 4.37 \text{ in.} \qquad r_y = 2.56 \text{ in.}$$

$$\frac{KL_x}{r_x} = \frac{9.5 \times 12}{4.37} = 26 \qquad \frac{KL}{r_y} = 44.5$$

with

$$\frac{P_u}{\phi_c P_n} = \frac{22}{465} = 0.05 < 0.2$$

Use the LRFD equation (H1-1b).
    From Table 8 of the *Specification* with $KL/r_x = 26$,

$$P_{e1}A_g = 423.4 \text{ kips}$$

$$B_1 = \frac{C_m}{1 - P_u/P_{e1}} = \frac{1.0}{1 - 22/423.4} = 1.055$$

$$M_{ux} = 1.055 \times 141 = 149 \text{ kip-ft} \qquad \text{From LRFD equation (C1-5)}$$

By applying the LRFD interaction equation (H1-1b) with

$$L_b = 9.5 \text{ ft} < L_b = 10.7$$

full plasticity is achieved and

$$\phi_b M_p = 180 \text{ kip-ft} \qquad \text{From LRFD selection table 4-19,}$$

$$\text{Part 4, } \textit{Specification}$$

$$\phi_c P_n = 465 \text{ kips}$$

Remember that $\phi_c P_n$ comes from the column design table 3-27 of the *Specification* but with a KL_x modified by the r_x / r_y factor for the major axis.

The beam is safe in carrying the factored gravity loads without support from the braces.

Magnitude of Postbuckling Effect

Figures 5.9 and 5.10 illustrate the design-detailing requirements for the out-of-plane buckling of the brace.

The next issue is the design of beams for inverted-V or chevron bracing. In this structural system the braces intersect at the midspan of the floor beam. There are two important aspects for consideration:

1. The 1997 UBC clearly reiterates C707.4.1 of the SEAOC "Recommended Lateral Force Requirements and Commentary" that the beam should be capable of supporting both dead and live loads without the help of the brace system "in the event of a loss in brace capacity." The 1997 UBC provisions for safeguarding such an event state: "The beam intersected by braces shall be continuous between columns" (CHAP. 22, DIV. IV, 9.4.a.2), "A beam intersected by V braces shall be capable of supporting all tributary dead and live loads assuming the bracing is not present" (CHAP. 22, DIV. IV, 9.4.a.3), and "The top and bottom flanges of the beam at the point of intersection of V braces shall be designed to support a lateral force equal to 1.5 percent of the nominal beam flange strength $(F_y b_c t_f)$" (CHAP. 22, DIV. IV, 9.4.a.4).

2. At the postbuckling stage, when the internal compressive force in the compression brace element will be smaller than the tensile force in the stiffer tensile brace, the unbalanced vertical component will impact on the horizontal connecting beam, which could cause a plastic hinge and large vertical beam deformation at the intersection.

   Unlike the 1994 UBC, the 1997 edition is silent about the magnitude of such unbalanced force and beam stiffness/strength requirement to counter such unbalanced force. In reviewing research data the issue appears complex.

Among important parameters impacting on one another are the *driving function* of the earthquake (frequency, magnitude, relationship between p and s waves), the relative stiffness of brace and beam elements, and the slenderness ratio of the braces.

A rough and somewhat simplified picture of the mechanism is as follows: A marked deterioration of compression brace member stiffness develops after a number of cycles. The stiffness/resistance of the compression brace can drop as low as 30% of the initial stiffness of the brace. A great number of researchers and the SEAOC *Recommendations* agree on this value (C708.4).[5]

There is a consensus among researchers and design engineers that the tensile brace retains, during the postbuckling phase, a stiffness and resistance greater than the compression brace. However, there seems to be less agreement over quantitative determination of the exact value of the tensile force in the tensile brace that is coupled with a reduced compression in the corresponding compressive brace.[6–9]

Some options for the analysis are as follows:

(a) Energy methods.
(b) Compatibility analysis based on inelastic deformations.
(c) Strength evaluation taking into account the maximum moment of resistance M_p of the beam at the brace intersection. What makes this option viable and relatively easy to apply in an actual design is that researchers and SEAOC agree on a 30% compression brace resistance as the lower bound.

We will apply method (c) to the floor beams of our project.

Assume the story drift is still small in relation to the dimensions of the structure. Then the unknown force in the diagonal tension brace and its vertical component can be calculated from the relationship

$$M_p = \tfrac{1}{4}(T \sin \alpha - 0.3P_n \sin \alpha)L = \tfrac{1}{4}[\sin \alpha(T - 0.3P_n)L]$$

where α is the angle measured from the horizontal to the centerline of the inclined brace, T is the axial force developed in the tension brace, and P_n is the nominal strength of the compression brace before postbuckling.

Applying the method for the second-floor beam of the SCBF prototype of our project,

$$\sin \alpha = \frac{128}{162.6} = 0.79$$

The nominal compression strength of the $7 \times 7 \times \tfrac{1}{2}$ structural tubing is derived as

$$P_n = \frac{P_u}{\phi_c} = \frac{375}{0.85} = 441 \text{ kips} \qquad \text{From column design table 3-27, Part 3,}$$

Specification

$$0.3 = 0.3(441) = 132.5 \text{ kips}$$

The nominal plastic moment of resistance M_p of the W10 × 54 beam (LRFD selection table 4-19, Part 4, *Specification*) is

$$M_p = \frac{\phi_b M_p}{\phi_b} = \frac{180}{0.9} = 200 \text{ kip-ft}$$

$$L = 18.94 \text{ ft} \quad \text{(clear span between columns)}$$

Therefore the maximum tensile force that can develop in the tensile chord simultaneous to postbuckling of the compression brace and full plastic yielding of the connecting beam is

$$T - 132.5 = \frac{4M_p}{(\sin \alpha)(L)} = \frac{4(200)}{(0.79)(18.94)} = 53.5$$

$$T = 53.5 + 132.5 = 186 \text{ kips}$$

and the vertical unbalanced component impacting at the midspan of the beam is given as

$$\sin \alpha \left(T - 0.3 \frac{P_u}{\phi_c} \right) = 0.79(186 - 132.5) = 42.2 \text{ kips}$$

The reaction by the relatively light dead and live loads on the still functioning braces was ignored.

It is worth noting that:

1. Full lateral support to beam flanges has been provided for the value of $0.015 \, F_y b_f t_f$ at the midspan of the beam where the CG of braces intersects the CG of the beam.

2. By proper choice of the beam the unsupported length of the W12 × 54 beam $L_b = 9.5 \text{ ft} < L_p = 10.7 \text{ ft}$, full development of plastic moment M_p can be expected at the beam–brace intersection, facilitating formation of a true plastic hinge.

   The results of research work aimed at identifying weaknesses in the conventional CBF design and developing a much improved SCBF adapted to strong ground motion seismic areas have been briefly described earlier in this chapter in Section 5.10.

Associated with the research activities, studies have also been conducted about the magnitude and nature of internal forces induced in the family of the conventional inverted-V braced frame under static loading.[10]

Six configurations were studied involving different brace arrangements plus the addition of vertical struts at intersecting points of the brace. Among these are the arrangements termed STG and its ZIP variant, a "zipper" type of configuration. (See Figures 5.15a, b, and c.) The magnitude of the internal forces was smaller than those carried by its cousins, in particular the V-braced frame, named VREG (conventional V CBF).

Despite modest savings achieved using the STG or ZIP, the problem of energy dissipation remains at the *anchor points* of the structure where an elastic medium, the steel structure, meets a virtually infinitely stiff medium— concrete footings or basement wall—demanding a high concentration of energy dissipation. The sudden change in stiffness—slender steel members meeting large masses of unyielding concrete—will cause even a well-designed structure to undergo brittle fracture failure.

Such was the case for the Oviatt Library at the California State University in Northridge, where 4-in.-thick steel base plates of the concentrically braced frame structure shattered in brittle fracture, in addition to other visible damage: bending of $1\frac{3}{4}$-in. anchor bolts and punching, shear-type failure cracks around the perimeter of columns solidly welded to base plates. This was most likely due to lack of sufficient energy dissipation at the crucial anchor points in an otherwise conservatively designed structure.

This brings us to a new field of development: *energy-dissipating mechanisms.* The purpose of these is to improve energy dissipation of the structure. Following the Loma Prieta and Northridge earthquakes it became evident that, contrary to general belief, steel structures possess relatively small inherent damping unless heavy—nonengineered architectural or engineered stiff elements such as concrete or masonry walls—help in reducing the prolonged swaying and time-dependent story drifts that occur during an earthquake[11]:

> During the Loma Prieta earthquake ($M = 7.1$), the East Wing of the 13-story Santa Clara Civic Center Office Building exhibited strong, prolonged building

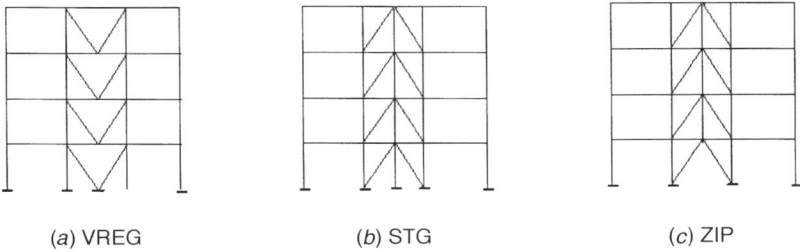

(a) VREG   (b) STG   (c) ZIP

Figure 5.15 Variations of concentric braced frame (CBF).

response. Results from a study conducted after the earthquake showed a lack of inherent damping to be the primary cause for the building's poor dynamic behavior. . . . researchers determined the building had very low inherent damping (<1%) and that bracing alone would not solve the problem experienced during the Loma Prieta Earthquake. One unusual feature was the long duration of strong vibration (available records measured response in excess of 80 seconds with little sign of decay and a torsional beat measured at 100 seconds). [From ref. 12.]

Although the consensus among engineers is still 2–3% inherent critical damping, researchers involved with the evaluation of the dynamic response of structures during an earthquake believe that the actual inherent damping of steel structures having only light curtain walls built in the last decades is considerably lower and seldom exceeds $\frac{1}{2}\%$. The general thought about the causes for such low actual damping values is the successful elimination of massive internal and external masonry and concrete architectural walls, stiff features to reduce the mass and impurity of structural response in modern earthquake engineering. It seems paradoxical that such commendable engineering effort leads to another issue: how to provide added damping to deprived structures. In Chapter 13 of this book we will discuss some of the new trends in engineering and project design.[13–15]

5.23   COLUMN BASE-PLATE DESIGN

The design discussed here involves two issues:

1. *Design the base plate for maximum axial compression.* The maximum axial design force is 690 kips [12], derived from formula (30-1), CHAP. 16, DIV. IV, 1630.1.1 of the 1997 UBC:

$$E = \rho E_h + E_v \qquad (30\text{-}1)$$

   Therefore it includes both lateral and vertical earthquake force components.

   The reader will recall that we had to multiply P_E, generated by E_h and E_v earthquake components, by $0.4R$ to obtain the maximum column design load. Because the column transfers the same axial load to the base plate, we are obliged to design the base plate for an increased earthquake design force $(0.4R)P_E$ added to factored dead and live loads, per formula (6-1), CHAP. 22, DIV. IV.

   Because the foundation is part of the load path system, it must safely transfer all gravity and seismic loads to the ground.

2. *Design the base plate–footing connection for maximum uplift.* The minimum column reaction will be the governing force for the column base-

plate design against uplift. The maximum uplift force is based on formula (30-1) and, instead of using the fundamental load combinations of 1612.2.1, the formula is derived from the special load combinations of formula (6-2), CHAP. 22, DIV. IV. Formula (6-2) also increases the seismic-caused uplift, $0.4R = 2.56$ times. Only the tributary dead load carried by the frame itself was entered in formula (6-2).

The computed uplift is

$$P_u = 634 \text{ kips} \qquad \text{Computer analysis load case [13]}$$

This is an extremely large uplift force demanding an equally large foundation capacity to counteract it, which might not prove practicable. However, CHAP. 22, DIV. IV, 6.1.c, states:

6.1.c. The axial Load Combinations 6-1 and 6-2 are not required to exceed either of the following:

1. The maximum loads transferred to the column, considering 1.25 times the design strengths of the connecting beam or brace elements of the structure.
2. The limit as determined by the foundation capacity to resist overturning up-lift.

The uplift limit for the foundation capacity for this project is $P_u = 500$ kip/ footing. We will use this value for maximum uplift in the column base-plate design.

Design of Base Plate for Compression

Design Parameters

Size of footing	10 ft $\times$ 10 ft
Strength of concrete	$f'_c = 3.0$ ksi
Steel plate	A36
Bolts	A490

Procedure   We will follow the procedure outlined in "Design of Axially Loaded Base Plates," Part 11, Volume II of the LRFD *Specification.* Following is a brief review of the basic issues.

The compressive strength of both concrete and soil is considerably smaller than the strength of steel. The f'_c of concrete used for footing is 3.0–4.0 ksi as compared to 36–50 ksi of structural steel, that is just about $\frac{1}{10}$th. The function of the base plate is to spread the highly concentrated column load over a sufficiently large area to keep the footing from being overstressed. Two types of concrete design strength apply depending on the relative geometric relationship of base plate and footing:

(a) Unconfined: LRFD equation (J9-1)
(b) Confined: LRFD equation (J9-2)

When the base plate covers the entire concrete support area, we refer to the uniaxial strength of concrete, and the resisting force is expressed as

$$P_u = \phi_c P_p = \phi_c(0.85 \ f'_c A_1) \qquad \text{From LRFD equation (J9-1)}$$

where A_1 = contact area
ϕ_c = 0.60 for bearing on concrete

from which

$$A_1 = \frac{P_u}{\phi_c 0.85 f'_c}$$

On the other hand, the confined strength of concrete is activated if the load-receiving concrete area—the top surface of the footing—is larger than the A_1 contact area. The gross concrete area $A_2 > A_1$ surrounding the contact area gives lateral confinement to the concrete in contact with the base plate. The correction factor to allow increase in bearing strength is given as

$$\sqrt{\frac{A_2}{A_1}} \text{ by the } \textit{Specification} \qquad \text{Limited to maximum value of 2}$$

The design strength includes the effect of confinement:

$$\phi_c P_p = \phi_c(0.85 \ f'_c A_1)\sqrt{\frac{A_2}{A_1}} \qquad \text{From LRFD equation (J9-2)}$$

Then

$$A_1 = P_u / \phi_c(0.85 \ f'_c A_1)\sqrt{\frac{A_2}{A_1}} \qquad \text{where } \sqrt{\frac{A_2}{A_1}} \leq 2$$

For the actual computation we refer to the procedure outlined by W. A. Thornton[16] and adopted by the *Specification:*

Notation

d, b_f	Depth and flange width of column
B	Width of base plate
N	Length of base plate
t	Required thickness of base plate
m, n	Cantilever *overhang* dimensions of base plate (Figure 5.16)
l	Maximum value of m, n, or $\lambda n'$
X	$\left[\dfrac{4db_f}{(d + b_f)^2} \right] \dfrac{P_u}{\phi_c P_p}$
P_p	Design strength of concrete bearing area
λ	$2 \dfrac{\sqrt{X}}{1 + \sqrt{1 - X}} \le 1$
$\lambda n'$	$\lambda\sqrt{db_f}/4$

The thickness of the plate is determined by

$$t = l\sqrt{\frac{2P_u}{0.9F_y BN}}$$

Steps to Design Column Base Plate of Braced Frame

1. Compute A_2.
2. Compute A_1 from LRFD equation (J9-2).
3. Select B and N.

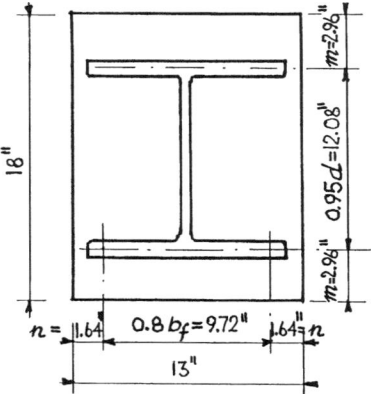

Figure 5.16 The LRFD design dimensions of the base plate.

4. Determine values of m, n, and $\lambda n'$.
5. Determine the required plate thickness:

$$A_2 = 120^2 = 14,400 \text{ in.}^2$$

$$A_1 = \frac{690}{0.6(0.85)(3.0)(2)} = 226 \text{ in.}^2$$

Note that $\sqrt{A_2/A_1} > 2$; we are using the value of 2.

The selection of B and N is also determined by geometric considerations to accommodate connection requirements for uplift:

$$B = 13.0 \text{ in.} \quad \text{and} \quad N = 18.0 \text{ in.}$$

are chosen with

$$A_1 = 234 \text{ in.}^2 > 226 \text{ in.}^2$$

Note that by selecting 13 in. for B, some conventional, although not sacred, rules of the construction industry which favor an even number of inches for plate dimension have been bypassed. However, plate sizes as chosen above can be ordered and delivered to the project site without difficulties.

$$m = \frac{N - 0.95d}{2} = \frac{18.0 - (0.95 \times 12.71)}{2} = 2.96 \text{ in}$$

$$n = \frac{B - 0.8b_f}{2} = \frac{13.0 - (0.80 \times 12.16)}{2} = 1.64 \text{ in.}$$

$$X = \left[\frac{4db_f}{(d + b_f)^2} \right] P_u/\phi_c P_p = \left[\frac{4(12.71)(12.16)}{(12.71 + 12.16)^2} \right] 690/716 = 0.963$$

$$\lambda = \frac{2\sqrt{X}}{1 + \sqrt{1 - X}} = \frac{2\sqrt{0.963}}{1 + \sqrt{1 - 0.963}} = 1.646$$

$$\lambda n' = \frac{\lambda\sqrt{db_f}}{4} = \frac{1.646\sqrt{(12.71)(12.16)}}{4} = 5.12 \text{ in} \leftarrow \textbf{GOVERNS}$$

$$t = l\sqrt{\frac{2P_u}{0.9F_y BN}} = 5.12\sqrt{\frac{2(690)}{0.9(36)(13.0)(18.0)}} = 2.18 \text{ in}$$

Provide a 13 × 18 × 2¼ base plate.

Design of Base Plate for Uplift

We will provide four A490 bolts. The design tensile strength of four $1\frac{1}{2}$ ϕ A490 bolts using Table 8-15, Volume II of the *Specification,* is

$$\phi_t P_u = 4(150) = 600 \text{ kips} > 500 \text{ kips applied}$$

Figure 5.17 shows the disposition of the bolts.

The required plate thickness to resist bending caused by the pair of bolts on each side of the column flange is determined as

$$\tfrac{1}{4}(0.9)(F_y)(B)(t)^2 = \tfrac{1}{2}P_u(I)$$

where I is the lever arm in inches. Then

$$t = \sqrt{\frac{2P_u I}{0.9 F_y B}} = \sqrt{\frac{2(500)(1.5)}{(0.9)(36)(13)}} = 1.887 \text{ in.} < 2.25 \text{ in.} \qquad \textbf{Provided}$$

Provide two L5 $\times$ 3 $\times$ $\frac{5}{8}$ $\times$ (4 in. long) to act as reinforcement welded to the base plate. This will supply the extra strength needed to improve the response of conventionally designed base plates (as experience has taught us) against stress reversals caused by strong ground motion. The base-plate dimensions, bolts, and connections are shown in Figure 5.17.

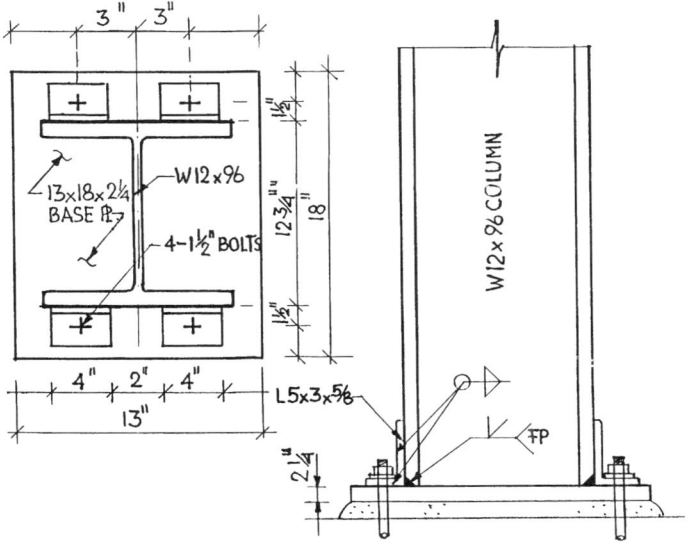

Figure 5.17 Base-plate dimensions, bolts, and connections.

5.24 SUMMARY OF DESIGN PROCEDURES

It would be appropriate to compare, once again, the results of the two different designs illustrated in Figures 5.7 and 5.8 to emphasize the importance of complying with all relevant provisions of the 1997 UBC. Figure 5.7 shows results of a design that, although applying the provisions for load combinations and lateral force analysis of CHAP. 16, did not incorporate the detailed systems design requirements of CHAP. 22, DIV. IV, for a 1997 UBC acceptable seismic design. The reader will note that the structural member sizes thus obtained are underdesigned as compared to the frame member sizes shown in Figure 5.8, synthesis of a structural design that takes into consideration the detailed systems design requirements of 1997.

5.25 SEAOC *BLUE BOOK* AND THE CODE

The SEAOC *Blue Books Recommendations* strongly influenced the UBC and IBC provisions and continued to be a determining factor in guiding the engineering community on the design of better and safer structures in California and other seismic zones. The *Recommendations* contain design principles based on lessons learned from Northridge as well as extensive empirical research. The results of experimental research on braced frames in particular were primarily based on over 20 years of tests on braced frames subjected to cyclic loading. The *Recommendations* also summarize the results of the FEMA/SAC joint venture work on steel moment frames reinitiated after Northridge.

Apart from the above considerations the reader is reminded that, while containing recommendations that will eventually be part of the IBC, the *Blue Book* is not per se a legally binding document. The IBC editions, in turn, become a binding document when adopted regionally by local legislators.

Steel is quoted in the *Recommendations* as one of the most efficient engineering solutions to counter earthquakes and with regard to CBFs: "Since their adoption into seismic design codes, improvements have been made to CBFs with emphasis on increasing brace strength and stiffness, primarily through the use of higher design forces that would minimize inelastic demand." The emphasis is on better brace performance and stiffness achieved by higher component design forces required by the special detailing requirements of the 1997 UBC (CHAP. 22, DIV. IV). The requirement for increased brace member stiffness goes well with research findings showing that insufficient brace stiffness promotes global buckling and leads to *pinching—* reduction—of the area of hysteresis envelope, a measure of the energy absorption of the system.

"More recently," the *Blue Book Commentary* adds, "ductility as an essential ingredient distinguishing lateral force resisting systems in seismically active areas, has been applied to the design of CBFs." *Ductility* equates to potential energy absorption of the structural system and is one of the most promising trends in engineering to improve structural response and minimize earthquake damage.

To maintain sufficient brace stiffness, the SEAOC *Recommendations* limit the brace slenderness of the SCBF to

$$\frac{KL}{r} \leq \frac{1000}{\sqrt{F_y}} \qquad\qquad \text{(C708.2)}$$

As a further refinement, the 1997 UBC subsequently restricted the brace slenderness to

$$\frac{KL}{r} \leq \frac{720}{\sqrt{F_y}} \qquad \text{From CHAP. 22, DIV. IV, 9.2.a}$$

The *Recommendations* truly reflect results of over 15 years of research work conducted chiefly at the University of Michigan as discussed earlier in this chapter. As stated in C708.1 of the *Recommendations:*

> Actual building earthquake damage (including the 1994 Northridge earthquake) as well as damage of CBFs (Concentrically Braced Frames) observed in laboratory tests have generally been caused by limited ductility and brittle failures. These brittle failures are most often observed as fracture of connection elements or brace members. Lack of compactness in braces results in severe local buckling of the brace, which leads to high concentrations of flexural strains at these locations, and reduces their ductility. . . . Extensive analytical and experimental work performed by Professor Subhash C. Goel and his collaborators has shown that improved design parameters, such as limiting width/thickness ratios (to prevent local buckling), closer spacing of stitches, and special design and detailing of end connections greatly improve post-buckling behavior of CBFs. A new system reflecting these developments, referred to as special concentrically braced frames (SCBFs), has been added to these requirements.

Indeed the 1997 UBC limits the width–thickness ratio of brace elements (CHAP. 22, DIV. IV, 9.2d):

> Braces shall be compact or non-compact, but not slender (i.e. $\lambda < \lambda r$). Circular sections shall have an outside diameter to wall thickness ratio not exceeding $1,300/F_y$, rectangular tubes shall have a flat-width to wall thickness not exceeding $\sqrt{110/F_y}$ unless the circular section or tube walls are stiffened.

Also reflecting test results, the SEAOC *Recommendations,* C708.3, "Bracing Connections," states:

> When brace buckling is in the plane of a gusset plate, the plate should provide flexural strength that exceeds that of the brace. This will ensure that plastic

hinges will form in the brace, rather than in the plate. . . . When brace buckling is out of plane of a gusset plate, the length needed to allow restraint-free plastic rotation is twice the plate thickness. This mode of buckling may lead to preferable inelastic behavior.[17]

REFERENCES

1. Black, R. G., Wenger, W. A., and Popov, E. P. 1980. "Inelastic Buckling of Steel Struts Under Cyclic Load Reversal," Report No. UCB/EERC-80/40. Berkeley: Earthquake Engineering Research Center, University of California.

2. Bruneau, M., and Mahin, S. A. 1990. "Ultimate Behavior of Heavy Steel Section Welded Splices and Design Implications." *Journal of Structural Engineering,* Vol. 116, No. 18, pp. 2214–2235.

3. Goel, S. C. 1992. "Cyclic Post Buckling Behavior of Steel Bracing Members." In *Stability and Ductility of Steel Structures under Cyclic Loading.* Boca Raton, FL: CRC Press.

4. Goel, S. C., and Hanson, R. D. 1987. "Behavior of Concentrically Braced Frames and Design of Bracing Members for Ductility." In SEAOC *Proceedings.* Sacramento, CA: Structural Engineers Association of California.

5. SEAOC. 1996. *Recommended Lateral Force Requirements and Commentary,* Seismology Committee, 6th ed. Sacramento, CA: Structural Engineers Association of California.

6. Goel, S. C. 1992. "Earthquake-Resistant Design of Ductile Braced Steel Structures." In *Stability and Ductility of Steel Structures Under Cyclic Loading.* Boca Raton, FL: CRC Press.

7. Hassan, O., and Goel, S. C. 1991. "Seismic Behavior and Design of Concentrically Braced Steel Structures," Report No. UMCE 91-1. Ann Arbor, MI: Department of Civil Engineering, The University of Michigan.

8. Tang, X., and Goel, S. C. 1987. "Seismic Analysis and Design Considerations of Braced Steel Structures," Report No. UMCE 87-4. Ann Arbor, MI: Department of Civil Engineering, The University of Michigan.

9. Whittaker, A. S., Uang, C. M., and Bertero, V. V. 1987. "Earthquake Simulation Tests and Associated Studies of a 0.3-Scale Model of a Six-Story Eccentrically Braced Steel Structure," Report No. EERC 87/02. Berkeley, CA: Earthquake Engineering Research Center, University of California.

10. Khatib, I. F., Mahin, S. A., and Pister, K. S. 1988. "Seismic Behavior of Concentrically Braced Steel Frames," Report No. UCB/EERC-88/01. Berkeley, CA: Earthquake Engineering Research Center, University of California.

11. Erdey, C. K. 1999. "Performance of Steel Structures in California Earthquakes." Paper presented at the Eurosteel '99 Symposium, Prague.

12. Crosby, P. 1994. "Seismically Retrofitting a Thirteen-Story Steel Frame Building." *SEF Structural Engineering Forum.*

13. Hanson, R. D. 1997. "Supplemental Energy Dissipation for Improved Earthquake Resistance." Ann Arbor, MI: Department of Civil & Environmental Engineering, University of Michigan, assignment with FEMA.

14. Kareem, A., and Tognarelli, M. 1994. "Passive & Hybrid Tuned Liquid Dampers." *SEF Structural Engineering Forum,* October.

15. Perry, C., and Fierro, E. A. 1994. "Seismically Upgrading a Wells Fargo Bank," *SEF Structural Engineering Forum,* October.

16. Thornton, W. A. 1990, "Design of Base Plates for Wide Flange Columns—A Concatenation of Methods." *Engineering Journal,* Vol. AISC 27, No. 4, pp. 173–174.

17. Astaneh, A., Goel, S. C., and Hanson, R. D. 1986. "Earthquake-Resistant Design of Double-Angle Bracing," *AISC Engineering Journal,* Fourth Quarter.

CHAPTER 6

IBC SEISMIC DESIGN OF SMRF STRUCTURES

6.1 IBC SETUP OF SEISMIC DESIGN FORCES

The procedure presented here demonstrates how to set up seismic design forces for a six-story SMRF according to IBC 2000. *This procedure is also applicable to the successor code 2003, which uses essentially the same method.* Once the seismic forces are set up, the design of the structure is very much the same as outlined in our seismic design example in compliance with the 1997 UBC. This is the reason why a detailed study of a SMRF by the 1997 UBC is so strongly recommended to the reader. Tables for IBC 2000 code requirements have been reproduced from the special structural design seminar held by the Arizona Building Officials, Fall Education Institute in Phoenix, Arizona, Fall 2000.

6.2 DESIGN EXAMPLE

Project Data

- Six-story moment-resisting frame
- Roof deck: 60 psf
- Floor deck: 80 psf; includes weights of suspended ceiling, ducts, external curtain wall
- Partitions: 10 psf; treated as dead load by IBC 2000
- Soil type: *D*
- Weight of roof: 650 kips

- Weight of floor: 973 kips
- Acceleration parameters:

$$S_s = 160\% \; g$$

$$F_a = 1.0$$

$$S_{MS} = F_a S_s = 1.0 \times 1.6 = 1.6 \; g \quad \text{From IBC 2000, Table 1615.1.2-1}$$

$$S_1 = 70\% \; g$$

$$F_a = 1.5$$

$$S_{M1} = F_a S_1 = 1.5 \times 0.7 = 1.05 \; g \quad \text{From IBC 2000, Table 1615.1.2-2}$$

- The 5% damped design spectral response acceleration:

$$S_{DS} = (\tfrac{2}{3}) S_{MS} = (\tfrac{2}{3})1.6 = 1.066 \; g$$

$$S_{D1} = (\tfrac{2}{3}) S_{M1} = (\tfrac{2}{3})1.05 = 0.70 \; g$$

- Period of structure (IBC 2000, 1617.4.2-2):

$$T_a = 0.1 \times 6 = 0.6 \; s$$

- Construct design response spectrum:

$$T_s = \frac{S_{D1}}{S_{DS}} = \frac{0.7}{1.066} = 0.656 \; s$$

- Design spectral response acceleration with $T_a < T_s$:

$$S_a = S_{DS} = 1.066 \; g \quad \text{From IBC 2000, 1615.1.4.2}$$

- Seismic response coefficient:

$$C_s = \frac{S_a I}{R} = \frac{S_{DS} I}{R} = \frac{(1.066)(1.0)}{8.0} = 0.13325$$

- Seismic base shear:

$$V = C_s W = 0.13325 \times 5515 = 735 \; \text{kips}$$

· Vertical distribution of seismic forces:

$$F_x = C_{vx}V \qquad \text{From IBC 2000, 1617.4.3-1}$$

where C_{vx} is the vertical distribution factor,

$$C_{vx} = \frac{w_x h_x^k}{\displaystyle\sum_{i=1}^{n} w_i h_i^k}$$

where $k = 1.1$ by interpolation (IBC 2000, 1617.4.3). Assume

$$C_{vR} = \frac{650 \times 81}{973(13.5^k + 27.0^k + 40.5^k + 67.5^k) + 650 \times 81^k} = 0.221$$

$$C_{v6} = \frac{973 \times 67.5^{1.1}}{370{,}702} = 0.270$$

$$C_{v5} = \frac{973 \times 54.0^{1.1}}{370{,}702} = 0.211$$

$$C_{v4} = \frac{973 \times 40.5^{1.1}}{370{,}702} = 0.154$$

$$C_{v3} = \frac{973 \times 27.0^{1.1}}{370{,}702} = 0.098$$

$$C_{v2} = \frac{973 \times 13.5^{1.1}}{370{,}702} = 0.946$$

· Vertical distribution of seismic forces:

$$F_{\text{roof}} = C_{vR}V = 0.221 \times 735 = 162 \text{ kips}$$

$$F_6 = C_{v6}V = 0.270 \times 735 = 198 \text{ kips}$$

$$F_5 = C_{v5}V = 0.211 \times 735 = 155 \text{ kips}$$

$$F_4 = C_{v4}V = 0.154 \times 735 = 114 \text{ kips}$$

$$F_3 = C_{v3}V = 0.098 \times 735 = 72 \text{ kips}$$

$$F_2 = C_{v2}V = 0.046 \times 735 = 34 \text{ kips}$$

Note the significant rate of increase in the vertical distribution of forces due to the exponent of h_x (IBC 2000, 1617.4.3).

The increase in the magnitude of applied lateral seismic forces with increased height is a significant change in the seismic force distribution as compared to the 1997 UBC seismic provisions. The contrast in the vertical distribution of seismic forces becomes even more emphasized as the structures become taller. Note the influence of height in the exponent as $k \to 2.0$.

6.3 IBC BUILDING CATEGORIES

At this stage we must determine the following requirements for the building*:

(a) Seismic use group (SUG)

(b) Occupancy importance factor

(c) Seismic design category

The 2000 IBC, 1616.2, states: "Each structure shall be assigned a Seismic Use Group and a corresponding Occupancy Importance Factor." Table 6.1 shows the SUGs in a simplified manner. Table 6.3 is based on short-period response accelerations. Table 6.4 is based on 1-s period response acceleration.

Our project is an office building that belongs to SUG I. The occupancy importance factor allocated to this group is $I_E = 1.00$, as noted in Table 6.2. The seismic design category is based on the *seismic use group* and the *Design spectral response acceleration* per IBC 2000, 1615.1.3. According to Tables 6.3 and 6.4 ($S_{DS} > 0.5$ g, $S_{DI} > 0.2$ g), the building fits into seismic design category D. The occupancy importance factor enters in the base-shear calculation (C_s is a function of I_E) and is influential in determining the magnitude of the base shear and earthquake design lateral forces applied to the structure. The seismic design category determines the type of structure and the height limitation that can be built on the site and, equally important, has overall control over structural component design and its connections. Table 6.5 provides a short list of structures pertaining to moment-resisting frame systems.

TABLE 6.1 Seismic Use Groups, IBC 2000, 1612.2

Use group I	Ordinary buildings: not assigned to SUG II or III
Use group II	High-occupancy buildings: substantial hazard to human life
Use group III	Essential facilities: emergency facilities/hospitals, police stations
Use group IV	Low hazard to human life: agricultural/temporary/storage

*"Seismic use group" has been replaced by "occupancy category" in IBC 2006, issued at the time of this writing.

TABLE 6.2   The Seismic Importance Factor I_E, IBC 2000, 1616.2 amplifies design forces to control damage and achieve enhanced performance in SUGs II and III

Seismic Use Group	I_E
I	1.00
II	1.25
III	1.50
IV	1.00

The first column in Table 6.5 indicates the type of structure, the second column the detailing reference section, and the third column the response modification coefficient R that enters in the C_s seismic response coefficient calculation. The coefficient is influential in determining the V base shear. The fourth column contains the Ω_0 system overstrength factor that controls element design. The fifth column is the deflection amplification factor C_D. If complied with, this value ensures that the interstory drift is within acceptable limits due to dynamic amplification, which is intended to rectify the static effect of the *assumed* "Equivalent (Static) Lateral Force Procedure for Seismic Design of Buildings" (IBC 2000, 1617.4). The last column gives system and building height limitations by seismic design category.

It can be seen that ordinary steel moment frames, for instance, cannot be used for our project because the height of the structure is over 80 ft and the height of the ordinary moment frame is limited to 35 ft in seismic design category D. Note that ordinary moment frames would have been allowed if the project were located in a zone of lesser earthquake spectral response acceleration. Note also that an intermediate steel moment frame could be used up to a maximum height of 160 ft. Our choice, however, was the special steel moment frame with $R = 8$. Should we have opted for an intermediate steel moment frame, the value of the response modification coefficient would have been less favorable: $R = 6$.

TABLE 6.3   Seismic Design Categories, IBC 2000, Table 1616.3-1[a]

Value of S_{DS}	Design Category		
	SUG I	SUG II	SUG III
$S_{DS} < 0.167\ g$	A	A	A
$0.167\ g \le S_{DS} < 0.33\ g$	B	B	C
$0.33\ g \le S_{DS} < 0.50\ g$	C	C	D
$0.50\ g \le S_{DS}$	D[a]	D[a]	D[a]

[a]For structures located on sites with spectral response acceleration at 1-second period, S_1, equal or greater than 0.75 g, IBC 2000 allocates Seismic Design Category E to SUG I and II structures, and Category F to SUG III structures.

TABLE 6.4   Seismic Design Categories, IBC 2000, Table 1616.3-2[a]

	Design Category		
Value of S_{DI}	SUG I	SUG II	SUG III
$S_{DI} < 0.067\ g$	A	A	A
$0.067\ g \le S_{DI} < 0.133\ g$	B	B	C
$0.133\ g \le S_{DI} < 0.20\ g$	C	C	D
$0.20\ g \le S_{DI}$	D[a]	D[a]	D[a]

[a]For structures located on sites with spectral response acceleraton at 1-second period, S_1, equal or greater than 0.75 g, IBC 2000 allocates Seismic Design Category E to SUG I and II structures, and Category F to SUG III structures.

The next major step for the design is to pass the checkpoint of the redundancy coefficient ρ. The purpose of the provision is to safeguard the structural system from a potential breakdown initiated by insufficient number of force-resisting components. The failure of such components, without a redundant backup system, could trigger the chain reaction of a progressive overall structural system failure. For instance, take a floor plan that contains only one pair of moment frames, one on each side of the building to counteract earthquake forces in their major direction. If one fails, a single frame would not be able to resist the seismic lateral load alone. Its reaction, now eccentric, would not be able to generate a reaction in line with the opposing earthquake forces. For this, not to mention torsional effect, a minimum of two frames is needed.

This simplified example proves the need for redundancy. The reliability/redundancy test is to ensure that there is enough structural redundancy for the task:

$$\rho_i = 2 - \frac{20}{r_{\text{max}_i}\ \sqrt{A_i}} \qquad \text{From IBC 2000, 1617.2-2a}$$

where r_{max_i} is the ratio of the design story shear resisted by the most heavily loaded single element in the story to the total story shear for a given direction of loading. Furthermore IBC 2000 (1617.2.2) states: "The redundancy coefficient ρ shall be taken as the largest of the values of ρ calculated at each story 'i' of the structure." The 2000 IBC sets a limit of

$$\rho = 1.25$$

for moment-resisting frames and

$$\rho = 1.5$$

for other systems in seismic design category D. If the limit is exceeded, it will warn the designer that the basic safety features for sufficient redundancy have been overlooked.

TABLE 6.5 Moment-Resisting Frame System Category

Basic Seismic-Force-Resisting System	Detailing Reference Section	Response Modification Coefficient, R	Over-Strength Factor, Ω_0	Deflection Amplification Factor, C_D	System and Building Height Limitations (ft) by Seismic Design Category per Section 1616.1				
					A or **B**	**C**	**D**	**E**	**F**
Moment-Resisting Frame Systems									
Special steel moment frames	Note k(9)	8	3	$5\frac{1}{2}$	NL	NL	NL	NL	NL
Special steel truss moment frames	Note k(12)	7	3	$5\frac{1}{2}$	NL	NL	160	100	NP
Intermediate steel moment frames	Note k(10)	6	3	5	NL	NL	160	100	NP
Ordinary steel moment frames	Note k(11)	4	3	$3\frac{1}{2}$	NL	NL	35	NP	NP
Dual Systems with Special Moment Frames									
Steel eccentrically braced frames, moment-resisting connections, at columns away from links	Note k(15)	8	$2\frac{1}{2}$	4	NL	NL	NL	NL	NL
Steel eccentrically braced frames, non-moment-resisting connections at columns away from links	Note k(15)	7	$2\frac{1}{2}$	4	NL	NL	NL	NL	NL
Special steel concentrically braced frames	Note k(13)	8	$2\frac{1}{2}$	$6\frac{1}{2}$	NL	NL	NL	NL	NL
Ordinary steel concentrically braced frames	Note k(14)	6	$2\frac{1}{2}$	5	NL	NL	NL	NL	NL

Note k. AISC Seismic Part II and Section number.
NL = not limited; NP = not permitted.
Source: IBC 2000, Table 1617.6

CHAPTER 7

MASONRY STRUCTURES

7.1 INTRODUCTION

How did masonry buildings fare in the Northridge earthquake? The City of Los Angeles assessment of building damage listed 350 tilt-ups/masonry that suffered moderate damage with 200 partial roof collapses and 213 unreinforced masonry with 6 partial roof collapses within the commercial group of structures. It was also noted that, while overall damage to reinforced masonry buildings was not as extensive as for tilt-ups, there was severe damage to many reinforced masonry buildings:

> The collapses that occurred in pre 1976 tilt-ups and reinforced masonry buildings were due in most cases to the lack of an adequate anchorage system of the walls to the roof and floor diaphragms, and due to the omission of an adequate pilaster-to-girder connection considering the wall panel between pilasters to be supported on 4 sides. The collapses in post 1976 of tilt-ups and reinforced masonry structures were primarily due to the following:
>
> 1. The failure of wall anchorage and sub diaphragm continuity tie connectors, where their ultimate capacity was about half of the amplified force level in the diaphragm;
> 2. Excessive elongations of the anchoring system that caused purlins/sub-purlins to lose their vertical support from hangars and plywood/strap anchor nails to lift out of the connecting members;
> 3. The failure of eccentric and twisted connectors with gross eccentricities.
>
> Also, quality control was a major factor in the failure of TU/RM structures, as it was in all low-rise construction.

The city implemented emergency measures that affected all building permits, including repair and retrofit permits issued after November 1994.

It is noteworthy that the City of Los Angeles acknowledged that part of the damage to masonry structures was the result of design shortcomings and inadequate quality control rather than inherent weaknesses in the type of structure.

The growth of reinforced masonry design as we know it in modern masonry times has been continuous from its 1933 Long Beach earthquake birth pains. . . . It devastated a jurisdiction of the Uniform Building Code, which had been organized earlier to disseminate Code improvements and uniformity in the western United States, along part of the Pacific Ring of Fire, a line of great seismic activity around the Pacific tectonic plate edge. The UBC was developed by the Pacific Coast Building Officials Conference, organized for building code uniformity, and was functioning well, with a regular pattern of review, improvement and updating in cooperation with professional building design profession. Hence, distribution of new methods was wide and prompt. The name was later changed to the International Conference of Building Officials (ICBO) as its influence spread internationally from the Pacific Coast jurisdictions.[1]

Figure 7.1   Unreinforced masonry buildings do not fare well in earthquakes. The photo shows the collapsed "House of Bread" in the city of Paso Robles, California during the 6.5-magnitude San Simeon earthquake of December 2003. The Old Clock Tower, symbol of Paso Robles, built in 1892 in the corner of the same building was also destroyed when the second story of the structure collapsed onto the street during the earthquake. (Courtesy of the University of California Berkeley Library. Photo: Janise E. Rodgers.)

7.2 CASE 1: RETAINING WALL SYSTEM

Seismic: Moderate Seismicity region

The engineering calculations for this project comply with the provisions of the 1997 UBC, Section 2107, "Working Stress Design of Masonry," applicable to moderate- to high-seismicity areas. The 1997 UBC specifically notes in 2107.1.3: "Elements of masonry structures located in Seismic Zones 3 and 4 shall be in accordance with this section." Soil bearing pressure is 2500 psf per soil report investigation. Soil profile: S_C. Because in our project the elevation of the retained ground varies, so does the height of the retaining wall system. (See Figures 7.2, 7.3, and 7.4.) We investigated the following clear heights of retained earth: 10 ft, 13 ft, and 15 ft. *Clear height = distance between top of wall and top of footings.*

ASSUMPTIONS

Assuming a 15-in.-deep toe and heel depth for the 10-ft wall and 18-in. depth for the 13- and 15-ft walls, the overall depth for overturning effect will be 11.25, 14.5, and 16.5 ft, respectively. The lateral active

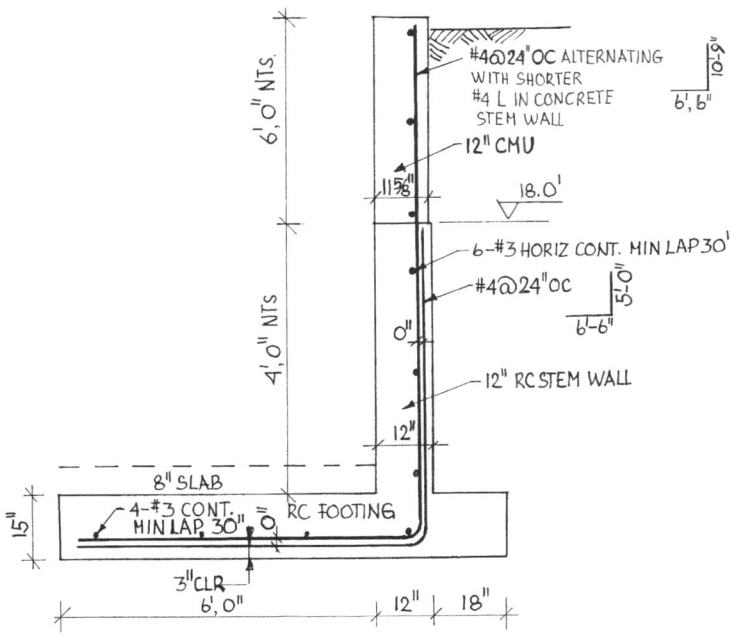

Figure 7.2 A–A 10-ft retaining wall.

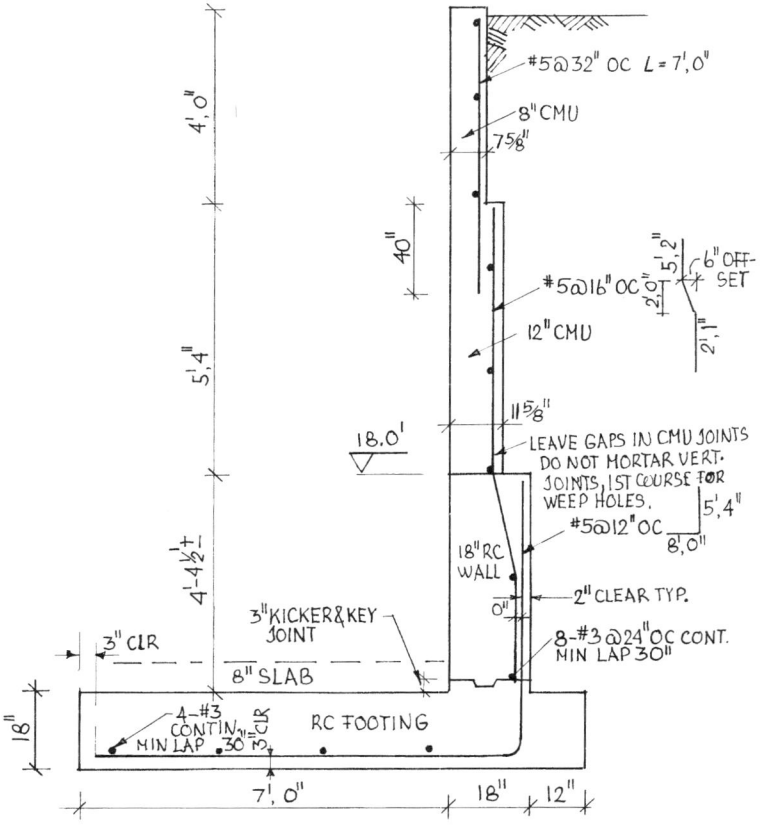

Figure 7.3 B–B 11–13-ft, 8-in. retaining wall.

soil pressure given by the soil report was 35 psf for each foot of depth acting as an equivalent lateral fluid pressure. The retaining walls consist of a top portion of reinforced-concrete masonry wall of variable thickness and a 4.0–6.0-ft-high, reinforced-concrete stem wall to which the top reinforced CMU (concrete masonry unit, type S mortar) portion is attached, sharing continuous vertical reinforcement. The engineering design parameters of the CMU are

$$f'_m = 1500 \text{ psf} \qquad E_m = 750 \times 1500 = 1,125,000 \text{ psi}$$

$$n = \frac{E_s}{E_m} = \frac{29,000}{1125} = 26$$

Allowable bending stress (no special inspection required) is

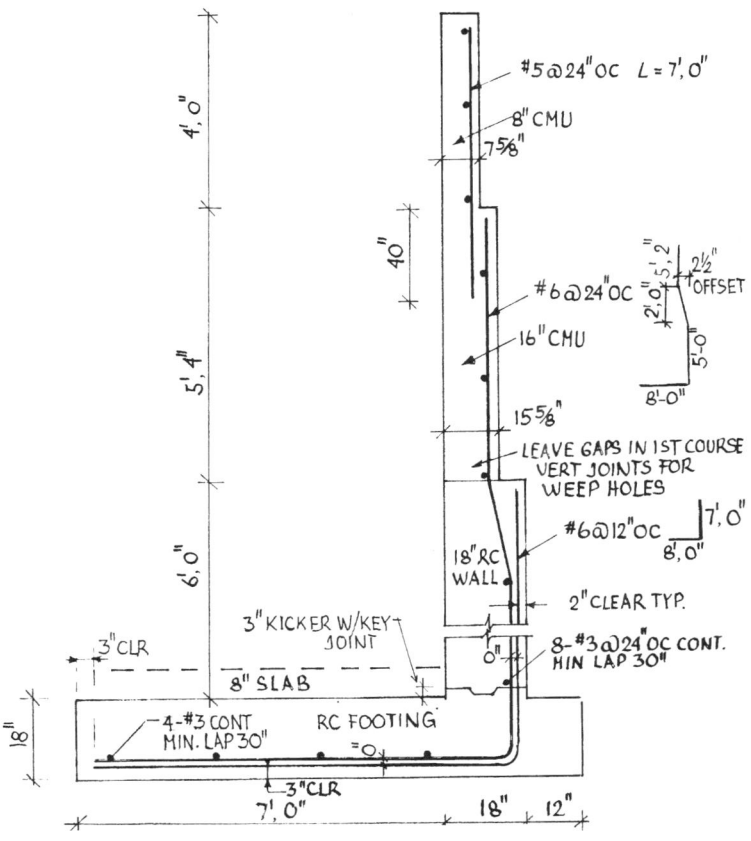

Figure 7.4 C–C 14–15-ft, 4-in. retaining wall.

$$F_b = \tfrac{1}{2}(0.33)f'_m = 250 \text{ psi}$$

Allowable tensile stress in steel reinforcement is

$$F_s = 0.5f_y \qquad \text{or 24,000 for deformed bars}$$
$$= 24,000 \text{ psi}$$

1997 UBC minimum reinforcing requirements for concrete masonry are found in 2106.1.12.4:

Minimum horizontal reinforcement: $0.0007 \times A_g$

$$0.0007 \times 7.625 \times 48 = 0.256 \text{ in.}^2 \qquad \text{\#4 @ 48 in. for 8 in. CMU}$$

$$0.0007 \times 11.625 \times 32 = 0.26 \text{ in.}^2 \qquad \text{\#4 @ 32 in. for 12 in. CMU}$$

DESIGN OF 10.0-FT-HIGH WALL

Design lateral pressures are derived as

$$p_{\text{max}} = 10.0 \times 35 = 350 \text{ psf}$$

$$H = 350 \times 10.0/2 = 1750 \text{ lb}$$

For the overturning moment about point A

$$M_o = 1750(4.58) = 8020 \text{ lb-ft} \qquad \text{at toe based on 1.0-ft strip of wall}$$

$$M_R = (300 + 600) \times 6.5 + 1.25 \times 150 \times 8.5^2/2 + 10.0 \times 110$$
$$\times 1.5 \times 7.75 = 25{,}410 \text{ lb-ft}$$

The factor of safety is

$$\text{FS} = \frac{25{,}410}{8020} = 3.17 > 1.5 \qquad \textbf{OK}$$

The retaining wall consists of a 4.0-ft-high reinforced-concrete stem and a 6.0-ft portion of 12-in. reinforced CMU (concrete masonry unit). The maximum moment acting on the masonry wall at the concrete–masonry interface is determined as

$$M_{12} = \tfrac{1}{6}(0.035 \times 6.0^3) = 1.26 \text{ k-ft} = 15.12 \text{ k-in.}/12\text{-in. wall length}$$

$$M = 15{,}120(\tfrac{32}{12}) = 40{,}320 \text{ lb-in. for 32-in. wall length}$$

Try #4 at 32 in. OC vertical reinforcement:

$$\rho = \frac{0.2}{32 \times (11.62 - 1.82)} = 0.0006377$$

$$n\rho = 26(0.0006377) = 0.01658$$

$$2n\rho = 0.03316 \qquad (n\rho)^2 = 0.000275$$

$$k = \sqrt{(n\rho)^2 + 2n\rho} - n\rho = 0.16627 \qquad (7\text{-}33)$$

$$j = 1 - \frac{k}{3} = 0.944576$$

$$jd = 0.944576 \times 9.8 = 9.26 \text{ in.}$$

The flexural compressive stress induced on the masonry is derived as

$$f_b = \left(\frac{M}{bd^2}\right)\left(\frac{2}{jk}\right) = \left(\frac{40{,}320}{32 \times 9.8^2}\right)\left(\frac{2}{0.94457 \times 0.16627}\right)$$

$$= 166 \text{ psi} < 250 \text{ psi} \qquad \textbf{OK}$$

$$f_s = \frac{M}{A_s jd} = \frac{40{,}320}{0.2 \times 9.26} = 21{,}770 \text{ psi} < 24{,}000 \text{ psi} \qquad \textbf{OK}$$

However, we will use #4 at 24 in. OC to match stem wall reinforcement.

DESIGN OF 4.0-FT-HIGH CONCRETE STEM WALL

Load factor for lateral earth pressure: $1.6H$ (1997 UBC, 1612.2.2)
Maximum bending moment at stem-to-footing fixed-end connection:

$$M_u = 1.6\left(\frac{0.035 \times 10.0^3}{6}\right) = 9.33 \text{ k-ft}$$

$$= 112.0 \text{ k-in./12-in. wall length}$$

Area of vertical reinforcement required with 2-in. cover (1997 UBC, 1907.7.1):

$$A_{st} = \frac{112.0}{0.9 \times 0.95 \times 9.75 \times 60} = 0.22 \text{ in.}^2$$

Use #4 @ 12 in. OC vertical.

Design Toe of Footing

The resultant of the vertical force from A without the earth pressure moment is

$$x = \frac{900 \times 6.5 + 1594 \times 8.5/2 + 1485 \times (8.5 - 0.75)}{900 + 1594 + 1486} = 6.06 \text{ ft}$$

The resultant will be pushed toward the toe by the eccentricity caused by earth-pressure-generated eccentricity:

$$e = \frac{8020}{900 + 1594 + 1485} = 2.00 \text{ ft}$$

The distance of the resultant from A due to vertical loads and earth pressure is

$$t = 6.06 - 2.0 = 4.06 \text{ ft} > \frac{8.5}{3} = 2.83 \text{ ft}$$

Therefore:

1. The resultant lies within the kern limit of the section.
2. The entire footprint of the footing will be under compression.
3. The classical formulas of elasticity hold.

The maximum bearing stress at A under service loads is determined by

$$f_B = \frac{P}{A} + \frac{M_o}{S} = \frac{3.98}{8.5} + \left(\frac{8020}{8.5^2}\right) \times 6 = 1.47 \text{ ksi}$$

1.47 ksi < 2.5 ksi allowed by soil report

Bearing pressure OK.

The bearing pressure at the stem wall with

$$I = \frac{8.5^3}{12} = 51.17 \text{ ft}^4$$

and distance from the section neutral axis of

$$y = \frac{8.5}{2} - 6.0 = -1.75 \text{ ft}$$

is

$$f_{B,\text{stem}} = \frac{3.98}{8.5} - \left(\frac{8.02}{51.17}\right) \times 1.75 = 0.194 \text{ ksf}$$

The strength-level bearing pressures are as follows

At the toe A

$$p_A = 1.6 \times 1.47 = 2.35 \text{ ksf}$$

At the stem wall

$$p_{ST} = 1.6 \times 0.194 = 0.31 \text{ ksf}$$

The bending moment at the footing-to-stem-wall connection at strength level is

$$M_u = \frac{0.31 \times 6.0^2}{2} + \frac{2.04 \times 6.0}{3} = 9.66 \text{ k-ft}$$

The area of steel required is

$$A_{st} = \frac{9.66 \times 12}{0.9} \times 0.95 \times 11.75 \times 60 = 0.196 \text{ in.}^2/\text{ft}$$

Use #4 @ 12 in. OC bottom.

DESIGN OF 13.0-FT-HIGH RETAINING WALL

The lateral force and its lever arm to A of the earth pressure are

$$H = \frac{0.035 \times 13.0^2}{2} = 2.95 \text{ k} \qquad L_e = 4.83 \text{ ft}$$

$$M_o = 2.95 \times 4.83 = 14.25 \text{ k-ft}$$

Assuming a 7.0-ft toe in front of the 12-in.-thick wall, the restoring moment is determined:

Weight of CMU	$5.33 \times 0.103 + 4.0 \times 0.058 = 0.78$ k $\times 7.5$	$= 5.85$
Weight of RC wall	$4.37 \times 0.15 = 0.66$ k $\times 7.5$	$= 4.95$
Weight of 9.5-ft-wide footing	$1.50 \times 9.5 \times 0.15 = 2.14$ k $\times 4.75 = 10.16$	
Weight of ground on heel	$1.5 \times 13.0 \times 0.11 = \underline{2.15}$ k $\times 8.75 = \underline{18.81}$	
	$R = 5.73$ k $\qquad M_R = 40.0$ k-ft	

The restoring moment is larger than the overturning moment. The factor of safety is

$$FS = \frac{40.0}{14.25} = 2.8 \qquad \textbf{OK}$$

The location of the vertical resultant from A is:

(a) Without lateral pressure $\qquad x = \dfrac{40.0}{5.73} = 6.98$ ft

(b) With lateral pressure $\qquad t = 6.98 - \dfrac{14.25}{5.73} = 4.50$ ft

$$4.50 \text{ ft} > \frac{9.5}{3} = 3.16 \text{ ft}$$

Formulas of elasticity apply. The maximum toe pressure under *service load* is

$$f_B = \frac{P}{A} + \frac{M}{S} = \frac{5.73(4.75 - 4.5) \times 6}{9.5^2} = 0.70 \text{ kip/ft}^2 \qquad \textbf{OK}$$

The factored toe pressure causing the bending moment is

$$p_t = 1.6 \times (0.70 - 0.225) = 0.76 \text{ ksf at } \textit{strength level}$$

The factored bearing pressure at the footing-to-wall connection causing the moment with

$$I = \frac{9.5^3}{12} = 71.4 \text{ ft}^4 \qquad \text{and} \qquad y_w = 4.75 - 7.0 = -2.25 \text{ ft}$$

is

$$p_w = 1.6\left(\frac{5.73}{9.5} - \frac{5.73(4.75 - 4.50) \times 2.25}{71.4} - 0.225\right)$$

$$= 0.53 \text{ ksf at strength level}$$

The moment is evaluated at the footing–wall connection:

$$M_u = \left(\frac{0.53 \times 7.0^2}{2} + \frac{(0.76 - 0.53) \times 7.0^2}{3}\right) = 18.62 \text{ k-ft}$$

$$= 223 \text{ k-in. at strength level}$$

$$A_s = \frac{223}{0.9 \times 0.95 \times 14.69 \times 60} = 0.296 \text{ in.}^2 < 0.31 \text{ in.}^2$$

Use #5 @ 12 in. OC bottom.

DESIGN OF RC STEM WALL

The maximum bending moment at the RC wall–footing connection due to lateral pressures is

$$M_u = 1.6\left(\frac{2.95 \times 13.0}{3}\right) = 20.4 \text{ k-ft} = 245 \text{ k-in.}$$

Provide an 18-in.-thick wall:

$$d = 18.0 - 2.0 - 0.30 = 15.7$$

$$A_s = \frac{245}{0.9 \times 0.95 \times 15.70 \times 60} = 0.30 \text{ in.}^2 < 0.31 \text{ in.}^2$$

Use #5 @ 12 in. OC

The maximum moment induced to the 12-in. CMU wall attached to the top of the RC wall for a 16-in.-long wall segment is

$$M = 1.33\left(\frac{0.035 \times (13.0 - 4.37)^3}{6}\right) = 5.0 \text{ k-ft}$$

$$= 60 \text{ k-in.}/16\text{-in. length}$$

Try #5 at 16 in. OC vertical reinforcement:

$$\rho = \frac{0.31}{16 \times (11.62 - 1.82)} = 0.001977$$

$$n\rho = 26(0.001977) = 0.0514$$

$$2n\rho = 0.1028 \qquad (n\rho)^2 = 0.0026423$$

$$k = \sqrt{(n\rho)^2 + 2n\rho} - n\rho = 0.27332 \qquad (7\text{-}33)$$

$$j = 1 - \frac{k}{3} = 0.90889$$

$$jd = 0.90889 \times 9.8 = 8.907 \text{ in.}$$

The flexural compressive stress induced on the masonry is given as

$$f_b = \left(\frac{M}{bd^2}\right)\left(\frac{2}{jk}\right) = \left(\frac{60,000}{24 \times 9.8^2}\right)\left(\frac{2}{0.90889 \times 0.27332}\right)$$
$$= 210 \text{ psi} < 250 \text{ psi} \qquad \textbf{OK}$$

$$f_s = \frac{M}{A_s jd} = \frac{60,000}{0.31 \times 0.90889 \times 9.8}$$
$$= 21,730 \text{ psi} < 24,000 \text{ psi} \qquad \textbf{OK}$$

12-in. CMU OK with #5 @ 16 in. OC vertical reinforcement

DESIGN OF 8-IN. CMU WALL ATOP THE 12-IN. CMU WALL

The height is $13.5 - 9.70 = 3.8$ ft, say 4.0 ft. The maximum moment for a 48-in.-long wall segment is

$$M = 4.0\left(\frac{0.035 \times 3.8^3}{6}\right) = 1.280 \text{ k-ft} = 15.4 \text{ k-in./48-in. length}$$

Assuming #5 at 48 in. OC vertical reinforcement,

$$\rho = \frac{0.31}{48 \times (7.62 - 1.82)} = 0.0011135$$

$$n\rho = 26(0.0011135) = 0.02895$$

$$2n\rho = 0.05790 \qquad (n\rho)^2 = 0.00083817$$

$$k = \sqrt{n\rho^2 + 2n\rho} - n\rho = 0.2134 \qquad (7\text{-}33)$$

$$j = 1 - \frac{k}{3} = 0.92886$$

$$jd = 0.9288 \times 5.8 = 5.3874 \text{ in.}$$

$$f_b = \left(\frac{M}{bd^2}\right)\left(\frac{2}{jk}\right) = \left(\frac{15,400}{48 \times 5.8^2}\right)\left(\frac{2}{0.9288 \times 0.2134}\right)$$

$$= 96.3 \text{ psi} < 150 \text{ psi} \quad \textbf{OK}$$

$$f_s = \frac{M}{A_s jd} = \frac{15,400}{0.31 \times 5.3874}$$

$$= 9222 \text{ psi} < 24,000 \text{ psi} \quad \text{Allowed} \quad \textbf{OK}$$

#5 @ 48 in. OC vertical OK, however, use #5 @ 32 in. OC.

DESIGN OF 15.0-FT-HIGH RETAINING WALL

The wall consists of three wall segments: The bottom portion is a 6-ft-high, 18-in.-thick reinforced concrete wall. The middle portion is a 5-ft-high 16-in.-thick CMU. The upper portion is a 4-ft-high, 8-in.-thick CMU. The footing is 9.5-ft-wide, 18-in.-thick reinforced concrete with a 7-ft toe and an 18-in. heel in the back. The overturning moment about A is

$$M_o = \left(\frac{0.035 \times 15.0^2}{2}\right) \times (5.0 + 1.5)$$

$$= 25.6 \text{ k-ft per 12-in. wall length.}$$

The weights of wall components and restoring moment are as follows:

8-in. CMU	$4 \times 0.058 = 0.23$ k $\times 7.5$	= 1.73
12-in. CMU	$5 \times 0.103 = 0.55$ k $\times 7.5$	= 4.12
18-in. RC wall	$6 \times 1.5 \times 0.15 = 1.35$ k $\times 7.75$	= 10.46
Footing	$9.5 \times 1.5 \times 0.15 = 2.14$ k $\times 4.75$	= 10.16

Weight of ground on heel

$$1.5 \times 15.0 \times 0.11 = \underline{2.47} \text{ k} \times 8.75 = \underline{21.66}$$
$$6.74 \text{ k} \qquad\qquad 48.13 \text{ k-ft}$$

The factor of safety for overturning is

$$FS = \frac{48.13}{25.6} = 1.90 \qquad \textbf{OK}$$

The resultant of the vertical forces measured from A is

$$t = \frac{48.13}{6.74} - \frac{25.6}{6.74} = 3.34 \text{ ft} > 3.16 \text{ ft}$$

The entire footprint is under compression. The bearing pressure is of triangular distribution. The toe pressure is

$$f_B = \frac{2 \times 6.74}{9.5} = 1.42 \text{ ksf} < 2.5 \text{ ksf} \qquad \textbf{Allowed} \qquad \textbf{OK}$$

The factored toe pressure is

$$p_T = 1.6 \times 1.42 = 2.27$$

The factored bearing pressure at the toe–wall interface is

$$p = 0.6 \text{ ksf}$$

The following maximum moments are applied to wall segments:

8-in. CMU: 21.4 k-ft. *See previous pages.*
12-in. CMU:

$$M = 1.33\left(\frac{0.035 \times 9.0^3}{6}\right) = 5.65 \text{ k-ft}$$

$$= 68.0 \text{ k-in. per 16-in. length of wall}$$

18-in. RC Wall:

$$M_u = 1.6\left(\frac{0.035 \times 15.0^3}{6}\right) = 31.5 \text{ k-ft} = 378 \text{ k-in.}$$

Reinforcement Design

8-in. CMU: #5 at 48 in. OC. *See previous pages.*

However, use #5 @ 24-in. OC vertical to match rest of wall reinforcement.

12-in. CMU: Try #6 at 16 in. OC:

$$\rho = \frac{0.44}{16 \times (11.62 - 1.82)} = 0.002806 \qquad n\rho = 0.072959$$

$$2n\rho = 0.0972 \qquad\qquad (n\rho)^2 = 0.005323$$

$$k = \sqrt{(n\rho)^2 + n\rho} - n\rho = 0.31594 \qquad\qquad (7\text{-}33)$$

$$j = 1 - \frac{k}{3} = 0.8947$$

$$jd = 0.911 \times 9.8 = 8.77 \text{ in.}$$

$$f_b = \left(\frac{68}{16 \times 9.8^2}\right)\left(\frac{2}{0.895 \times 0.3159}\right) = 0.31 \text{ ksi} > 250 \text{ psi}$$

Allowed

Must use 16-in.-thick CMU.

The effective depth of reinforcement is

$$d = 13.8 \text{ in.}$$

Try #6 at 24 in. OC:

$$\rho = \frac{0.44}{24 \times 13.8} = 0.001328 \qquad \eta\rho = 0.03454$$

$$2n\rho = 0.06908 \qquad\qquad (n\rho)^2 = 0.001193$$

$$k = \sqrt{(n\rho)^2 + 2n\rho} - n\rho = 0.2306 \qquad\qquad (7\text{-}33)$$

$$j = 1 - \frac{k}{3} = 0.923$$

$$jd = 12.74 \text{ in.}$$

$$f_b = \left(\frac{102}{24 \times 13.8^2}\right)\left(\frac{2}{0.923 \times 0.2306}\right) = 0.21 \text{ ksi}$$

$$= 210 \text{ psi} < 250 \text{ psi} \qquad \textbf{OK}$$

$$f_s = \frac{M}{A_s jd} = \frac{102}{0.44 \times 12.74} = 18.2 \text{ ksi}$$

$$= 18,200 \text{ psi} < 24,000 \text{ psi} \qquad \textbf{OK}$$

Design of 18-in. RC Wall

$$M_u = 378 \text{ k-in.}$$

Try #6 at 12 in. plus #6 at 24 in. OC alternating:

$$A_s = \tfrac{1}{2}(0.6 + 0.44) = 0.52 \text{ in.}^2$$

$$\phi M_n = 0.9 \times 0.52 \times 60 \times (0.875 \times 15.62)$$

$$= 392.0 \text{ k-in.} > 378 \text{ k-in.} \quad \textbf{OK}$$

Design the footing at *strength level:*

$$M_u = \frac{0.6 \times 7.0^2}{2} + \frac{(2.27 - 0.6) \times 7.0^2}{3} = 42.0 \text{ k-ft} = 503 \text{ k-in.}$$

The amount of reinforcement provided is as follows:

From concrete stem wall: #6 @ 12 in. OC	0.6 in.^{2}
From 16-in. CMU: #6 @ 24 in. OC	0.3 in.^{2}
	0.9 in.^{2}

$$\phi M_n = 0.9 \times (0.9 \times 60) \times (0.875 \times 14.62)$$

$$= 622.0 \text{ k-in.} > 503 \text{ k-in.} \quad \textbf{OK}$$

7.3   CASE 2: SEISMIC VERSUS WIND

Seismic: Low to Moderate Seismicity area

The following engineering calculations were done for a real project. Despite comparatively high winds in a relatively low seismicity area, *seismic governed over wind* in this case study.

PROJECT DESCRIPTION

The subject structure is a *reinforced masonry hangar* (Figures 7.5 and 7.6).

Figure 7.5 Partial view of the hangar during construction.

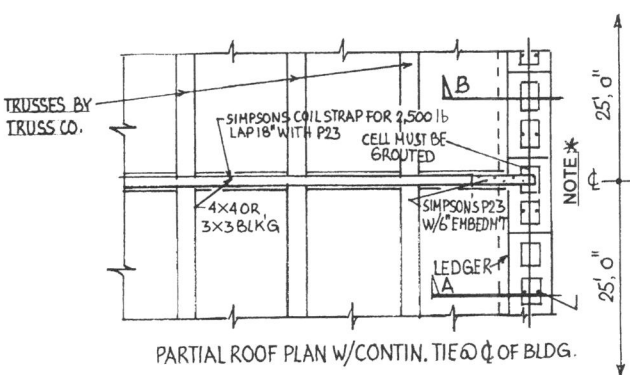

PARTIAL ROOF PLAN W/CONTIN. TIE @ ℄ OF BLDG.

✳ NOTE: 8-IN. CMU W/2–#5 @ 24 IN. OC VERTICAL, #3 @ 32 IN. OC CONT.
HORIZONTAL. MIN. LAP 40-IN. GROUT EVERY ALTERNATE CELL.

Figure 7.6 Partial roof plan of the masonry hangar with continuous tie at centerline of building.

DESIGN CRITERIA

Wind 70 mph; exposure C; 1997 UBC

Soil profile: S_C

Allowable bearing pressure 1000 psf

Dimensions: 50 × 50-ft, 8-in.-thick walls, 12-in. piers at front entrance

Plate height 14 ft

Ridge height 20 ft

Roof: wood trusses provided by truss manufacturer

Mortar: type S, $f'_m = 1500$ psi, $n = 26$, $F_b = 250$ psi

The reinforced-concrete masonry walls are supported top and bottom with a 14.0-ft simply supported clear span:

$$f'_m = 1500 \text{ psi} \qquad E_m = 750 \times 1500 = 1,125,000 \text{ psi}$$

$$n = \frac{E_s}{E_m} = \frac{29,000}{1125} = 26$$

The allowable bending stress (*no special inspection required*) is

$$F_b = \frac{1}{2} \times 0.33 \ f'_m = 250 \text{ psi}$$

The allowable compressive stress with partial fixity at bottom and effective height of

$$h' = 0.70(168) = 118 \text{ in.}$$

$$\frac{h'}{r} = \frac{118}{2.53} = 46 < 99 \qquad\qquad (7\text{-}11)$$

is

$$F_a = \left(\frac{1}{2}\right)(0.25)f'_m \left[1 - \left(\frac{h'}{140r}\right)^2\right] = 167 \text{ psi}$$

The allowable *shear stress* in shear walls without shear reinforcement, for $M/Vd < 1.0$, is

$$F_v = 1.5\sqrt{f'_m} < 75 \text{ psi}$$

The allowable *tensile stress* in steel reinforcement is

$$F_s = 0.5f_y \text{ or } 24{,}000 \text{ deformed bars}$$

$$= 24{,}000 \text{ psi}$$

REINFORCEMENT REQUIREMENTS

Minimum vertical reinforcement

0.2 in.^{2} or #4 bar continuous at each side of opening and at each corner [2106.1.12.3 (2)]

8-in. walls, $0.0015A_g = 0.0015 \times 7.625 \times 24.0 = 0.275$ in.^{2} < 0.31 in^2 @ 24 in. OC

Provided by #5 vertical @ 24 in. OC.

Horizontal reinforcement

$0.0015 + 0.0006 > 0.002$ [2106.1.12.4 (2.3)]

Provided by #3 @ 32 in. OC.

(1) #4 horizontal to be placed at top and bottom of wall as well as in lintels, top and bottom, minimum embedment 24 in., $L_b = 24$ in.

MASONRY WALLS

Wind and Seismic Loads Acting on Wall

1. **Wind** (projected area method), UBC 1997, (20-1), Tables 16-F, 16-G, and 16-H:

$$p = C_e C_q q_s I$$

$$q_s = 12.6 \text{ psf}$$

$$C_e = 1.06 \qquad \text{Exposure C}$$

$$C_q = 1.3 \qquad \text{Structures 40 ft or less in height}$$

$$p = 12.6(1.06)(1.3)(1.0) = \textbf{17.4 psf}$$

2. **Seismic**—acting out of plane of wall. The lateral load for the 12-ft-wide wall is

$$F_p = \frac{a_p C_a I_p}{R_p (1 + 3h_x/h_r) W_p}$$

that is,

$$\frac{(1.0)(0.24)}{3.0[1 + 3(1.0)]56} = 17.9 \text{ psf}$$

Seismic GOVERNS.

Next, we complete our calculations for the design of the hangar structure. The maximum moment acting at midheight on a 24-in.-wide strip of wall is

$$M = \frac{wL^2}{8} = 2 \times 17.9(14.0^2) \times 1.5 = 10,525 \text{ lb/in.}$$

Try #5 at 24 in. vertical reinforcement:

$$\rho = 0.31/24(7.62 - 1.82) = 0.00223$$

$$n\rho = 26(0.00223) = 0.058$$

$$2n\rho = 0.1158$$

$$(n\rho)^2 = 0.00335$$

$$k = \sqrt{(n\rho)^2 + 2n\rho} - n\rho = 0.287 \qquad (7\text{-}33)$$

$$j = 1 - \frac{k}{3} = 0.9$$

$$jd = 0.9(5.8) = 5.22 \text{ in.}$$

$$f_b = \left(\frac{M}{bd^2}\right)\left(\frac{2}{jk}\right) = \frac{10,525}{24 \times 5.8^2(2/0.9 \times 0.287)}$$

$$= 101.0 \text{ psi} < 250 \text{ psi} \qquad \textbf{OK}$$

$$f_s = \frac{M}{jdA_s} = \frac{10,525}{0.31 \times 5.22}$$

$$= 6500 \text{ psi} < 24,000 \text{ psi} \qquad \textbf{OK}$$

COMBINED AXIAL AND BENDING ON 24-IN.-WIDE WALL SUPPORTING TRUSSES AT 2.0 FT OC

The reaction of the main roof on the 8-in. CMU wall will be derived from the following loads:

DL of roof—two-ply roofing felt, $\frac{1}{2}$-in. OSB/trusses @ 24 in. OC

$$DL = 10.0 \text{ psf}$$

$$LL = 16.0 \text{ psf} \qquad \text{Total } 26.0 \text{ psf}$$

With a 50-ft roof span the end reaction for a 2.0-ft slice of roof is given as

$$\frac{2 \times 26 \times 50.0}{2} = 1300 \text{ lb}$$

The reaction of a roof truss on a 24-in.-wide strip of wall is

$$P = 1300 \text{ lb}/24 \text{ in.}$$

The weight of the wall is derived as

$$2 \times 14.0 \times 56 = 1568 \text{ lb}/24 \text{ in.}$$

$$f_a = 1300 + \frac{1568}{24.0 \times 5.8} = 21.0 \text{ psi}$$

$$\frac{f_a}{F_a} + \frac{f_b}{F_b} = \frac{21.0}{167} + \frac{101.0}{250} = 0.53 < 1.0$$

Combined bending and axial force. **OK.**

DESIGN OF MASONRY PIER AT ENTRANCE

Total Seismic Load/Building

Weight of roof (concrete tiles): $22 \times 50 \times 50 = 55,000$ lb
Weight of side wall: $2 \times 14.0 \times 50 \times 56 = 78,400$ lb
Weight of rear wall: $14 \times 50.0 \times 56 = 39,000$ lb
Weight of 12-in. front piers: $2 \times 3.0 \times 14.0 \times 90 = 7600$ lb

Total weight of building: 180,000 lb

Using the 1997 UBC static force procedure (1630.2) gives

$$V = \left(\frac{2.5\ C_a I}{R}\right) W = \left(\frac{2.5 \times 0.24 \times 1.0}{4.5}\right) \times 180 = 24.0\ \text{k}$$

$C_a = 0.24$ (soil profile S_C)

$R = 4.5$

The horizontal force acting at the top of each pier is

$$H_u = \frac{24.0}{4} = 6.0\ \text{k} \qquad \text{at } \textit{strength} \text{ level}$$

Using working stress design, divide H by 1.4 per UBC 1997, 1612.3, "Allowable Stress Design,"

$$H = \frac{6.0}{1.4} = 4.28\ \text{k/pier}$$

The moment caused by horizontal force H applied at plate elevation is

$$M = 4.28 \times 14.0 = 60.0\ \text{k/ft} = 720\ \text{k-in.}$$

The reaction of the front roof truss above the entrance is as follows:

$$P_T = \frac{2.0 \times 42 \times 46.0}{2} = 1940\ \text{lb} \qquad \text{For roof } D + L$$

$$P_D = \frac{20 \times 46.0}{2} = \underline{\ \ 460\ \text{lb}} \qquad \text{For weight of door}$$

$$\text{Total}\quad 2400\ \text{lb} \qquad \text{Each end}$$

The weight of the pier at midheight is

$$3.0 \times 7.0 \times 90 = 1890\ \text{lb} \qquad \text{say 1900 lb}$$

The total vertical load on the pier is 4300 lb.

Design Pier for Combined Axial and Bending

Assume

$$F_a = 167 \text{ psi} \qquad f_b = 250 \text{ psi}$$

The actual working compressive stress is derived as follows:

$$\frac{P}{A} = \frac{4300}{11.625 \times 36.0} = 10.3 \text{ psi}$$

$$\frac{f_a}{F_a} + \frac{f_b}{F_b} = \frac{10.3}{167} + \frac{25.5}{250} = 0.163 < 1.0$$

Pier OK in compression

$$f_s = \frac{M}{jdA_s} = \frac{720}{(30.7)(0.88)} - \frac{4.3}{2 \times 0.88} = 24.2 \text{ ksi}$$

Pier reinforcement OK

Design Continuous Footing for Side and Rear Walls

Roof reaction	$\frac{1}{2}(10.0 + 16.0)50.0 =$ 650 lb/ft
Weight of 8-in. CMU	$14.0 \times 56 =$ 785 lb/ft
	Total $=$ 1435 lb/ft

Bearing pressure

$$P = \frac{1435}{1.5} = 956 \text{ psf} < 1000 \text{ psf}$$

Footing OK.

Design Footing for 12-in. × 3.0-ft Piers at Front Entrance

From the pier calculations the maximum moment transferred to the footing by the pier is

$$M = 60.0 \text{ k/ft}$$

The moment is resisted by the 18-in.^{2} *grade beam* connecting the footings of the two individual piers. The reinforcement of the grade beam is comprised of 3 − #7 continuous top and bottom bars. The factored applied moment is

$$M_u = 1.5 \times 60 = 90.0 \text{ k/ft} = 1080 \text{ k/in.}$$

The resistance of the 18-in.^{2} beam is supplied by the $3 - \#7$ main reinforcement:

Effective depth: $18.0 - 3.0 - 0.44 = 14.56$ in.

$$M_u = 0.9 \times 3 \times 0.6 \times 60 \times (0.8 \times 14.56)$$

$$= 1132 \text{ k/in.} > 1080$$

Therefore a uniformly distributed bearing pressure can be expected between the footing and soil. The pressure with 4300 lb total vertical load per pier is

$$P = \frac{4300}{3.0 \times 6.0} = 240 \text{ psf} < 1000 \text{ psf allowed} \quad \textbf{OK}$$

7.4   CASE 3: DESIGN OF CMU WALL AND PRECAST CONCRETE PLATE

Seismic: Moderate- to High-Seismicity Area

PROJECT DESCRIPTION

Precast concrete circular plate for a new sewer well and a reinforced masonry, 6-ft wall.

ANALYSIS OF STRESSES AND STRAINS

The following analysis is based on R. J. Roark's[2] *Formulas for Stress and Strains.*

$$\text{LL} = 150 \text{ psf} \qquad \text{DL} = 75 \text{ psf}$$

The maximum load acting on top of the circular concrete lid at *strength level* is

$$[1.4(0.15) + 1.7(0.075)] = 0.3375 \text{ ksft} \qquad \text{say } 0.34 \text{ ksft}$$

$$q = 0.034 \text{ ksft}$$

The thickness of the plate is $t = 6$ in. The modulus of elasticity is

$$E = 3 \times 10^6 \text{ psi}$$

Poisson's ratio v for concrete is 0.32 and

$$D = \frac{3(10^6)(6.0^3)}{12(1 - 0.32^2)} = 60 \times 10^6$$

$$a = \frac{15.0}{2} = 7.5$$

$$M_{max} = \frac{qa^2 (3 + 0.32)}{16} = 3.97 \text{ k-ft} = 47.6 \text{ k-in.}$$

With $1\frac{1}{2}$-in. bottom cover the effective depth is

$$6.0 - 1.75 = 4.25 \text{ in.}$$

and

$$A_{st/radial} = \frac{47.6}{0.9 \times 4.25 (0.85) \times 60} = 0.244 \text{ in.}^2/\text{ft width}$$

$$0.244 < 0.31 \text{ in.}^2$$

Provided by #5 @ 12 in. OC.

To compensate for the local, weakening effect of two 3.0-ft manhole openings, we provided a 5 − #5 bar bundle to be placed in the diagonal direction at the bottom next to each opening and 3 − #5 bottom in the cross direction.

FOOTING DESIGN

Total DL + LL acting on the lid, *working stress design:*

$$P = r^2 \pi q = 7.5^2 \pi (0.15 + 0.075) = 39.76 \text{ kips} \qquad \text{say } 40.0 \text{ kips}$$

Linear reaction of lid on footing:

$$w = \frac{r^2 \pi q}{2r\pi} = 0.85 \text{ k/ft}$$

Required footing width with 1.0 ksf allowable soil bearing pressure:

$$b = \frac{0.85}{1.0} = 0.85 < 1.0 \text{ ft}$$

Provided by a ring foundation with a minimum footing depth of 18 in.   OK.

SEISMIC DESIGN OF 6-FT-HIGH, 8-IN. CMU WALL

Using the UBC 1997 seismic provisions for cantilevered walls with a unit weight of 60 psf, fully grouted:

$$a_p = 2.5 \qquad R_p = 3.0 \qquad I_p = 1.0$$

$$C_a = 0.33, \text{ seismic zone 3, soil profile } S_C$$

$$W_p = 6.0 \times 60 = 360 \text{ lb}$$

$$F_p = 2.5(0.33)(1.0)\left[1 + 3\left(\frac{0}{6.0}\right)\right]\left(\frac{360}{3.0}\right)$$

$$= 99 \text{ lb/linear foot of wall} \qquad \text{say } 100 \text{ lb}$$

The maximum overturning moment is

$$M_o = 0.1\left(\frac{6.0}{2} + 1.5\right) = 0.45 \text{ k-ft}$$

The restoring moment with a 2.0-ft-wide × 12-in-deep footing is

$$M_r = \frac{0.15(2.0^2)}{2} + 0.36(1.66) = 0.9 \text{ k-ft} > 0.45 \text{ k-ft} \qquad \textbf{OK}$$

The factor of safety is

$$\text{FS} = \frac{0.9}{0.45} = 2.0 \qquad \textbf{OK}$$

Reinforcement Design: Working Stress Design

Lateral force $F = \dfrac{0.099}{1.4} = 0.0707$

$$M = 0.0707 \times \left(\frac{6.0}{2} + 0.5\right) = 0.247 \text{ k-ft} = 2.96 \text{ k-in.}$$

$$A_{st} = \frac{2.96}{0.85 \times 3.75 \times 20.0} = 0.046 \text{ in.}^2 < 0.23 \text{ in.}^2$$

Provided by #5 @ 16 in. OC vertical.

7.5 CASE 4: RETAIL STORE, MASONRY AND STEEL

Seismic: Low-to-Moderate Seismicity area

Our task was the analysis and engineering design of a reinforced masonry building in Mesquite, Nevada. Despite being a low- to moderate-seismicity area, **seismic forces governed** in this case study, as shown in the structural calculations.

PROJECT DESCRIPTION

Structural configuration of building: rectangular with approximately 17,500 total square footage (Figure 7.7)

Reinforced masonry walls, split-face CMU

Soil profile S_D

Metal roof joists: VULCRAFT 22K 10 at 6.0 ft OC

Steel columns: HSS (Hollow Structural Section) $8 \times 8 \times \frac{5}{16}$ with 12-in.$^2 \times \frac{5}{8}$ base plates and column head plate at centerline of store sales room

Figure 7.7 Front elevation of structure.

Pad footing for HSS 8×8 column: $6.0 \text{ ft}^2 \times 18$ in. with $7 - \#5$ reinforcement bars bottom each way

Architectural design by Thistle Architecture, Henderson, Nevada

WIND ANALYSIS

Basic formula:

$$P = C_e C_q q_s I \qquad \text{(UBC 1997 20-1)}$$

75 mph, exposure C

$q_s = \frac{1}{2}(12.6 + 16.4) = 14.5$ psf (UBC 1997 16-F)

C_e (UBC 1997 16-G):

    At 15.0 ft: 1.06

    At 25.0 ft (average 20–25): 1.16

C_q:

    Below parapet/roof connection: 0.8, method 1, 1997 UBC, 16-H

    For parapet: $0.8 + 0.5 = 1.3$, method 1

Wind pressures on component, service level (load factor 1.0):

    0–15 ft $p = 14.5 \times 0.8 \times 1.06 \times 1.0 = 12.3$ psf External wall

    15–24.5 ft $p = 14.5 \times 0.8 \times 1.16 \times 1.0 = 13.5$ psf External wall

    Parapet $p = 14.5 \times 1.3 \times 1.16 \times 1.0 = 21.9$ psf Parapet wall

SEISMIC DESIGN FORCES

Basic formula:

$$F_p = \left(\frac{a_p C_a I_p}{R_p}\right)\left(1 + 3\frac{h_x}{h_r}\right) W_p \qquad (32\text{-}2)$$

where

$$C_a = 0.28 \qquad (16\text{-Q})$$

External wall:

$$a_p = 1.0 \qquad R_p = 3.0 \qquad (16\text{-O})$$

Unbraced parapet:

$$a_p = 2.5 \qquad R_p = 3.0 \qquad I_p = 1.0$$

First term of Equation (32-2):
   External wall:

$$\frac{1.0 \times 0.28 \times 1.0}{3.0} = 0.0933 \qquad \text{But } 0.7 \times C_a$$

$$= 0.196 \qquad \textbf{GOVERNS}$$

   Parapet:

$$\frac{2.5 \times 0.28 \times 1.0}{3.0} = 0.2333$$

Second term of Equation (32-2):
   External wall:

$$1 + \left(\frac{12.25}{24.5}\right)3 = 1.0 + 1.5 = 2.5$$

   Parapet:

$$1 + \left(\frac{25.75}{24.5}\right)3 = 1.0 + 3.15 = 4.15$$

Seismic pressure on component:

$$W_p = 68 \text{ psf} \qquad \text{8-in. CMU, } \textbf{fully grouted}$$

   External wall:

$$F_p = 0.196 \times 2.5 \times 68 = 33.3 \text{ psf} \qquad \text{at } \textit{strength level}$$

   Parapet:

$$F_p = 0.233 \times 4.15 \times 68 = 65.7 \text{ psf} \qquad \text{at } \textit{strength level}$$

Seismic pressure on component:
   External wall: $F_p = 33.3/1.4 = 23.8$ psf at service level (ASD)
   Parapet wall: $F_p = 65.7/1.4 = 46.9$ psf at service level

Seismic GOVERNS.

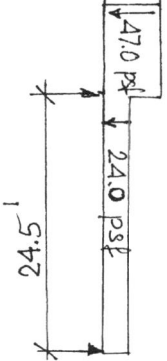

Figure 7.8 Seismic pressure on external wall and parapet.

Vertical roof reaction on wall, span = 88.0 ft:

DL: $w_D = \frac{1}{2}(14.0 \times 88.0) = 616$ lb/ft at service level

DL + LL: $w_{D+L} = \frac{1}{2}(14.0 + 16.0)88.0 = 1320$ lb/ft at service level

Total axial loads include the weight of the wall:

    Parapet: 1320 + 180   = 1500 lb/ft   DL + LL

    Wall (midheight):   1320 + 1000 = 2320 lb/ft   DL + LL

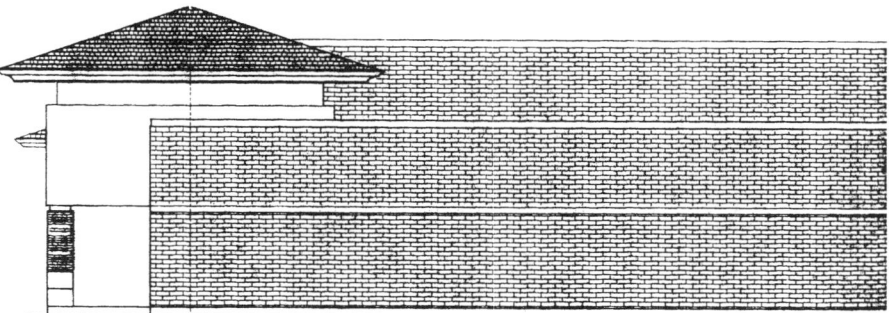

Figure 7.9 View of the masonry wall on the east elevation of the building.

Maximum bending moment:

$$0.024 \times 24.5^2 \times 1.5 = 21.6 \text{ k-in./ft}$$

Check stresses in CMU wall:

$$m = 28.8 \text{ k-in./16-in. strip}$$

Try 8-in. CMU with #5 at 16 in. OC:

$$\rho = \frac{0.31}{16 \times (7.62 - 1.82)} = 0.00334$$

$$n\rho = 26(0.00334) = 0.08685$$

$$2n\rho = 0.1737 \qquad (n\rho)^2 = 0.007543$$

$$k = \sqrt{(n\rho)^2 + 2n\rho} - n\rho = 0.3389$$

$$j = 1 - \frac{k}{3} = 0.887$$

$$jd = 0.887 \times 5.8 = 5.1448$$

$$jk = 0.30$$

$$f_b = \frac{(m/bd^2)2}{jk} = \frac{2 \times 28.8}{16 \times 5.8^2 \times 0.3} = 357 \text{ psi} > 250 \text{ psi}$$

The stress is *too high* for CMU *without special inspection.* Try #5 at 8 in. OC:

$$\rho = 0.00668$$

$$n\rho = 0.1737$$

$$2n\rho = 0.34736 \qquad (n\rho)^2 = 0.03016$$

$$k = \sqrt{(n\rho)^2 + 2n\rho} - n\rho = 0.4407$$

$$j = 0.853$$

$$jd = 4.947$$

$$jk = 0.376$$

$$f_b = \frac{2(10.8)}{8 \times 5.8^2 \times 0.376} = 0.213 \text{ ksi} < 250 \text{ psi}$$

OK without special inspection.

$$f_s = \frac{m}{A_s jd} = \frac{10.8}{0.31 \times 4.947} = 7.04 \text{ ksi}$$

Try #5 at 24 in. OC for 8-in. CMU with special inspection, $f'_m = 1500$ psi:

$$\rho = \frac{0.31}{24 \times 5.8} = 0.002227$$

$$n\rho = 0.0579$$

$$2n\rho = 0.1158 \qquad (n\rho)^2 = 0.003353$$

$$k = 0.28729$$

$$j = 0.904$$

$$jd = 5.243$$

$$jk = 0.256$$

$$f_b = \frac{2(43.2)}{24 \times 5.8^2 \times 0.256} = 0.418 \text{ ksi} = 418 \text{ psi} < 0.33$$

$$f'_m = 500 \text{ psi}$$

$$f_s = \frac{2 \times 21.6}{0.31 \times 5.243} = 26.58 \text{ ksi} \qquad \text{Bending component only}$$

The axial stress in the bar caused by roof DL + weight of wall at midheight—each half of wall receives one half axial force:

$$26\left(\frac{616 + 1000}{2.9 \times 24}\right) = 600 \text{ psi}$$

$$f_{stotal} = 26,580 - 600 = 25,980 \text{ psi} > 24,000 \text{ psi allowed}$$

Reinforcement is overstressed when #5 is placed at 24 in. OC. Try #5 at 16-in. OC:

$$f_s = \frac{28.8}{0.31 \times 5.145}$$

$$= 18.0 \text{ ksi} < 24.0 \text{ ksi allowed} \qquad \text{Bending only}$$

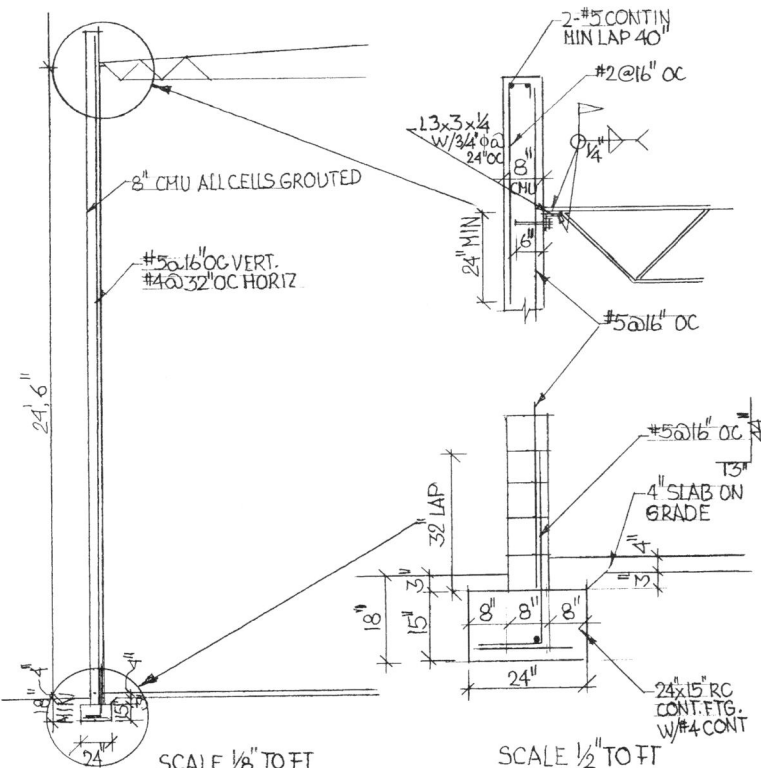

Figure 7.10   Structural details of the project.

VULCRAFT JOISTS

Maximum span 44.0 ft

Maximum spacing 6 ft, 4 in.

Maximum DL + LL:

$$12 + 3.0 \text{ (joist)} + 20.0 = 35 \text{ psf} \quad w_{D+L} = 6.33 \times 35 = 221 \text{ lb/ft}$$

From the SJI load table, page 11, supplied by NUCOR/VULCRAFT, Phoenix, Arizona, a 22K10 has a safe load-carrying capacity of 272 lb/ft of the K series with a depth of 20 in.

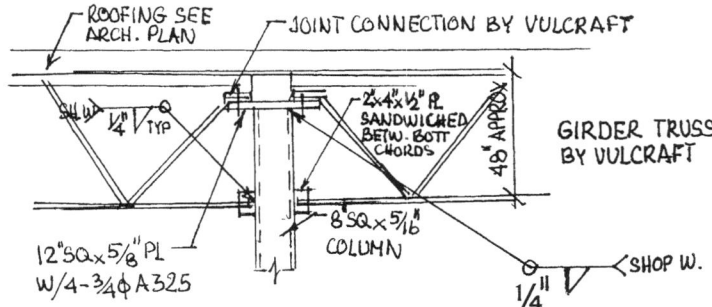

Figure 7.11   Joist girder-to-column connection.

HSS STEEL COLUMN

Carries a tributary roof area of $44.0 \times 48.0 = 2112$ ft^2

Maximum DL + LL on column, maximum column spacing 48 ft OC:

$$0.035 \times 2112 + 0.04 \text{ (girder weight)} \times 48.0$$

$$= 75.8 \text{ k} \qquad \text{say } 76.0 \qquad \textbf{OK}$$

From the AISC load tables, an HSS $8 \times 8 \times \frac{5}{16}$ has a safe load-carrying capacity of 141 k $>$ 76.0 k with $KL = 25$ ft.

$$\textbf{HSS } 8 \times 8 \times \tfrac{5}{16} \qquad \textbf{OK.}$$

BASE PLATE

Assume 12-in.^{2} $\times \frac{5}{8}$-in. plate.

Maximum bearing pressure under plate:

$$\frac{76,000}{12.0^2} = 528 \text{ psi}$$

Effective plate span: $3.0 - 0.625 = 2.355$ in.

Maximum bending moment per inch of slice:

$$\tfrac{1}{2}(0.528 \times 2.355^2) = 1.463 \text{ k-in.}$$

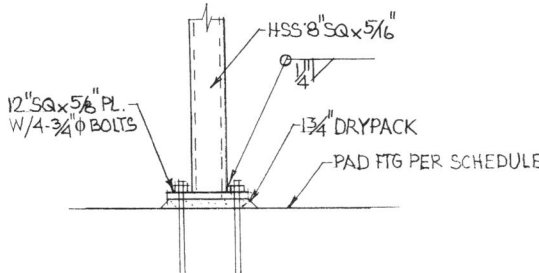

Figure 7.12 Base plate and column-to-footing connection.

$$S_{req.} = \frac{1.463}{27.0} = 0.054 \text{ in.}^3$$

or

$$t = \sqrt{6 \times 0.054} = 0.57 \text{ in.} < 0.625 \text{ in.}$$

Provided by $\frac{5}{8}$-in. plate.

A 12-in.^{2} × $\frac{5}{8}$-in. base plate with $4\frac{3}{4}$-in. ϕ A307, 12-in. embedment.   OK.

FOOTING DESIGN

Strip Footing

To support external CMU walls
Maximum DL + LL vertical load at top of footing:

From roof	$(14.0 + 0.6 \times 20)88.0/2 =$	1144
Weight of wall	$26.5 \times 68 =$	1802
		2946 say 2950 lb/ft

Minimum required footing width using 2000 psf allowable soil bearing pressure (per soil report):

$$b = \frac{2950}{2000} = 1.48 \text{ ft} \approx 1.5 \text{ ft}$$

However, use 24-in. × 15-in. continuous footing.

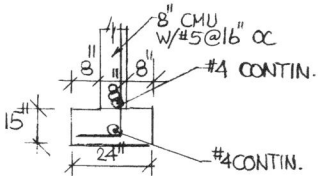

Figure 7.13 CMU walls continuous footing.

Pad Footings

Under the most loaded column, carrying 40-ft × 46-ft tributary area,

$$\text{Maximum DL + LL} = 72.0 \text{ k column load}$$

Using square footings, the minimum width b with 2.0 ksf allowable bearing pressure is

$$b = \sqrt{\frac{72.0}{2}} = 6.0 \text{ ft}$$

From the CRSI footing design tables, the minimum footing thickness with 10-in. × 10-in. plate is 14 in. and a reinforcement 7 − #5 each way, grade 60.

Use 6.0-ft^2 × 18-in. deep footing with 7 − #5 bottom each way.

REFERENCES

1. Schneider, R. R., and Dickey, W. L. 1994. *Reinforced Masonry Design.* Upper Saddle River, NJ: Prentice-Hall.
2. Roark, R. J. 1975. *Formulas for Stress and Strains,* 5th ed., McGraw-Hill, Koga-kusha: Tokyo.

CHAPTER 8

WOOD-FRAMED BUILDINGS

8.1 INTRODUCTION

Historically, wood-framed buildings (type-V timber design) emerged as one of the best structures to resist earthquakes.

> Past performance of wood structures indicates wood to be a safe, durable and economical building material when it is used properly [ref. 1, p. xiv].

> Wood frame dwellings and similar small wooden structures performed excellently [*in the Alaskan earthquake of 1964*] as a class of construction when located in an area not subjected to land movement. Also, in many cases, wood frame dwellings behaved remarkably well in areas that were subject to land movement [ref. 2, p. 212].

The design of wood-framed buildings was based on sound principles: a stiff concrete main wall attached to a lightwood framing so that the rigidity of the concrete was combined with the resilience of the wood framing—a sound concept, an ideal formula. How such a formula was put into practice can be appreciated by the damage suffered by residential wood buildings during the Northridge earthquake of 1994.

At the time of the Sylmar (San Fernando) earthquake of 1971 the general public and the engineering and construction communities in highly populated areas had enjoyed a "seismic break" and were under the impression that much time would elapse until another major event would happen. Lessons from the San Fernando earthquake were not fully utilized. Although relatively not a major earthquake, it had seriously damaged buildings, bridges, underground pipes, and other installations.

The seismic chasm was shorter than expected. Sixteen years later we were hit by Whittier Narrows, Loma Prieta, Upland, Sierra Madre, Big Bear, Landers, and Bishop and less than 25 years after San Fernando by a more destructive earthquake roughly in the same populated area.

8.2 NORTHRIDGE LESSON

Before Northridge it was presumed that the construction industry knew enough to meet the earthquake challenge. The information available had been mostly gathered from the Long Beach earthquake of 1933, and the experience was mainly based on the behavior of unreinforced masonry buildings, leaving other type of structures virtually without scrutiny.

The City of Los Angeles alone reported 2280 failures of wood-framed buildings during the 1994 Northridge earthquake. Of these, 1650 were homes that suffered moderate to severe damage, including partial or total collapse. The remaining 630 were apartment houses, hotels, and condos, 40 with severe damage that involved leaning or partially collapsed structures.

The 2003 IBC contains provisions for general design requirements for lateral-force-resisting systems in Section 2305: "Structures using wood shear walls and diaphragms to resist wind, seismic and other lateral loads shall be designed and constructed in accordance with the provisions of this section."

Next we analyze the type of failures found in wood-framed structures damaged during the Northridge earthquake. The author gathered the information and took the photographs during his inspection of homes located near or relatively close to the epicenter of the earthquake.

Crawl Spaces

Crawl spaces are most revealing to show the response of wood-framed buildings under seismic loads. Unlike the rest of the type-V structure, concealed under architectural rendering, structural wood members in the crawl space are fully exposed to visual inspection.

Many contemporary residential buildings, either single or tract houses, are built on a concrete slab, but a very large number of older properties in California were constructed over a crawl space, generally a very shallow "mini-basement" where a number of posts and piers support the interior of the timber structure. Thus the crawl space plays a pretty important role.

A view of a typical crawl space would show supports consisting of 4×4 posts (some even 2×4) resting on relatively small pyramid-shape piers. In some cases the truncated pyramid piers were replaced by equally small cylindrical footings, typically cast in place. The piers in general were not wide enough or did not go deep enough into the ground; thus they were rather unstable to resist an earthquake; prone to wobble, the piers denied a good firm support to the floor above.

Typically all elements of the assembly were separated, that is, individual pieces were not connected by structural strong-tie or other metal straps, anchors, or saddles. Before the January 1994 Northridge earthquake, little attention was paid to these internal supports regardless of the vital role they play in holding up the building. Their adequacy was seldom questioned, taking for granted that the entire assembly would respond to satisfaction.

Among numerous buildings inspected after the Northridge earthquake none appeared to be engineered; apart from architectural design and layouts, the structure was unresolved. The entire responsibility for putting the building together rested on the framer. Common earthquake-caused damage found in crawl spaces consisted of leaning and tilted posts, displacement of posts relative to the support, some to the point of collapse, beams and girders pulled off the support, and footings too small to support the structure (Figures 8.1, 8.2, and 8.3). During inspections of earthquake-damaged buildings the author found posts squashed against the beam as a consequence of bearing failure of an overstressed structure. Some posts were merely 2 × 4 nailed together in a T assembly that could have buckled at any time without a seismic event. It made one wonder what would have happened had the earthquake lasted 30 seconds longer.

Framing Connections

Engineered metal connectors are a must to join individual wood elements into a structural unit. Although engineered metal connectors had been around for many years, only a few buildings had metal straps which, in most cases, were merely *plumber straps* (Figures 8.4 and 8.5). What was then used to connect the individual wood elements to meet a violent earthquake is called *toenailing*—several nails driven in at the edge of the wood: joist, beam, or girder (Figure 8.6). Generally the nails end up at the edge of the post and often splinter the wood.

Stucco and Wood Frame Assembly

Destructive inspections to expose damages concealed under architectural rendering—stucco and drywalls—revealed that the severe vibrations caused by the earthquake had separated the stucco from the wood frame. This is a serious impairment to the seismic resistance of the structure because the two major components, the stucco and wood frame, should work together to resist seismic forces that are mostly lateral.

Cripple Walls

A 16-ft-high wall in the living room of a residential building rocked severely during the Northridge earthquake and continued to do so afterward (Figure 8.7). Destructive inspection revealed that the wall was constructed of two

Figure 8.1   Wooden block in a crawl space crushed by the 1994 Northridge earth-quake.

panels nailed together at joining plates in what is known as *cripple wall* construction. A cripple wall—two panels nailed together—is commonly used where extra ceiling height is needed (higher than the regular 9-ft-tall wall). Because in a cripple wall the upper panel does not form a monolithic unit with the rest of the assembly, the assembly tends to perform badly against seismic forces.

Cripple wall construction is an accepted practice *under the umbrella of the building codes* but not from an engineering standpoint. The assembly is not sufficiently stable to resist lateral forces caused by winds or earthquakes. The joint created by nailing two separate panels tends to pivot and fall apart even under moderate seismic activity. In structural engineering it would constitute a *hinged* joint that provides poor lateral resistance to the wall.

Figure 8.2 Dangerously unstable support to the floor above.

Figure 8.3 Rough-and-ready workmanship. Wood wedges inserted under the support assembly.

Figure 8.4 Inspection of damaged buildings found plumber straps in lieu of engineered metal connectors to join individual wood elements.

Figure 8.5 Properly engineered metal connector used to join individual wood elements in the retrofitting of an earthquake-damaged residential building.

Figure 8.6 Toenailing, an unsafe practice to face a violent earthquake, which pulled out the nails.

Figure 8.7 Cripple wall damaged during the Northridge earthquake.

Garages

Type-V garages were perhaps one of the most neglected structures. With very few exceptions, those inspected after the Northridge earthquake appeared to lack sound engineering design and displayed poor workmanship (Figure 8.8). In most cases garages showed a downgrade in workmanship as compared to the dwelling portion of the building, which received better attention from the framer.

Most garages share a major weakness: a relatively large opening created in the front that leaves no wall surface to resist earthquakes. In the absence of any structural provision to stabilize the front, garages are at the mercy of earthquakes and high winds. One of the solutions the author adopted for the retrofitting of damaged garages was to provide a steel portal frame along with adequate footings to resist lateral seismic loads.

Vaulted Ceilings

Most property owners love space and extra story-height. The trouble with such a choice is lack of workmanship to support it and, above all, lack of engineering to back up the style with a sound structure. When inspecting some earthquake-damaged *cathedral*-type ceilings the author found ridge beams with only a thin metal anchor to connect beam and post, which resulted

Figure 8.8 Common sight in earthquake-damaged garages coupled with poor workmanship.

in a girder twisted by the earthquake. Obviously, the connection was entirely insufficient to hold the heavy girder that was holding most of the roof in place.

Chimneys and Fireplaces

Earthquake-damaged chimneys became a familiar sight in the San Fernando Valley area of California after the Northridge earthquake (Figures 8.9 and 8.10). A general belief is that, if an unreinforced chimney and fireplace can be made stronger against seismic forces, the problem is solved. In the author's opinion, masonry chimneys, reinforced or unreinforced, are incompatible with the relatively frail framework of a wood construction building.

Construction Defects and Earthquake Damage

Consideration of construction defects versus earthquake damage is not always clear-cut:

Case 1 It had to be established whether sagging of the roof was the result of earthquake-caused damage. Inspection of the loft found large wood knots at a critical location in two adjacent roof rafters. The wood defect had reduced by about 40% the load-carrying capacity of the rafters.

Figure 8.9 Two-story stone fireplace that collapsed during the Northridge earthquake.

Figure 8.10 One of the numerous poorly built fireplaces damaged by the Northridge earthquake.

Being adjacent to each other, no load transfer could take place between the rafters. This was a true construction defect.

Case 2 The insured claimed that his vaulted ceiling was sagging due to earthquake damage. The problem appeared similar to case 1. When the author examined the exposed structure it turned out that a minor horizontal crack caused by the earthquake had split the rafter lengthwise. The crack had propagated from a sizable wood knot at the critical point of support, attracting maximum shear force and stresses. The insurance company ruled that this was earthquake damage because the earthquake had aggravated an initial construction defect. The roof was of the asphalt shingle type and had to be entirely replaced—a costly operation to replace one faulty structural element.

In case 1 a trained engineering eye helped avoid expensive reroofing. The entire matter is complex—a lot depends on individual judgment and the skill, knowledge, and experience of the inspector to make the difference between a less costly repair and an expensive replacement.

Revisiting the crawl spaces, images of leaning posts, and disintegrating support elements, one would ask: Are these truly earthquake-caused damages or could they have been avoided by better construction? As mentioned earlier, had the connections been strapped by *proper metal hardware,* the entire assembly would have performed far better in the earthquake. Leaving connections loose without proper metal straps is a *serious construction deficiency.* Depending on the knowledge and the experience of forensic engineers and inspectors in the matter of safety issues, some may not rule such cases as construction defects. The 1994 UBC was not specific on the issue of strapping of wood elements, leaving it to the discretion of plan examiners and city inspectors. Next we analyze design examples of wood-framed structures in various seismic zones.

8.3   CASE 1: STEEL-REINFORCED WOOD-FRAMED BUILDING

Seismic: Low- to Moderate-Seismicity Area

The structure is located in a relatively moderate seismicity zone near a higher seismicity region. We provided steel frames *to resist seismic and wind forces* (Figures 8.11–8.15).

PROJECT DESCRIPTION

The project is an 8400-ft^2 wood-framed residential building consisting of a partial underground basement of reinforced masonry construction and two-garage-and-shop, 1900-ft^2 attached structure.

WIND ACTING ON THREE-STORY STEEL FRAME ON GRIDLINE 1 (EAST–WEST DIRECTION)

At roof plate level

$$H_r = \frac{20.0(8.0 + 8.0/2) \times 14.0}{2} = 1680 \text{ lb} = 1.68 \text{ k}$$

At second-floor level

$$H_2 = \frac{19.0 \times 3.0 \times (14.0 + 8.0)}{2} = 630 \text{ lb} = 0.63 \text{ k}$$

At first-floor level

$$H_1 = \frac{17.4 \times 3.0 \times 14.0}{2} = 580 \text{ lb} = 0.58 \text{ k}$$

These loads were input into a computer analysis, "Two-Story Front M Frame," combined with roof and floor dead loads,

Figure 8.11 View of the three-story north elevation. As the terrain slopes down sharply toward the riverfront, the north elevation became three stories high. (*Architectural design by SM Designs, Inc.*)

$$w_r = 0.50 \text{ k/ft} \qquad \text{and} \qquad w_2 = w_1 = 0.15 \text{ k/ft}$$

respectively, and roof and floor live loads,

$$w_{rL} = 0.44 \text{ k/ft} \qquad \text{and} \qquad w_{2L} = w_{1L} = 0.50 \text{ k/ft}$$

WIND ACTING ON STEEL FRAME, GRIDLINE 2

$$H_r = 20.0\left[3.0 \times 7.0 + \left(8.5 + \frac{8.0}{2}\right)\left(\frac{21.0}{2}\right)\right] = 3046 \text{ lb} = 3.046 \text{ k}$$

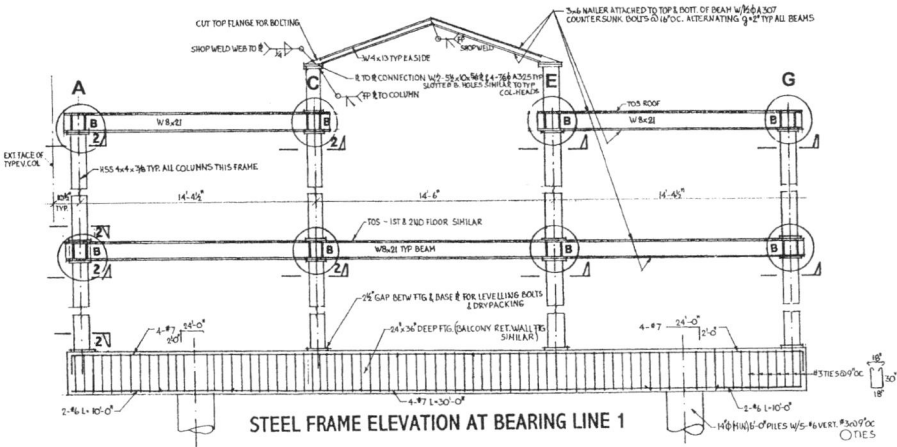

STEEL FRAME ELEVATION AT BEARING LINE 1

Figure 8.12   Plan view of the three-story steel frame at the north elevation showing the horizontal steel beams and the HSS $6 \times 4 \times \frac{3}{8}$ steel columns.

Figure 8.13   First stages of erection of the steel frames that will provide additional support against wind and seismic forces to the type-V building.

Figure 8.14 View of the second and third floors of the north elevation with the steel frame in place.

$$H_2 = 19.0\left[3.0 \times 7.0 + \left(\frac{14.0 + 8.0}{2}\right)\left(\frac{21.0}{2}\right)\right] = 2600 \text{ lb} = 2.60 \text{ k}$$

combined with floor dead and live loads as

$$w_{2D} = 0.25 \text{ k/ft} \qquad w_{2L} = 0.86 \text{ k/ft}$$

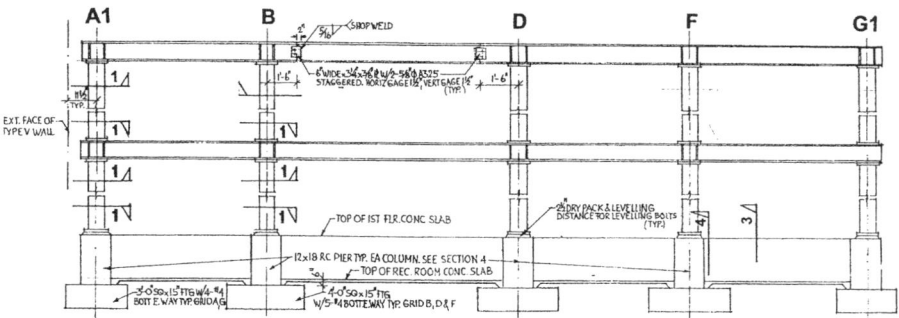

Figure 8.15 Plan view of the steel frame that supports the two-story structure at the south elevation.

STEEL FRAME ANALYSIS AND DESIGN, GRIDLINE 1

1. *Design of Beams*
   Largest beam moment for the entire structural system:

   16.55 k-ft = 198.6 k-in. for beam 16, loading case 5

   Allowable stress for the W8 × 24:
   with

   $$L_u = 15.2 \text{ ft} \qquad F_b = 22 \text{ ksi}$$

   Required section modulus:

   $$S_{req.} = \frac{198.6}{2} = 9.02 \text{ in.}^3 < 18.2 \text{ in.}^3$$

   Provided by W8 × 24.   OK.

2. *Design of Columns*
   6 × 4 × $\frac{3}{8}$ HSS—structural tubing.
   The worst load combinations are for column 10.
   The maximum moment is caused by dead plus 0.75 live plus 0.75 wind, 1997 UBC load combination (12-11), our computer load combination 5. See Appendix at the end of this chapter.

   $$M_{max} = 12.32 \text{ k-ft} = 147.8 \text{ k-in.}$$

   Axial: $P = 6.30$ k.

   Design 6×4×$\frac{3}{8}$ HSS (AISC ASD design):

   $$L = 10.0 \times 12 = 120$$

   Sectional properties, Table C-50, AISC *Manual*:

   $$A = 6.58 \text{ in.}^2$$

   $$S_x = 9.90 \text{ in.}^3$$

   $$\frac{KL}{r_y} = \frac{1.0 \times 120}{1.54} = 78$$

   $$r_y = 1.54$$

   $$F_a = 15.6 \text{ ksi}$$

Maximum allowable bending stress:

$$F_b = 0.75 \times F_y = 0.75 \times 46 = 34.0 \text{ ksi}$$

$$f_b = \frac{147.8}{9.9} = 14.92 \text{ ksi}$$

$$f_c = \frac{6.3}{6.58} = 0.96 \text{ ksi}$$

$$\frac{f_b}{F_b} + \frac{f_c}{F_c} = \frac{14.92}{34} + \frac{0.96}{15.6} = 0.5 < 1.0 \qquad \textbf{OK}$$

Design Beam-to-Column Plate and Bolts

Maximum axial tension in the two $\frac{7}{8}$-in. ϕ bolts:

$V = M_{max}$/lever arm to centerline of compression edge of HSS.

Lever arm $= 6.0 + 1.5 = 7.5$ in.:

$$V = \frac{147.8}{7.5} = 19.7 \text{ k} < 2 \times 26.5 = 53.0 \text{ k}$$

53.0 k is the capacity of two $\frac{7}{8}$-in. ϕ A325 bolts (AISC ASD Table I-A, Part 4). OK.

Maximum moment acting on the beam-to-column plate at the face of HSS 6×4 column:

$$19.7 \times 1.25 = 24.6 \text{ k-in.}$$

Attach 2-in-wide, 2-in-deep triangular haunch to support beam-to-column bearing plate. Static moment about the top surface of the T assembly:

$$4.5 \times 0.5 \times 0.25 = 0.562$$
$$2.5 \times 0.5 \times 1.00 = \underline{1.250}$$
$$1.812 \text{ in}^3$$

Area $= 4.5 \times 0.5 + 2.5 \times 0.5 = 3.50 \text{ in.}^2$:

$y_T = 1.812/3.50 = 0.518$ in.

$I_{1\text{-}1} = \frac{1}{3}(4.5 \times 0.5^3 + 0.5 \times 2.5^3) = 2.792$ in.^{4}

$I_{CG} = 2.792 - (3.50 \times 0.518^2) = 2.792$ in.^{4}

$S_{\text{bott}} = 2.792/(2.5 - 0.52) = 1.41$ in.^{3}

Maximum stress in bracket:

$$f_b = \frac{24.6}{1.41} = 17.5 \text{ ksi} < 27.0 \text{ ksi} \qquad \textbf{OK}$$

DESIGN OF BEAMS, GRIDLINE 2

The largest moment occurs in beam 9, from #5 computer load combination:

$$M_{\text{max}} = 14.75 \text{ k-ft} = 177 \text{ k-in.}$$

$$S_{\text{req.}} = \frac{177.0}{22.0} = 8.0 \text{ in.}^3 < 18.2 \text{ in.}^3$$

Provided by W8 × 21.   OK.

LATERAL ANALYSIS AND DESIGN: GARAGE

Wind is acting in the east–west direction.

Forces on Front Shear Walls

Total length: $3 \times 4 = 12.0$ ft
Tributary wind area $(47.0/2) \times (18.0 - 9.5) = 223$ sqft
Wind force on three-shear-wall system: $V = 17.4 \times 223 = 3884$ lb
Linear shear: $v = 3884/12.0 = 324$ lb/ft < 380 lb/ft

Provided by $\frac{3}{8}$-in. OSB with 8d nails at 3, 3, 12-in. OC.
(1997 UBC, Table 23-II-I-1)

Note. OSB: oriented strand board wood panel.

Overturning/Hold-Down Bolts

$$M_o = \left(\frac{3884}{3}\right) \times 18.0 = 23,000 \text{ lb-ft} = 23.0 \text{ k-ft}$$

Weight of roof and wall on shear wall:

$$w = \frac{8 \times 2.0 \times 40.0}{2} + 4 \times 18.0 \times 12 = 1184 \text{ lb}$$

Net uplift:

$$T = \frac{23.0}{3.75} - 1.18 = 4.95 \text{ k} < 5.51 \text{ k}$$

Provided by Simpson's HD6A   OK.

Footing Design

Each shear wall pier will be:

1. Provided by 2.5 × 6.0 × 2.0-ft deep footing
2. Connected by an 18-in.² continuous grade beam

The grade beam serves to counteract the base moment caused by the wind. The factored base moment per shear wall is

$$M_u = \frac{1.3 \times 23.0}{3} = 10.0 \text{ k-ft} = 120.0 \text{ k-in.}$$

The resistance of the 18-in.² beam with two #6 longitudinal reinforcements is

$$\phi M_n = 0.9 \times (2 \times 0.44) \times 60.0 \times 0.80 \times (18.0 - 3.375) = 556 \text{ k-in.}$$

Larger than 120 k-in. applied.   OK

Glued Laminated Beams (Glulam, GLBs) to Support Garage Roof (Figure 8.16)

Maximum span 22.0 ft
Unit weights: DL = 8.0 psf
                       LL = 20.0 psf. Total 28.0 psf
$M_{max} = 0.56 \times 22.5^2 \times 1.5 = 283.5$ k-in.
$w = 0.028 \times 20 = 0.56$

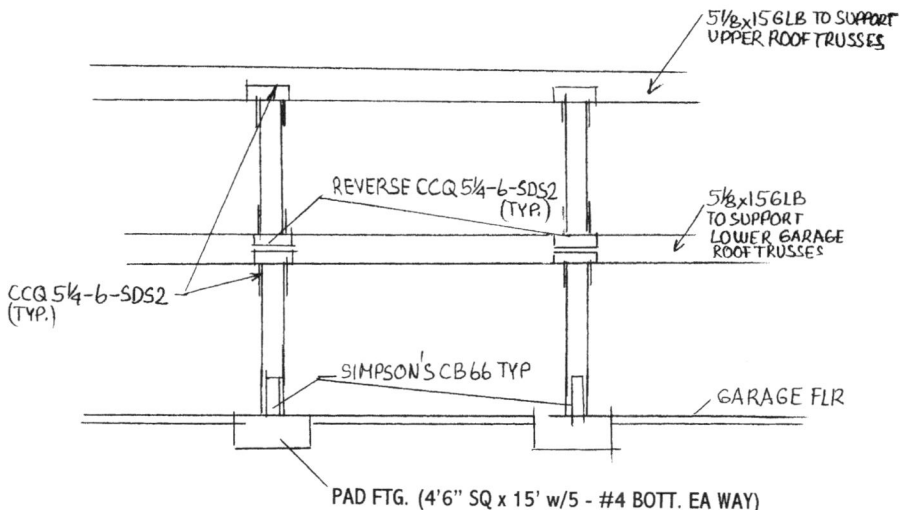

Figure 8.16 Glued laminated beams to support garage roof and door.

Tributary width = 20.0 ft using 24 DF/DF (Douglas fir) grade

$$S_{req.} = \frac{283.5}{2.4} = 118 \text{ in.}^3 < 192 \text{ in.}^3$$

Provided by $5\frac{1}{8} \times 15$ GLB.   OK.

GLB Header Over 16.0-ft Garage Door, Span 16.5 ft.

$$DL = 2.0 \times 8 + 6 \times 10 + 15 = 91 \text{ lb/ft}$$

$$LL = 2.0 \times 20 = 40 \text{ lb/ft. Total } 131 \text{ lb/ft} = 0.131 \text{ k/ft}$$

$$M_{max} = 0.131 \times 16.5^2 \times 1.5 = 53.4 \text{ k-in.}$$

$$S_{req.} = \frac{53.4}{2.4} = 22.3 \text{ in.}^3 < 123 \text{ in.}^3$$

Provided by $5\frac{1}{8} \times 12$ GLB.   OK

GLB to Support Second Floor Over Kitchen (Figure 8.17)

Tributary width: $(21.0 + 7.0)/2 = 14.0$ ft. Maximum span = 14.5 ft
DL = 12 psf
LL = 40 psf, Total 52 psf

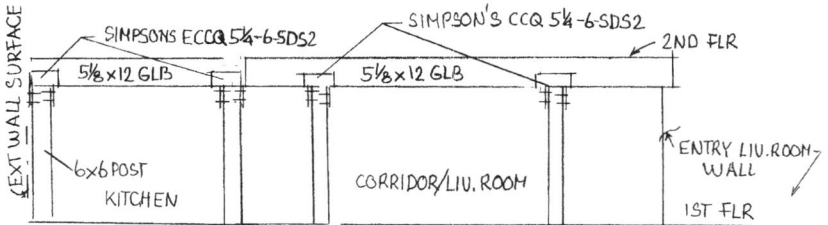

Figure 8.17   GLB to support second floor over kitchen.

$w_{D+L} = 0.052 \times 14.0 = 0.73$ k/ft   Say 0.75 k/ft
$M_{max} = 0.75 \times 14.5^2 \times 1.5 = 236.5$ k-in.
$S_{req.} = 236.5/2.4 = 98.6$ in.$^3 < 123$ in.^{3}

Provided by $5\frac{1}{8} \times 12$ GLB.   OK.

$\Delta_{max} = (5/384)(0.75 \times 14.5) \times 14.5^3 \times 12^3/738 \times 1800 = 0.56$-in. $< L/240 = 0.72$ in.

Provided by $5\frac{1}{8} \times 12$ GLB.   OK.

6 × 6 Posts

Consider two maximum loads:

1. Garage: $P_{max} = 12.0 \times 20.0 \times 0.032 = 7.8$ k/9.0 ft high
2. Kitchen/living room: $P_{max} = 9.0 \times 14.0 \times 0.052 = 6.6$ k with 13.0-ft height

GOVERNS

$F_c = 1.2$ ksi
Effective length factor: $L_e/d = 13.0 \times 12/5.5 = 28 > 26$
Slender, use $F'_c = 0.3E/(L_e/d)^2 = 0.3 \times 1600/28^2 = 0.61$ ksi
Capacity of column: $P = 0.61 \times 5.5^2 = 18.5$ k > 6.6 k applied   **OK**

6 × 6 post OK.

Garage Foundation

Footing under 6 × 6 post supporting roof GLB
Tributary area: $12.0 \times 20 = 240$ sqft

$P_{max} = (0.012 + 0.02) \times 240 = 7.68$ k

Minimum footing size: $\sqrt{7.68/1.0} = 2.77$ ft

**Provide 4-ft, 6-in.^{2} × 15 in. with 5 #4
bottom each way reinforcement.**

Footing for 6 × 6 Post in Kitchen

$P_{max} = 6.6$ k

Minimum footing size: $b = \sqrt{6.6/1.0} = 2.57$ ft

**But provide 3-ft, 6-in.^{2} × 15 with 4 #4
bottom each way reinforcement.**

8.4   CASE 2: WOOD-FRAMED TWO-STORY HOME

Seismic: Moderate- to High-Seismicity Area

PROJECT DESCRIPTION

The structure comprises living quarters at ground level and a game room
and portion of a balcony at the second level. Building total length is
94.0 ft measured along its centerline. The footprint of the building forms
a V shape; the lower roof and second floor act as a continuous diaphragm that activates all the shear walls resisting wind or seismic. A
three-car garage attached to the wing of the building will be treated as
separate entity for lateral forces.

Roof LL 20 psf; floor LL 40 psf; external balcony LL 60 psf
Soil profile S_C
Allowable bearing pressure 1000 psf
Wind 75 mph, exposure C
UBC 1997 design provisions were followed.

SEISMIC ANALYSIS

Mass of First-Story Structure

Second-floor and balcony weight
    $10.00(18 \times 20.5 + 18 \times 30) + 8.0[12.0(19.5 + 14.5)] \approx 12,600$ lb
Lower roof

$$22.0(46 \times 7.33 + 30 \times 17 + 9 \times 30 + 10 \times 8 + 15 \times 16 + 25 \times 20) = 42,600 \text{ lb}$$

Upper roof

$$22.0(18.5 \times 50) + 6 (12 \times 33) = 22,800 \text{ lb}$$

Weight of walls impacting lower roof/second-floor diaphragm

First-story walls	8,000 lb
Second-story walls	3,000 lb
Second-story walls impacting upper roof	3,000 lb

Mass impacting second-floor/lower roof diaphragm	66,200 lb
Mass impacting upper roof diaphragm	25,800 lb
Total mass	92,000 lb

Longitudinal Analysis

$$T = C_t(h_n)^{3/4} = 0.02(19.0)^{0.75} = 0.182 \text{ s}$$

$$C_a = 0.33 \qquad C_v = 0.45$$

$$T_s = \frac{C_v}{2.5C_a} = \frac{0.45}{2.5 \times 0.33} = 0.545 \text{ s} > 0.182 \text{ s}$$

$$T_o = 0.2T_s = 0.110 \text{ s}$$

Use upper plateau of Fig. 16.3, UBC 1997, Vol. 2 (Figure 8.18)

$$V = \left[2.5(0.33)\left(\frac{1.0}{5.5}\right) \right]92 = 0.15(92) = 13.8 \text{ k} \qquad \textbf{GOVERNS}$$

Wind is *not* a determining factor in the longitudinal direction at its base shear:

$$(1.4) \times (5.9) = 8.3 \text{ k} < 13.8 \text{ k}$$

Between the two longitudinal shear walls the inner (rear) shear wall carries a larger unit shear:

$$v = \frac{13,800}{2 \times 26.67} = 258 \text{ lb/ft} < 260 \text{ lb/ft allowed}$$

By $\frac{3}{8}$-in. OSB/Plywood with/8d nails at 6, 6, 12 in. OC, UBC Table 23-II-I-1.

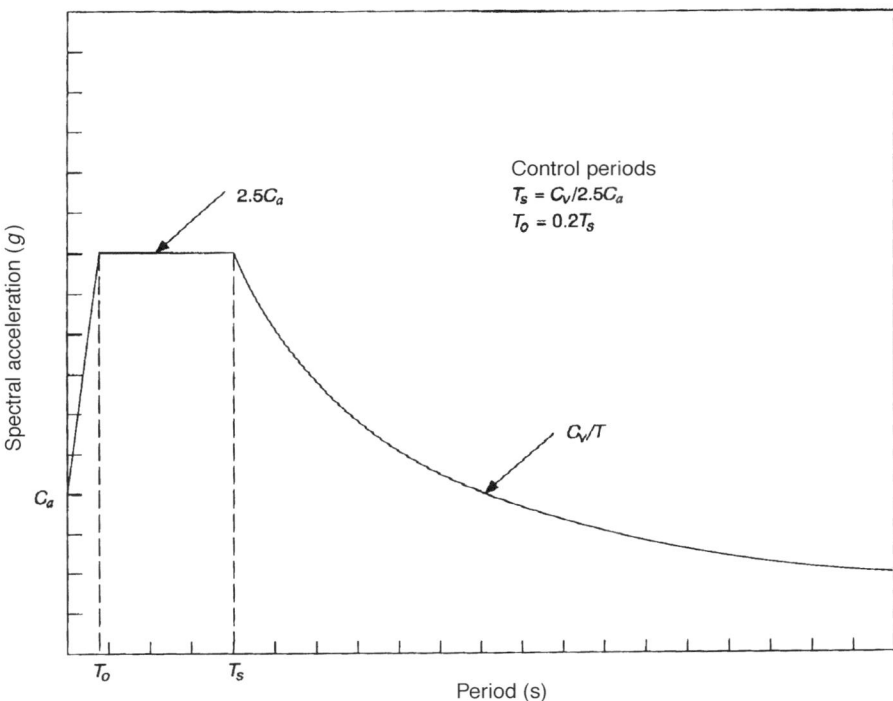

Figure 8.18   Design response spectra.

Hold-Down/Uplift

Among the shear wall panels the 4.0-ft shear wall panel will receive the largest of the hold-down (HD) uplifting force. The shear wall will share $4.0/2 \times 26.67 = 0.075$ of the total base shear.

Uplift
$13.8(0.075) \times 10.0/3.75 =$ -2.67 k
Less roof and second-floor DL reaction
$7.25(9.5 \times 22 + 6.0 \times 10)/1000 =$ $\underline{2.00 \text{ k}}$
Net uplift 0.67 k < 1.95 k

Provided by Simpson's STHD8 or HD2A with/$\frac{5}{8}\phi$ HD bolts.

SEISMIC IN TRANSVERSE DIRECTION

Seismic in the transverse direction is resisted by several walls on the left side but lesser shear walls on the right-hand side of the building (kitchen/family room). The share of lateral force on each wall will be calculated on the basis of tributary areas. As the width of the building, effective to counter lateral forces, is constant, the exercise is reduced to length issues proportional to tributary areas. Because the total length of the building is 94.0 ft measured along its centerline, the central wall adjacent to the stairs shares

$$31.5/94(13.8) = 4.60 \text{ k}$$

The total length of the shear wall including the dining room wall aligned within 2.0 ft of the stair wall is 31.0 ft:

$$v = \frac{4600}{31.0} = 150 \text{ lb/ft} < 260 \text{ lb/ft}$$

Allowed by $\frac{3}{8}$-in. OSB/plywood with 8d at 6, 6, 12 in. OC, UBC 1997, Table 23-II-I-1.

Family Room/Bedroom 2 and 3 Shear Wall

Total length 26.5 ft
Lateral force shared by this wall: $(17.5/94)(13.8) = 2.57$ k
$v = 2570/26.5 = 97$ lb/ft < 100 lb/ft

Allowed by $\frac{1}{2}$-in. gypsum drywall with 5d cooler nails at 7 in. OC.

End Wall, Right Elevation

Load share of this wall: $(9.0/94)(13.8) = 1.32$ k
Length of OSB shear wall panels: $3 \times 4.0 = 12.0$ ft
$v = 1320/12.0 = 110$ lb/ft < 260 lb/ft

Provided by $\frac{3}{8}$-in. OSB with 8d at 6, 6, 12 in. OC.

VERTICAL LOAD ANALYSIS

Master Bedroom Door Header

Span 8.25 ft
DL $= 8.4 \times 22 + 6.0 \times 6 = 223$ lb/ft
LL $= 8.4 \times 14 + 6.0 \times 14 = 200$ lb/ft
Total DL + LL $= 0.42$ k/ft

$$S_{req.} = \frac{0.42 \times 8.25^2 \times 1.5}{1.45} = 29.6 \text{ in.}^3 < 30.6 \text{ in.}^3$$

Provided by 4 × 8.

GLB to Support Second-Floor Patio

Span 16.0 ft
LL $= 6.0 \times 60 = 360$ lb/ft
DL $= 6.0 \times 6 = \underline{36 \text{ lb/ft}}$
$ 396$ lb/ft   Say 0.4 k/ft

$$S_{req.} = \frac{0.4 \times 16.0^2 \times 1.5}{2.4} = 64.0 \text{ in.}^3 < 117 \text{ in.}^3$$

Provided by $3\frac{1}{8}$ × 15-in. GLB.   OK.

Deflection

$$\Delta = \left(\frac{5}{384}\right)(0.4 \times 16)(16^3)(12^3)879 \times 1.5 \times 10^3$$

$$= 0.44 \text{ in.} < 0.53 \text{ in.} = \frac{L}{360} \quad \textbf{OK}$$

Patio Column Footings

Pad footings will be provided.
Two-story portion: $\frac{1}{2}(14.5 + 16.5)(17.0) = 263$ sqft

Maximum floor tributary area: $\dfrac{(14.5 + 16.5) \times 12.0}{4} = 93.0 \text{ ft}^2$

DL roof $= 0.022 \times 263 = 5.79$ k
LL roof $= 0.4 \times 0.02 \times 263 = 2.11$ k
DL floor $= 0.008 \times 93 = 0.74$ k
LL floor $= 0.4 \times 0.06 \times 93 = \underline{2.23 \text{ k}}$
$ $Total 10.87 k

$$b_{req.} = (10.87/1.0)^{1/2} = 3.3 \text{ ft} = 3 \text{ ft, 4 in. square footing dim.}$$

Strip Footing for Main Front Bearing Wall

In front of but not at the entrance door:
Tributary width of roof: $34.0/2 = 17.0$ ft
Tributary width of floor: $10.0 + 5.0 = 15.0$ ft

Maximum linear loads received from
  Roof: $(22 + 0.4 \times 20)\ 17.0 = 510$ lb/ft
  Floor: $(8 + 0.4 \times 40)\ 15.0\ = 360$ lb/ft
  Weight of wall: $10.0 \times 20\ \ = \underline{\ 200\ \text{lb/ft}}$
             Total load   1070 lb/ft
Required footing width:

$$\frac{1.07}{1.0} = 1.07 \text{ ft} < 1.3 \text{ ft} = 15 \text{ in. provided} \qquad \textbf{OK}$$

8.5 CASE 3: STEEL-REINFORCED TWO-STORY DUPLEX

Seismic: Moderate- to High-Seismicity Area

PROJECT DESCRIPTION

This is a wood-framed construction, two-story residential duplex at the riverfront in a moderate- to high-seismicity area (Figure 8.19). The engineering analysis took into consideration the seismic region as well as the 75-mph winds in the area. The soil profile is S_C. The allowable bearing pressure is 1500 psf per the geotechnical report. The front frame is reinforced with HSS steel columns (Figure 8.20) and rolled wide-flange steel beams at roof level and GLBs at the second-floor balcony level.

WIND LATERAL ANALYSIS

Riverfront Frame

The riverfront frame receives roof reaction because wind impacts on the front frame blowing downstream in the north–south direction. The major impact is at eaves level. Using the 1997 UBC projected-area method, the lateral wind pressure on the gable end at 23.0 ft CG height, exposure C, is

$$p = 12.6 \times (1.166) \times (1.3)(1.0) = 19.0 \text{ psf}$$

The tributary area is $\frac{1}{2}(9.6 \times 22) = 106$ sqft. The total lateral wind load acting at eaves level is

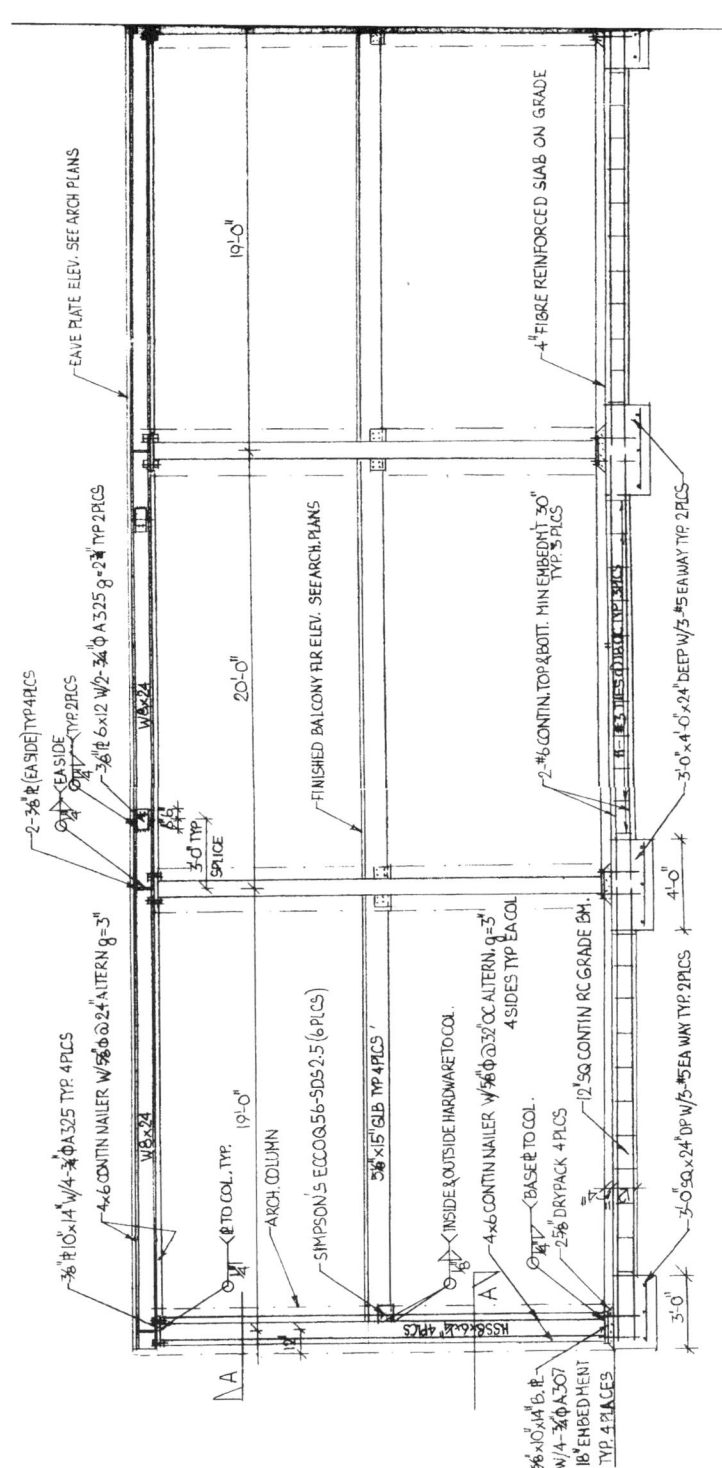

Figure 8.19 Front plan view of structural elevation showing steel frame.

253

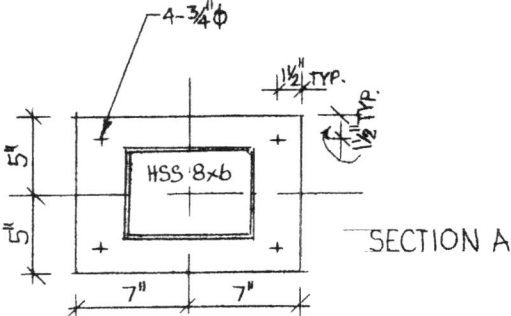

Figure 8.20 Detail of HSS tubing in section A of riverfront structural elevation.

$$H_r = 0.019 \times 106 = 2.0 \text{ k horizontal}$$

The wind load will be shared by four HSS steel columns, i.e., 0.5 k per column. The columns are 20.0 ft high and

$$M_{\text{max}} = 0.5 \times 20.0 = 10.0 \text{ k-ft} = 120 \text{ k-in. at fixed base}$$

$$S_{\text{req.}} = \frac{120}{32} = 3.75 \text{ in.}^3 < 8.5 \text{ in.}^3$$

Provided by HSS $7 \times 5 \times \frac{3}{16}$ structural steel tube.　　**OK.**

However, a $\frac{1}{4}$-in.-thick tube will be used for better deflection control.

Design Steel Beam

$$S_{\text{req.}} = \frac{120}{22} = 5.45 \text{ in.}^3 < 13.4 \text{ in.}^3$$

Provided by W6 × 20 grade A36.　　**OK.**

Note: A36 vs. A992 A992 is noted as the preferred structural steel in the AISC *Manual of Steel Construction,* LRFD, 3rd edition, 2005. A concern about the welding properties of A992 as opposed to the old A36 was addressed by AISC Technical Assistance Director Kurt Gustafson, S.E.

　　There has been a lot of research on the subject of connection details for seismic conditions since the Northridge E.Q. Much of this involves the ductility of the connections. In addition to the base material, the weld rod and the specific details of seismic connections were studied. Stipulations were incorporated in the AISC Seismic Provisions to enhance the ductility of connections. ASTM A992 was added to the AWS D1.1-2000 in Table 3.1. The steel mills did considerable testing to get it approved. On the material question inquired, it was known that almost all ASTM A36 material of recent production actually had yield stresses closer to 50 ksi than the 36 ksi required minimum. This placed the yield level very close to the tensile strength resulting in diminished ductility from material with lower yields. The ASTM A992 Specification now has a Carbon Equivalent (CE) of 0.45 for shapes, and places a maximum yield-to-tensile ratio of 0.85 on the material to assure some level of ductility. The AISC 2002 Seismic Provisions stipulate a design requirement in the form of a R_y Factor to assure this level of ductility. ASTM A36 still can be used, but the associated R_y Factor is greater than that for ASTM A992 material.

Riverfront Shear Walls Each Side of Bay Windows

Total length: $5.0 + 2 \times 5.0 + 5.0 = 20.0$ ft

Total lateral force acting on shear wall system: $(10.0 \times 40.0/2)(19.0) = 3800$ lb

Linear shear induced to shear walls $v = 3800/20.0 = 190$ lb/ft $<$ 260 lb/ft

Provided by $\frac{3}{8}$-in. OSB with 8d at 6, 6, 12 in.   OK,
1997 UBC, Table 23-II-I-1.

Shear Walls, Garage Front

Total north-south wind impacting on each door shear wall: $19.0 \times (10.5 \times 30.0/2) = 2990$ lb

Total shear wall length: $2 \times 40 = 8.0$ ft. The linear shear: $v = 2990/8.0 = 374$ lb/ft $<$ 380 lb/ft

Provided by $\frac{3}{8}$-in. OSB w/8d at 4, 4, 12 OC   OK.

Overturning effect on each shear wall panel:

Vertical uplift at each end: $T_o = (2.99/2) \times 9.0/4.0 = -3.36$ k

Less weight of roof: $0.022 \times 7.0 \times 26.0/2 = \underline{2.00\ k}$

Net uplift $-1.36 < 4.0$ k

Provided by HD5A on 6 × 6 post   OK.

Second-Floor Balcony GLB

Carries balcony reaction

Spans 19.4 ft between HSS steel columns

Tributary width: $14.0/2 = 7.0$ ft

$W_{D+L} = 7.0 (60.0 + 10.0) + 10.0 = 500$ lb/ft

Maximum vertical reaction:

$9.4(0.5) = 4.7$ k

$$S_{req.} = 0.5 \times 19.4^2 \times \left(\frac{1.5}{2.4}\right) = 117.6 \text{ in.}^3 < 123 \text{ in.}^3$$

Provided by $5\frac{1}{8}$-in. × 12 GLB.   OK.

Maximum horizontal shear:

$$f_h = \frac{4700}{61.5} = 76.0 \text{ psi} < 200 \text{ psi}   \textbf{OK}$$

GLB over Riverfront Bay Windows

Supports balcony and second floor

Span 20.0 ft

Tributary width: $(30.0 + 14.0)/2 = 22.0$ ft

$W_{D+L} = 22.0 (40.0 \times 0.6 + 10) = 750$ lb/ft $= 0.75$ k/ft

$S_{req.} = 0.75 \times 22.0^2 \times 1.5/2.4 = 226 \text{ in.}^3 < 232 \text{ in.}^3$

Provided by $5\frac{1}{8}$ × $16\frac{1}{2}$ GLB.   OK.

8.6   CASE 4: WOOD-FRAMED COMMERCIAL BUILDING

Seismic: High-Seismicity Area

The calculations will show that, even though the structure is in a high-seismicity region, seismic *does not govern* in this case study.

PROJECT DESCRIPTION

This is a 50 × 80-ft commercial building on a 56,000-sqft lot in Victorville, California, with a retail area of 4000 sq ft, in a high-seismicity area (*Developers: Glen & Pearl Ludwig. Architect: Steve Shover*). Plate height is 14.25 ft, parapet height is 16.0 ft, and roof structure is open web trusses. See Figure 8.21.

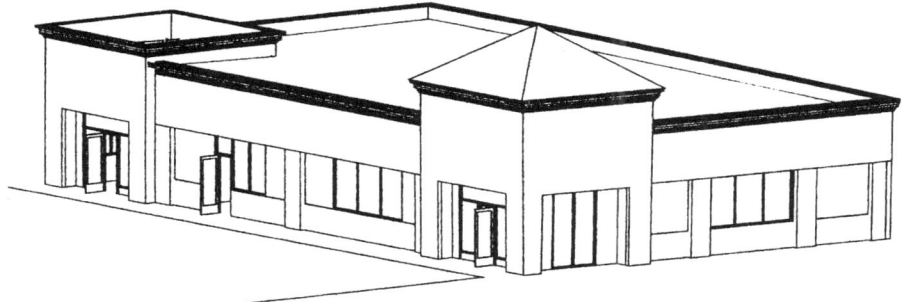

Figure 8.21 Architectural rendering of case 4 building.

DESIGN CRITERIA

The building was designed and engineered in compliance with the 1997 UBC, the 1997 Fire Code, and the City of Victorville Development Code. See Figure 8.22 for the building plan.

Wind 70 mph, exposure C, seismic zone 4
Soil profile S_C
Allowable bearing pressure 1000 psf
Maximum wind pressure by projected-area method:

$$p = C_e C_q q_s I_w = 12.6(1.07)(1.3)(1.0) = 17.5 \text{ psf}$$

WIND ANALYSIS

First east–west wind analyzed
Total horizontal wind force acting on larger (80-ft-long) *west wall* reacting at plate level:

$$V = 17.5 \times (80.0) \times \left(4.0 + \frac{12.0}{2}\right) = 14,000 \text{ lb}$$

Lateral wind force reacting on northern external shear wall at plate level:

$$H = \frac{14,000}{2} = 7000 \text{ lb}$$

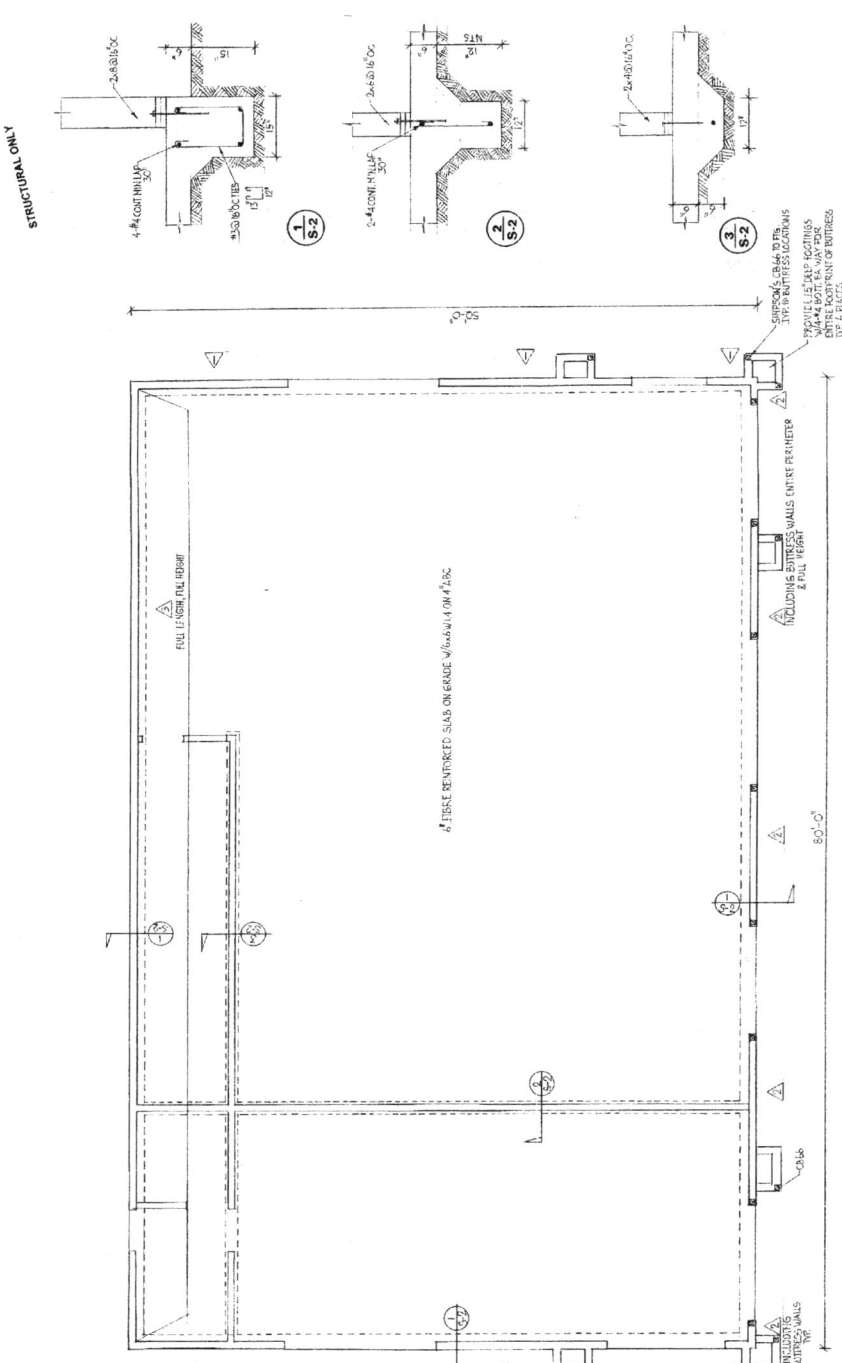

Figure 8.22 Foundation plan showing the layout of the building. On the right: details of perimeter footings and steel reinforcement shown encircled on the layout as 1/S-2, 2/S-2, and 3/S-2.

258

North External Shear Wall

Length: $L = 13.5 + 12.5 = 26.0$ ft

Unit shear at base of wall: $v = 7000/26.0 = 260$ lb/ft

Provided by $\frac{3}{8}$-in. Plywood/OSB with 8d at 6, 6, 12 in., (UBC Table 23-II-I-1.)

Overturning effect on 12.5-ft-long shear wall

Shear force shared: $H_{12.5} = (12.5/26.0)\ 7000 = 3365$ lb

Overturning effect: $M_o = 3365 \times 12.0 = 40.4$ k-ft

Uplift at the HD bolt is

$$V = \frac{40.4}{12.5} = 3.23 \text{ k}$$

less

$$6.25 \times 16.0 \times 10 + (6.25 + 6.0)(1.0)(10) = 1130 \text{ lb} = 1.13 \text{ k}$$

Net uplift at HD $= 2.00$ k < 2.03 k

Provided by HPAHD22 (Simpson's Catalog).

Right South Shear Wall

Length: $L = 12.5 + 15.3 + 4.0 = 32.0$ ft.

The base shear on this wall consists of 7000 lb from the main structure plus the wind reaction on the tower.

Reaction of wind on the tower: $V_T = 20 \times (18.0) \times (6.0 + 3.5) = 3400$ lb.

Total wind force reaction on right (south) shear wall at base:

$$H = 7000 + 3400 = 10,400 \text{ lb}$$

$$v = \frac{10,400}{32.0} = 325 \text{ lb/ft} < 380 \text{ lb/ft}$$

Provided by $\frac{3}{8}$-in. plywood/OSB with 8d at 4, 4, 12 in., UBC Table 23-II-I-1.

West Shear Wall

Wind acting in north–south direction

Wind reaction on west wall including wind on tower:

$$H_w = 17.5 \left(\frac{52.66}{2}\right)(6.0 + 4.0) + 3400 = 8.000 \text{ lb}$$

Total shear wall length: $L = 2.25 + 14.0 + 11.75 + 10.0 + 2.25 = 40.25$ ft

$v = 8000/40.25 = 198$ lb/ft < 260 lb/ft

> **Provided by $\frac{3}{8}$-in. plywood/OSB with 8d at 6, 6, 12 in.,**
> **UBC Table 23-II-I-1.**

East Shear Wall

Total length of wall 80.0 ft

Total shear force acting at its base:

$$H_E = 4600 \text{ lb}$$

$$v = \frac{4600}{80} = 58 \text{ lb/ft} < 125 \text{ lb/ft}$$

Provided by $\frac{5}{8}$-in. drywall with 6d cooler nails at 7 in. OC, 1997
UBC, Vol. I.

EARTHQUAKE ANALYSIS, ZONE 4, SOIL PROFILE S_C

East–west E.Q. direction.

Seismic Force on North External Wall

Weight of east–west external wall reaction as mass at plate height
$(80.0/2)(12.0/2 + 4.0) \times 8.0$ $= 6,400$ lb
Weight of office/storage roof $(40.0)(8.0)(6.0)$ $= 1,910$ lb
Weight of east–west partition wall $(51.167)(12.0/2)(6.0)$ $= 1,842$ lb
Weight of main roof $(51.167)(80.0/2)$ $= \underline{2,048}$ lb
$12,200$ lb

Using the UBC 1997 seismic provisions:

$$C_t = 0.2$$

$$R = 5.5 \qquad \text{(Table 16-H)}$$

$$I = 1.0$$

$$T = C_t\,(h)^{3/4} = 0.02(16.0)^{3/4} = 0.16 \text{ s}$$

$$Z = 0.4$$

$$N_a = 1.0 \qquad \text{(10 km, Table 16-S)}$$

$$C_a = 0.4\,N_a = 0.4$$

$$N_v = 1.0 \qquad \text{(Table 16-T)}$$

$$C_v = 0.56 \qquad N_v = 0.56$$

The base shear for the wall is

$$V = \frac{C_v I}{R\,T}\,W = \frac{0.56(1.0)}{5.5(0.16)}\,W = 0.636W \qquad (30\text{-}4)$$

$$V = 0.11 C_a I\,W = 0.044\,W \qquad (30\text{-}6)$$

$$= \frac{0.8 Z N_v I}{R}\,W = \frac{0.8(0.4)(1.0)(1.0)}{5.5}\,W = 0.058\,W \qquad (30\text{-}7)$$

But the total design base shear need not exceed the following

$$V = \frac{2.5 C_a I}{R}\,W \qquad \text{UBC 1997 (30-5)}$$

$$= \frac{2.5 C_a I}{R}\,W = 0.1818W$$

Therefore the maximum base shear on the north external shear wall is derived as

$$V = 0.1818\,(12.2) = 2.218 \text{ kips}$$

corresponding to the linear shear

$$v = 2218/26.0 = 85 \text{ lb/ft}$$

Seismic DOES NOT GOVERN.

Seismic Design Force on South Wall Acting East–West

Total weight acting at plate level including weight of tower:

Main east and west walls		6,400 lb
Main roof		2,048 lb
Buttress walls		2,302 lb
	Total	10,750 lb

$$V = 0.1818 \, (10.75) = 1.95 \text{ k} = 1900 \text{ lb}$$

$$v = \frac{1900}{32.8} = 60 \text{ lb/ft}$$

Seismic DOES NOT GOVERN.

VERTICAL LOAD SUPPORT SYSTEM

GLB to Support Main Roof Reaction over Opening

Maximum span 12.5 ft (Figure 8.23)
The GLB carries:

Main roof LL	$20.0 \times 50.0/2 = 500$ lb/ft
Main roof DL	$10.0 \times 50.0/2 = 250$ lb/ft
Girder and miscellaneous	$= 150$ lb/ft
Total	900 lb/ft

$$M_{max} = 0.9 \times 12.5^2 \times 1.5 = 211.0 \text{ k-in.}$$

$$S_{req.} = \frac{211.0}{2.4} = 88.0 \text{ in.}^3 < 123.0 \text{ in.}^3$$

Provided by $5\frac{1}{8}$-in. × 12 GLB.

Shear

Maximum shear span 12.0 ft
$$v_{max} = [(3/2) \; 12.0(0.9)]/2 \; \times \; 61.5 \; = \; 0.132 \; \text{ksi} \; < \; 155 \; \text{psi}$$
allowed   OK

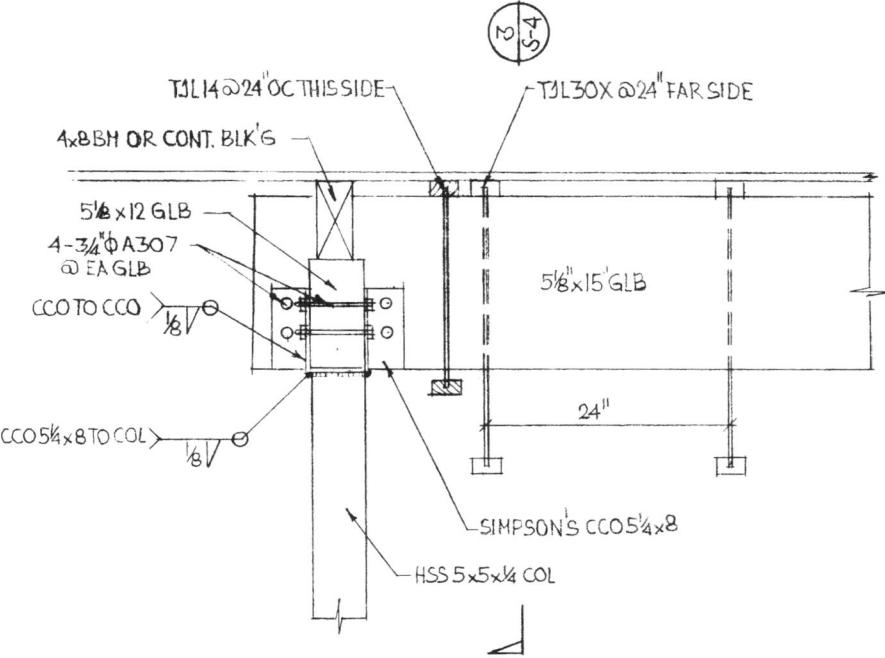

Figure 8.23   Glulam beam-to-column connection.

Maximum deflection:

$$\Delta_{\max} = \frac{(5/384)(0.9 \times 12.5)(12.5)^3(12)^3}{738 \times 1.8 \times 10^3} = 0.37 \text{ in.} < 0.42 \text{ in.}$$

$$\frac{L}{360} = \frac{150}{360} = 0.42 \text{ in.}$$

Deflection OK.

FOUNDATIONS: PERIMETER FOOTINGS

The worst loading condition occurs for the longitudinal external walls:

Maximum roof reaction (DL + LL) (37)(50.0/2) =	925 lb/ft
Weight of external wall	200 lb/ft
Total	1125 lb/ft

The required minimum footing width is

$$b = \frac{1.125}{1.0} = 1.125 \text{ ft} < 1.25 \text{ ft} \qquad \textbf{Provided}$$

8.7 CASE 5: WOOD-FRAMED RESIDENTIAL BUILDING

Seismic: High-Seismicity Area.

PROJECT DESCRIPTION

This is a one-story residence, including three-car garage, in the County of Riverside, California (*Architecture by Dondelinger Designs*). See Figures 8.24 and 8.25.

> Dimensions: dwelling 3026 sqft; garage area 816 sqft; patios and porches 246 sqft
> Wind 75 mph, exposure C
> Soil profile S_D
> Allowable bearing pressure 2500 per geotechnical report

The building was designed and engineered in March of 2003 in compliance with the 1997 UBC. Following are the engineering calculations, which apply to *lateral analysis and design* and *gravity load analysis and design*.

LATERAL ANALYSIS AND DESIGN

The lateral analysis covered *seismic analysis* and *wind analysis*.

Figure 8.24 Architectural rendering of the building.

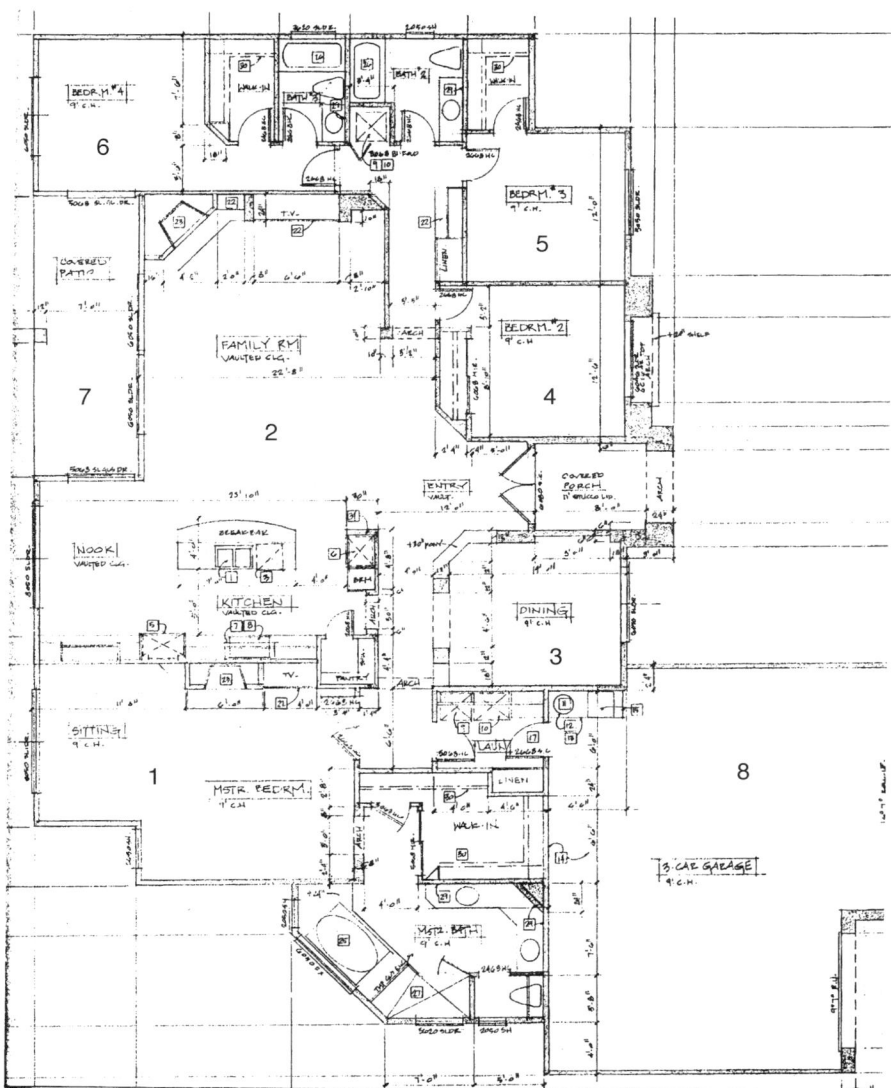

Figure 8.25   Floor plan of the building. 1. Master Bdrm, 2. Family Room, 3. Dining area, 4. Bdrm. 2, 5. Bdrm. 3, 6. Bdrm. 4, 7. Covered patio, 8. 3-car garage.

Seismic Analysis

We used UBC 1997 seismic provisions in conjunction with the recommendations of the geotechnical report. The seismic input data are as follows:

High-seismicity area
Soil profile S_D with median shear wave velocity 270 m/s
Fault type B at 7.0 km
$N_a = 1.0$ (16-S)
$N_v = 1.12$ (16-T)
$C_a = 0.44N_a = 0.44$
$C_v = 0.64N_v = 0.72$
$T = C_t (h^{3/4}) = 0.02 (18.0^{0.75}) = 0.175$ s
$T_s = C_v/2.5 C_a = 0.655$ sec > 0.175 s

Therefore equations (30-5) and (30-6) will be compared and selected:

$$V = \frac{2.5C_a I}{R} W = \left(1.1 \times \frac{1.0}{5.5}\right)W = 0.2W \qquad \textbf{GOVERNS} \qquad (30\text{-}5)$$

$$V = (0.11)(0.44)(1.0)W = 0.048W \qquad\qquad (30\text{-}6)$$

Adding the individual unit weights of external and internal walls plus the weight of the roof and V_{seismic} of 18 k lateral, we reached a total of 90,000 lb for the design mass of the building.

Wind Analysis

The worst scenario is wind acting on one of the long sides of the building, that is, the entrance side. With 75 mph, exposure C,

$$P = (1.09)(1.3)(15.8)(1.0) = 22.4 \text{ psf} \qquad \text{for roof}$$

$$P = (1.06)(1.3)(15.8)(1.0) = 21.8 \text{ psf} \qquad \text{for walls}$$

The total wind at *working load level* is

$$82 [(9.0/2) \times 21.8 + (7.0 \times 22.4)] = 20,900 \text{ lb}$$

At the *limit-state level*

$$V_W = 1.4(20.9) = 29.2 \text{ k}$$

Wind GOVERNS acting on the long side.

and

$$V_W = 6.6[(9.0/2) \times 21.8 + (6.0 \times 22.4)](1.4) = 21,500 \text{ lb} = 21.5 \text{ k}$$

Wind also governs acting on the short direction.

Thus the governing base shears to fit UBC 1997, Chapter 23, are

$$V = 20.9 \text{ k} \qquad \text{for minor axis}$$

$$V = 15.4 \qquad \text{for major axis}$$

Shear Wall Design

Shear walls will share lateral force in proportion to their tributary width. The *external garage shear wall* shares

$$H = \left(\frac{7.5}{82}\right)(20.9) = 1.91 \text{ k}$$

The linear shear is

$$v = \frac{1910}{23.5} = 81 \text{ lb/ft}$$

using two 4-ft-wide shear wall panels and

$$v = \frac{1910}{8.0} = 238 \text{ lb/ft} < 260 \text{ lb/ft}$$

Provided by $\frac{3}{8}$-in. plywood with 8d at 6, 6, 12 in. OC.

Hold-down requirements are as follows:

Uplift: entire garage wall resists uplift with roof sitting on it
$V_o = 1.91 \times 9.0/2.30 = 0.75 \text{ k}$
   Less weight of roof = $\underline{0.97 \text{ k}}$
$\qquad\qquad\qquad 0.22 \text{ k} < 3.80 \text{ k compression}$

No theoretical uplift occurs. Nonetheless one Simpson's STHD8 will be provided at each end.

External Master Bedroom Wall

Length 17.0 ft

$H = (15.0/82)(20.9) = 3.82$ k

$v = 3820/17.0 = 225$ lb/ft < 260 lb/ft

Provided by $\frac{3}{8}$-in. plywood with 8d at 6, 6, 12 in. OC.

Hold-down requirements are as follows

$V_o = 3.82 \times 9.0/16.75 = 2.05$ k

Less weight of roof and wall $= 2.14$ k

No theoretical hold-down is needed. Nevertheless one STHD8 will be provided at each end of the external surface.

Garage (Right), Dining, Kitchen Shear Wall Complex

Total lateral load on this wall system, tributary width 17.5 ft, is determined as

$$H = (17.5/82)(20.9) = 4.46 \text{ k}$$

$$L = 21.5 + 14.5 + 22.0 = 58.0 \text{ ft}$$

$$V = \frac{4460}{58} = 77 \text{ lb/ft}$$

However, only 21.0 ft wall width needs to be shear paneled using $\frac{3}{8}$-in. plywood:

$v = 4460/21 = 212$ lb/ft < 260 lb/ft

Provided by $\frac{3}{8}$-in. plywood with 8d at 6, 6, 12 in. OC, UBC Table 23-II-I-1.

$V_o = 4.46 \times (9.0/21.0) = 1.91$ k

Less weight of roof and wall $= 1.99$ k

No theoretical uplift occurs. Nevertheless one STHD8 will be provided at each end.

Porch/Bedroom 2 Wall

Length 13.5 ft
Tributary width 20.0 ft
$H = (20.0/82)\ 20.9 = 5.1$ k
$v = 5100/13.5 = 377$ lb/ft < 380 lb/ft

> **Provided by $\frac{3}{8}$-in. plywood with 8d at 4, 4, 12 in. OC,**
> **UBC Table 23-II-I-1.**

Hold-down requirements are

$$V_o = 5.1 \times \left(\frac{9.0}{13.0}\right) = 3.5\ \text{k} < 4.0$$

> **Provided by HD5A (Simpson's Catalog)   OK.**

Family Room/Bedroom Wall

Length 17.5 ft
Tributary width 16.0 ft
$H = (16.0/82)(20.9) = 4.08$ ft
$v = 4080/17.5 = 233$ lb/ft < 380 lb/ft

> **Provided by $\frac{3}{8}$-in. plywood with 8d at 4, 4, 12 in. OC.**

$V_o = 4.08(9.0/17.0) = 2.1$ k < 3.7 k

> **Provided by HD5A on 2–2x wall stud.   OK.**

Right External Wall

Length 38.25 ft
$H = (6.0/82)\ 20.9 = 1.53$ k
$v = (1530/38.25) = 40$ lb/ft < 260 lb/ft

> **Provided by $\frac{3}{8}$-in. plywood with 8d at 6, 6, 12 in. OC.**

$V_o = 1530 \times 9.0/38.0 = 360$ lb
Less weight of roof and wall $= 2600$ lb

No uplift occurs; nevertheless one STHD8 strap will be provided at each end of the wall.

Longitudinal Shear Wall Design

Only the external walls of dwelling and garage front entrance will be utilized to resist lateral forces. The total length of the left (rear) shear wall system is

$$L = 2 \times 2.5 + 6.5 + 3.75 + 6.5 + 3.25 + 3.5 = 28.5 \text{ ft}$$

The total length of the right (front) shear wall system excluding the front wall of the garage is

$$L = (30.25 - 2.75) + 4.0 + 2.75 + 3.5$$
$$+ 3.25 + (3.0 \times 2) + 7.75 = 54.75 \text{ ft}$$

The garage being treated as a separate structure, the total wind force acting along the longitudinal axis is

$$V = [\tfrac{1}{2}(22.4 + 21.8)][\tfrac{1}{2}(9.0) + 8.0)][\tfrac{1}{2}(38.7 + 45.3)] = 11,600 \text{ lb}$$

or

5800 lb acting on each, left or right, shear wall system

Left (Rear) Shear Wall System

The nominal linear shear on the left (rear) wall system is

$$v = \frac{5800}{28.5} = 204 \text{ lb/ft} < 260 \text{ lb/ft}$$

Provided by $\tfrac{3}{8}$-in. plywood with 8d at 6, 6, 12 in. OC.

Uplift

The worst scenario for uplift is the 3.25-ft shear wall panel at bedroom 4. The load shared by the panel is derived as follows:

$$H = (3.25/28.5) \times 5.8 = 0.66 \text{ k}$$
$$V_o = 0.66 \times (9.0/3.0) = 1.98 \text{ k}$$
Less weight of portion of roof & wall = 2.96 k

No uplift can occur; nevertheless one STHD8 will be provided at each end of the bedroom wall. The front shear wall system is less stressed than the rear. The same $\tfrac{3}{8}$-in. plywood with 8d at 6, 6, 12 in. OC will be used with STHD8 straps at shear wall extremities.

Garage Front Wall

The total wind load impacting the garage front shear walls is

$$V = 22.1 \times \left(\frac{27.5}{2}\right) \times 13.0 = 3950 \text{ lb}$$

acting on four 24-in.-wide piers. The linear shear for each pier is

$$v = \frac{3950}{4 \times 2} = 490 \text{ lb/ft} < 510 \text{ lb/ft}$$

**Provided by $\frac{19}{32}$-in. plywood with 10d at 4, 4, 12 in. OC,
UBC Table 23-II-I-1.**

Hold-down requirements are as follows:

$$V_o = 1.0 \times (9.0/1.75) = 5.14 \text{ k}$$

Less weight of roof $(12.0 \times 23.5)(22/4 \times 1000) = \underline{1.54 \text{ k}}$

Net uplift $= 3.60 \text{ k} < 4.0 \text{ k}$

Provided by HD5A on 6 × 6 lumber.   OK.

GRAVITY LOAD ANALYSIS AND DESIGN

Headers

8-ft Opening, Nook

Roof tributary width: $(45.0/2) + 2.0 = 24.5$ ft
$w_{D+L} = 24.5(22.0 + 16.0) = 930$ lb/ft, say 1.0 k/ft
With allowable stress $F_b = 1350$ psi for DF (Douglas Fir) 1
$S_{\text{req.}} = 1.0 \times 8.5^2 \times 1.5/1.35 = 80.3 \text{ in.}^3 < 121.2 \text{ in.}^3$

Provided by 6 × 12.

Deflection

D + L presents the worst scenario:

$$\Delta_{D+L} = \frac{1.0(8.5)(8.5^3 \times 12^3)\,5}{384 \times 1.6 \times 697 \times 10^3} = 0.10 \text{ in.} < 0.4 \text{ in. allowed} \qquad \textbf{OK}$$

Use the same 6 × 12 type of header for the 7.0-ft master bedroom opening.

6.0-ft Openings

Dining room is the worst scenario with 1.0 k/ft uniformly distributed
D + L

$S_{req.} = 1.0 \times 6.25^2 \times 1.5/1.35 = 43.4$ in.$^3 < 82.7$ in.^{3}

Provided by 6 × 10.

Glulam Beam (GLB) Design

9.0-ft Single-Garage Door Opening

Receives concentrated load at 3.0 ft from support

Average tributary width from girder truss causing point load: 18.0/2
= 9.0 ft

Point load:

$$P = 0.038(9.0)\left(\frac{24.0}{2}\right) = 4.1 \text{ k}$$

$$M_{max} = 4.1 \times \left(\frac{6.0}{9.0}\right) \times 3.0 = 8.2 \text{ k-ft} = 98.4 \text{ k-in.}$$

With $F_b = 2.4$ ksi for 24-V7 DF/DF GLB

$S_{req.} = 98.4/2.4 = 41.0$ in.$^3 < 123.0$ in.^{3}

Provided by $5\frac{1}{8}$ × 12 GLB.

16.0-ft Double-Garage Opening

Carries 4.5-ft-wide strip of roof D+L

$w = 0.04 \times 4.5 = 0.18$ k/ft, say 0.20 k/ft

$S_{req.} = 0.2 \times 16.5^2 \times 1.5/2.4 = 34.0$ in.$^3 < 123.0$ in.^{3}

Provided by $5\frac{1}{8}$ × 12 GLB.

FOUNDATION DESIGN

Allowable soil pressure is 2500 psi with one-third increase for wind and
seismic per geotechnical report.

External Bearing Wall Footing

Roof span 45.0 ft

$w_{D+L} = 0.038(45.0/2 + 2.0) = 0.93$ k/ft roof reaction.

Total load on strip footing, including weight of wall: 1.0 k/ft
Required footing width: $b = 1.0/2.5 = 0.4$ ft < 1.0 ft

Provided OK.

Pad Footing to Support Girder Truss over Master Bedroom

Girder carries $(22.0/2) = 11.0$-ft-wide roof load. The point load reaction caused by the girder truss is

$$P = 0.038 \times 11.0 \times (39.0/2 + 2.0) = 9.0 \text{ k}$$

The required footing width is

$$b = \sqrt{\frac{9.0}{2.5}} = 1.9 \text{ ft} < 2.5 \text{ ft}$$

Provided OK.

Using an overall depth of 18 in., an effective depth of 14 in., and ultimate actual bearing pressure of 2.2 kips/ft^2,

$$A_{st} = \frac{33.0}{0.9 \times 0.85 \times 14 \times 60} = 0.05\text{-in.}^2 \text{ footing} < 0.8 \text{ in.}^2$$

**Provided by 4-#4 bars bottom each way.
2 ft, 6 in.^{2} × 18-in. footing.   OK.**

Footing to Support Front of Garage

Worst scenario: 16.0-ft opening
Roof reaction: $0.2 \times 16/2 = 1.6$ k
Force resisted by 1-ft-wide bearing area
Width of continuous footing and length of the pier 2.0 ft
Actual bearing pressure:

$$f_B = \frac{1.6}{1.0 \times 2.0} = 0.8 \text{ ksf} < 2.5 \text{ ksf allowed}   \textbf{OK}$$

8.8   CASE 6: WOOD-FRAMED GARAGE AND WORKSHOP

Seismic: Low- to Moderate-Seismicity Area

We have seen that seismic forces governed over wind in a low- to moderate-seismicity region for a masonry design project. In this case wind is prevalent and therefore the calculations are adjusted accordingly for this wood-framed structure where a garage is combined with a work-shop. See Figure 8.26.

PROJECT DESCRIPTION

Soil profile S_C
Allowable bearing pressure 1000 psf
Code used for analysis and design: 1997 UBC
Wind 70 mph

Garage
    One-story height, footprint 18 × 36 ft, length of building 36.0 ft
    Plate height 14.0 ft.
    Maximum roof height 16.0 ft
    Maximum tributary width for longitudinal shear walls: 48.0/4 = 12.0 ft for wind acting on front

Workshop: Attached to the garage is a 48-ft-wide workshop, self-supporting against lateral forces such as wind.

WIND ANALYSIS

Basic *wind stagnation* pressure 12.6 psf
Combined height, exposure coefficient for 0–15.0 ft height: $C_e = 1.06$

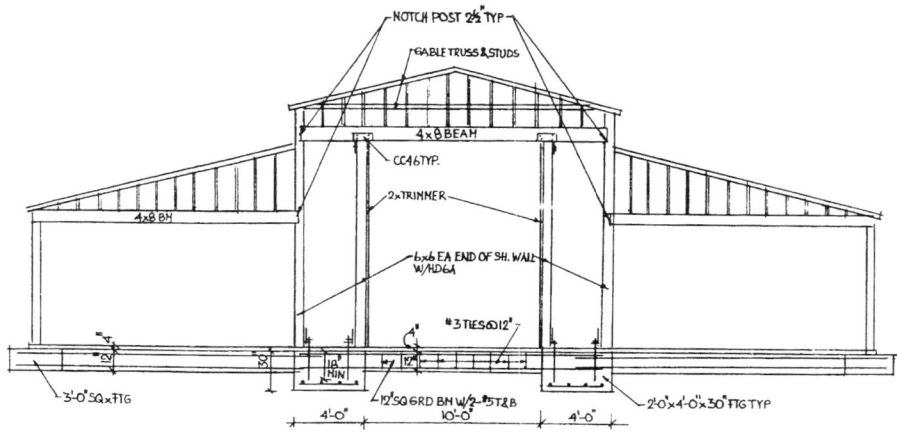

Figure 8.26 Front structural elevation of the building.

Use Method 2, *projected-area method.*

Importance factor: $I = 1.0$

Wind pressure acting on walls and roof: $p = 1.06 \times 1.3 \times 12.6 \times 1.0 = 17.4$ psf

Wind pressure on half of left façade: $H = (48.0/2)(7.0 + 2.00) \times 17.4 \approx 3860$ lb

Longitudinal Analysis

Total length of shear walls: $2 \times 4.0 = 8.0$ ft

Resistance along longitudinal front wall of building

Linear shear: $v = 3860/8.0 = 483$ lb/ft < 490 lb/ft

Provided by $\frac{3}{8}$-in. OSB with 8d at 3, 3, 12 in. OC.

DESIGN OF FRONT TRANSVERSE TWIN SHEAR WALLS

Maximum wind on the twin walls 3860 lb

Maximum overturning on each wall: $(3.86/2)(14.0) = 27.0$ k-ft

Maximum uplift at HD of 6×6 posts:

$$27.0/4 = 6.75 \text{ k}$$

$$\text{Less weight of roof and wall} = \underline{0.85 \text{ k}}$$

$$\text{Net uplift} \quad 5.90 \text{ k} < 7910 \text{ lb}$$

Provided by HD8A (Simpson's Catalog p. 21).

DESIGN OF $6 \times$ BEAMS FOR WORKSHOP ROOF

Span 24.0 ft.

Tributary width 12.0 ft

Total uniform roof load acting on beam including weight of beam:

$$w_{D+L} = (14.0 + 6.0) \times 12.0 + 14.0 = 254 \text{ lb/ft}$$

$$M_{max} = 0.254 \times 24.0^2 \times 1.5 = 219 \text{ k-in.}$$

$$S_{req.} = \frac{219}{1.33} = 165 \text{ in.}^3 < 167 \text{ in.}^3$$

Provided by 6×14 beam. **OK.**

DESIGN OF 2 × ROOF RAFTERS

Span 12.0 ft
Try 2.0 ft OC:

$$w_{D+L} = (6.0 + 20.0) \times 2.0 \approx 56 \text{ lb/ft}$$

Say 58 lb/ft including weight of roof rafter

$$M_{\max} = 0.058 \times 12.0^2 \times 1.5 = 12.53 \text{ k-in.}$$

$$S_{\text{req.}} = \frac{12.53}{1.4} = 8.9 \text{ in.}^3 > 7.56 \text{ in.}^3$$

NG on 2.0 ft OC.

2 × 6 rafters must be at 16 in. OC.

FOUNDATION DESIGN

The foundation will be designed for an allowable soil pressure at 12 in. minimum depth, 1000 psf at service loads.

External Footing Maximum gravity load:

$$\frac{24.0 (24/2) + 10.0 \times 8.0}{2} \approx 350 \text{ lb/ft}^2$$

$$\frac{350}{1.0} = 350 \text{ psf} < 1000 \text{ psf} \quad \textbf{OK}$$

Individual Pad Footing for 6 × 6 Posts
Maximum vertical load:

$$12.0 \times 24.0 (14.0 + 6.0) = 5760 \text{ lb} = 5.76 \text{ k}$$

Required footing size:

$$b = \sqrt{\frac{5.76}{1.0}} = 2.40 \text{ ft} < 3.00 \text{ ft provided} \quad \textbf{OK}$$

8.9   LIGHT-GAUGE STEEL AS ALTERNATIVE TO WOOD FRAMING

Light-gauge steel is gaining increased acceptance in the commercial and hous-ing markets. Some reasons are the environmental concerns over deforestation to meet the pressing demands for housing, the cost of lumber, and the need to import from foreign sources (ref. 3 p. 5):

> The wood industry looked toward imports from Canada to satisfy construction needs. They assured builders that the supply would meet demand, especially through a combination of imports and managed forests. They described timber products as a "sustainable" industry, with new growth satisfying demand. How-ever, over the past decade overall lumber prices have both increased and become more volatile.

The interest in light-gauge steel for residential construction is based on the strength and uniform quality of studs and joists as compared to their lumber counterparts. Steel framing offers fire safety and termite protection. In addi-tion, the strength and ductility of steel are a good prospect to resist wind and seismic loads. An added advantage is that about 25% of the content of man-ufactured steel today is derived from recycled sources. So why not build more steel houses? According to Timothy J. Waite, P.E., of the National Association of Home Builders (NAHB) Research Center, there are several reasons: While steel framing material costs are now competitive with lumber, labor and en-gineering costs tend to be higher. There is a lack of skilled framing labor in steel framing. There is an inherent higher connection cost using screws and screw guns versus nails and nail guns. Also, because there are no standards in the energy codes, steel suffers a penalty in extra insulation costs in colder climates. Finally, consumers are not aware of all the benefits of steel.

Building Code Provisions

As another example of the interaction between building codes and other stan-dards, IBC 2003 adopted the *North American Specification for the Design of Cold-Formed Steel Structural Members* (AISI-NASPEC) for the design and construction of cold-formed stainless steel structural members in Section 2209, "Cold-Formed Steel." Section 2210 of the code addresses cold-formed steel light-framed construction specifically for the design, installation, and construction of cold-formed carbon or low-alloy steel, structural and nonstruc-tural steel framing. In this case the code adopts the *Standard for Cold-Formed Steel Framing—General Provisions* of the American Iron and Steel Institute (AISI—General) as well as AISI-NASPEC. Section 2211 of IBC 2003 de-voted several pages to cold-formed steel light-framed shear walls, including detailed tables for shear values for wind and seismic forces.

The 2006 IBC, Sections 2209 and 2210, very much abridged as compared to IBC 2003, prescribe that the design of cold-formed carbon and low-alloy steel structural members should comply with AISI-NAS. The design of cold-formed stainless steel structural members, on the other hand, must follow the provisions of ASCE 8. Cold-formed steel light-framed construction should comply with IBC 2006, Section 2210. Section 2210 of IBC 2006 specifically addresses steel light-framed construction, which in general should adhere to the recommendations of AISI—General, AISI-NAS, and in each specific case AISI-Header, AISI-Truss, AISI-WSD for wall stud design, and AISI—Lateral for lateral design for light-framed, cold-formed steel walls and diaphragms to resist wind and seismic loads.

8.10 CASE 7: LIGHT-GAUGE STEEL IN MULTISTORY PROJECT

Seismic: High-Seismicity Area

PROJECT DESCRIPTION

Light-gauge steel curtains and partition walls for two nine-story concrete towers in Long Beach, California

Clear curtain wall stud height 9 ft. 4 in.

Maximum hollow pier/fin stud height 19 ft, 4 in.

Design by 1997 UBC, zone 4

SCOPE OF WORK

The following calculations cover Light-Gauge Steel Curtain Walls and Internal Partition Walls

The light-gauge steel components in the project fall into two categories: curtain walls and fins and other architectural (LGS) façade features (Figure 8.28):

(a) *Curtain Walls* The span is 9.33 ft for 10.0 ft floor-to-floor height. These walls will be designed for wind pressures corresponding to the highest floor elevation, that is, LVL 9 with an average of 93 ft above ground. Exposure is type C of the 1997 UBC, Table 16-G, method 1, Normal Force Method, and 70 mph basic wind speed:

$$q_s = 12.6 \text{ psf} \qquad (16\text{-F})$$

$$C_e = 1.58 \qquad (16\text{-G})$$

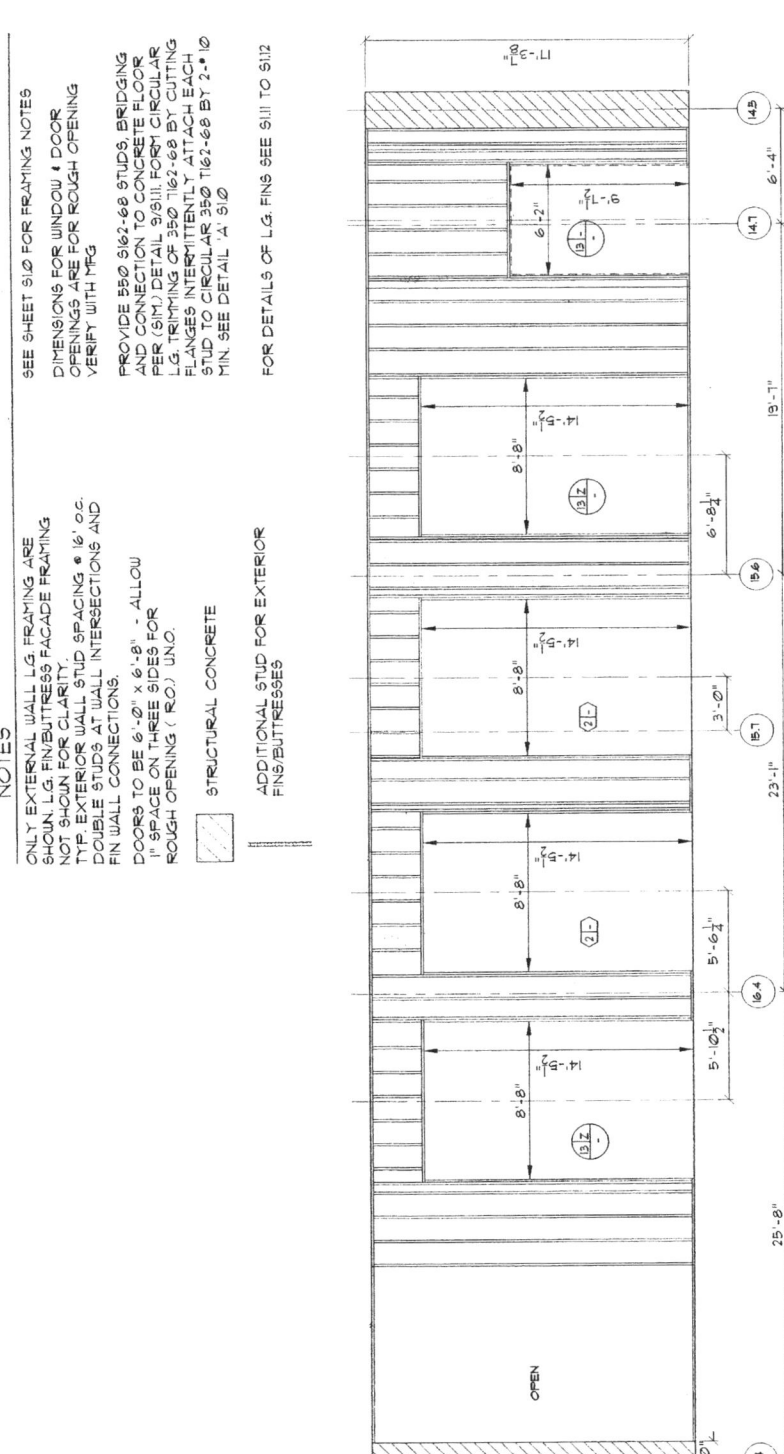

Figure 8.27 One of the exterior walls at first level showing the use of cold-formed steel. The shaded area is a structural concrete bearing wall.

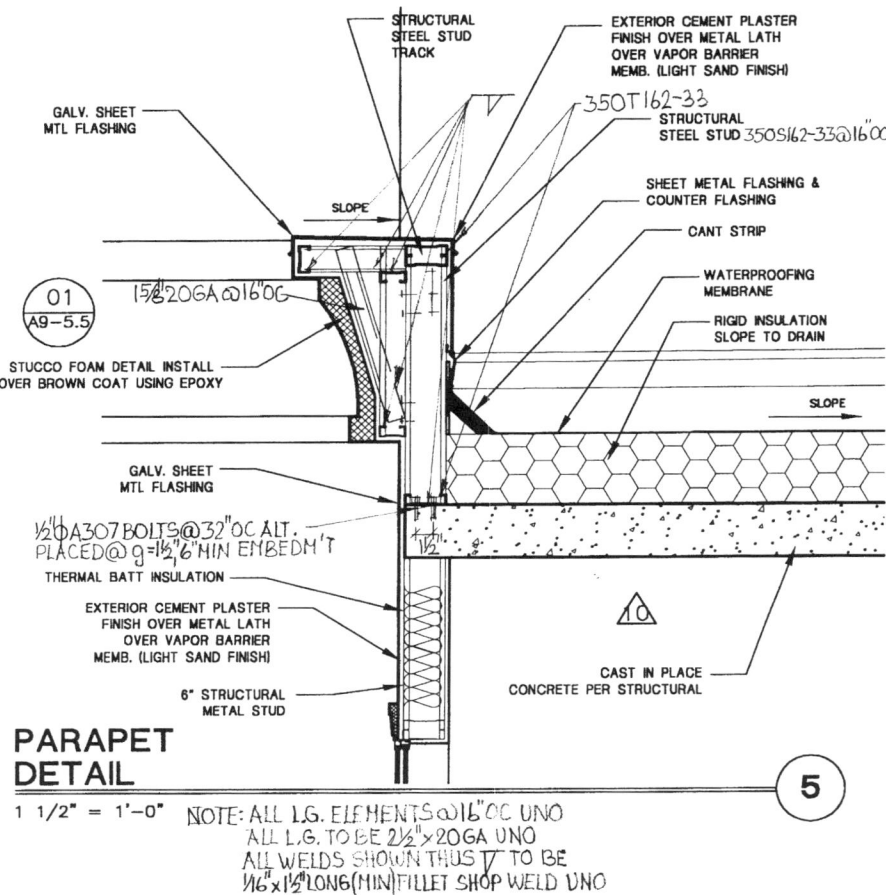

Figure 8.28 Detail of a parapet above the ninth floor of the towers showing application of light-gauge steel elements.

$$C_q = 0.80 \text{ inward} \tag{16-H}$$

$$I_w = 1.0 \tag{16-K}$$

$$p = 1.58(0.8)(12.6)(1.0) = 15.93 \text{ psf}$$

for highest floor elevation

From Table 6.6 of the NASFA (North American Steel Framing Alliance) "Prescriptive Method for Cold-Formed Steel Framing," 2000 edition, with a stud spacing of 16 in. and a 10.0-ft span, the 550S162-33 stud (33 mils, 20 gauge), **the wall is good for 110 mph wind with exposure C.**

Deflection.   The 550S162-33 stud is allowed to span 12 ft, 1 in. for an L/360 and 15.0 psf wind pressure. (Reference: CEMCO Design Tables, Table 5, p. 34 "Allowable Curtain Wall Spans.") The maximum shear force at each end, top and bottom of stud, is

$$\left(\frac{16}{12}\right)(15.93)\left(\frac{9.33}{2}\right) = 99 \text{ lb} < 2 \times 177 = 354 \text{ lb}$$

$$\text{allowable capacity of two \#10 screws}$$

(Reference: NASFA, "Prescriptive Method for Residential Cold-Formed Steel Framing," 2000 edition, p. 160.)

(b) *Hollow Piers and Fins*   Maximum span of 19.33 ft is at first and second stories, 550S162-68 (14 gauge) studs at 16 in. OC. Maximum wind pressure with $C_e = 1.23$, exposure C, method 1, is

$$p = 1.23(0.8)(12.6)(1.0) = 12.4 \text{ psf}$$

The hollow piers/fins have four side walls of 2 × 6, 14-gauge (68-mil) studs cross-braced and bridged at 32 in. OC. Therefore a sound load sharing exists between the walls—the side and rear walls each sharing at least 33% with the front wall. Thus the applied uniformly distributed wind load on an external stud at 16 in. OC is

$$w = (0.33)(\tfrac{16}{12})(12.4) = 5.5 \text{ psf}$$

From Table 1, p. 5, of CEMCO, "CS-Punched C Studs & Joists, $1\frac{5}{8}$-in. Flanges," the inertia of 550S162-68, 14-gauge, $5\frac{1}{2}$-in.-deep studs is

$$I_x = 2.922 \text{ in.}^4$$

The maximum deflection under the above uniformly distributed wind load with a 19.33-ft span is

$$\Delta_{\text{max}} = \frac{(5/384)wL^4}{EI} = \frac{(5/384)5.5(19.33)(19.33^3) \, (12^3)}{2.922 \, (29)(10)^6}$$

$$= 0.20 \text{ in.} < \frac{L}{360} = 0.644 \text{ in.}$$

Deflection OK.

2. *Partition Walls*

350S125-18 (gauge 25) studs at 16 in. OC

Floor-to-floor height 10.0 ft, giving clear stud height of 9 ft, 4 in.

Maximum allowable stud height for non-load-bearing partition walls for 350S125-18 (gauge 25) stud at 16 in. OC: 10 ft, 8-in.

10 ft, 8 in. > 9 ft, 4 in.   OK.

(Reference Table 7.1, NASFA, "Prescriptive Method for Residential Cold-Formed Steel Framing," 2000 edition).

SPECIFICATIONS

Following is a summary of the light-gauge steel specifications for all walls of the two towers of the project, levels 2–9:

- Materials should comply with the American Iron and Steel Institute (AISI) specifications for the design of cold-formed steel structural members as adopted by the NASFA (North American Steel Framing Alliance), ASTM A 653, grades 33 and 50.
- Minimum nominal yield strength for all members should be $F_y =$ 33 ksi for 20-gauge steel and $F_y = 50$ ksi for 14-gauge steel.
- All studs for external walls and firewalls should be 550 S162-33 (gauge 20) cold-formed steel.
- All tracks for external walls should be 550-T162-33 (gauge 20).
- All studs for internal partition walls should be 350-S162-18 (gauge 25) cold-formed steel.
- All tracks for internal partition walls should be 350-T162-18 (gauge 25) cold-formed steel.
- Tracks of external walls should be anchored to concrete by $\frac{1}{2}$-in.-diameter A307 at 32 in. OC with a minimum embedment of 6 in.
- Tracks and studs at anchor boltholes should have a 6-in.-long, 20-gauge stud or plate reinforcement at each bolthole, thus doubling the metal at the anchor bolthole.
- Boltholes for tracks and studs secured to concrete floor, ceiling wall, or column should be $\frac{1}{32}-\frac{1}{16}$ in. oversize. Holes over $\frac{1}{16}$ in. require replacement of sheet metal.
- All external light-gauge stud walls should be anchored at top and bottom to concrete floor by anchor bolts as stated above.
- Clips and all other components should be minimum 20-gauge cold-formed steel.
- All screws connecting metal to metal should be self-drilling #8 hex head or low-profile head if drywall, OSB, or plywood sheathing covers connection.

- All interior walls should be anchored to concrete with 0.145-in.-diameter approved power drive pins at 24 in. OC maximum.
- Slotted-hole tracks should be provided at top of steel studs (bottom of structural slab). Centerline of holes should coincide with bolts specified per engineering plans. Length of slots should be 1 in. ≤ *L* ≤ 3 in.

APPENDIX

COMPUTER ANALYSIS OF STEEL FRAME

```
 1. STAAD PLANE FRAME
 2. TWO STORY M-FRAME3, LINE2
 3. UNIT KIP FEET
 4. JOINT COORDINATES
 5. 1 0. 0.; 2 11. 0.; 3 22. 0.
 6. 4 0. 11.5; 5 11. 11.5; 6 22. 11.5
 7. 7 0. 22.5; 8 11. 22.5; 9 22. 22.5
 8. MEMBER INCIDENCE
 9. 1 1 4; 2 2 5; 3 3 6
10. 4 4 7; 5 5 8; 6 6 9
11. 7 7 8; 8 8 9; 9 4 5; 10 5 6
12. MEMBER PROPERTY AMERICAN
13. 1 3 4 6 TABLE ST TUB60406
14. 2 5 TABLE ST TUB60406
15. 7 8 9 10 TABLE ST W8X21
16. UNIT INCHES
17. CONSTANTS
18. E 29000.0
19. MEMBER OFFSET
20. 7 8 9 10 START 4.25 0.0
21. 7 8 9 10 END -4.25 0.0
22. PRINT MEMBER INFORMATION
```

MEMBER INFORMATION2

MEMBER	START JOINT	END JOINT	LENGTH (INCH)	BETA (DEG)	RELEASES
1	1	4	138.000	0.00	
2	2	5	138.000	0.00	
3	3	6	138.000	0.00	
4	4	7	132.000	0.00	
5	5	8	132.000	0.00	
6	6	9	132.000	0.00	
7	7	8	123.500	0.00	
8	8	9	123.500	0.00	
9	4	5	123.500	0.00	
10	5	6	123.500	0.00	

***************** END OF DATA FROM INTERNAL STORAGE ***************

```
23.SUPPORT
24.1 2 3 FIXED
25.DRAW MEMBER JOINT PROPERTY SUPPORT
```

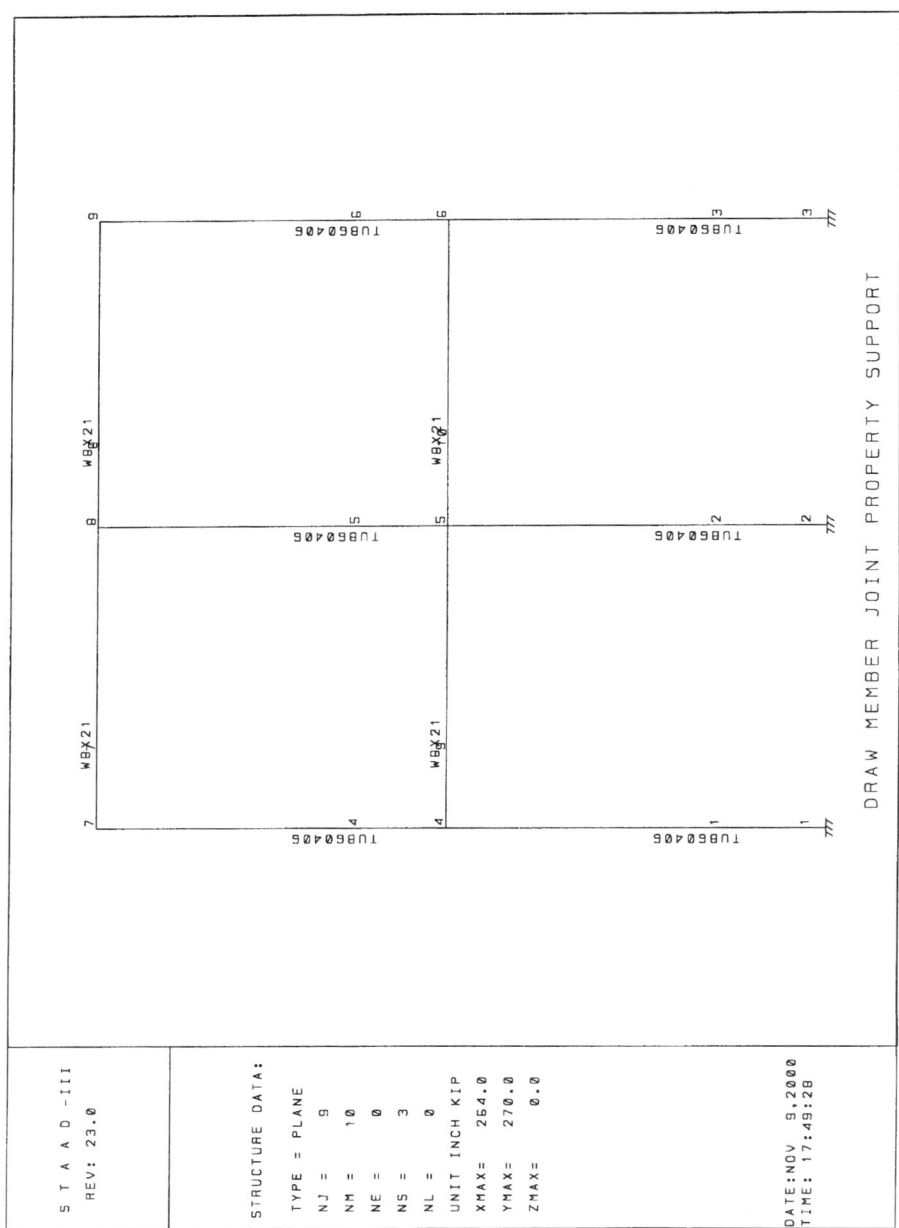

```
TWO STORY M-FRAME3, LINE2
26.UNIT FT
27.LOADING 1 W FROM LEFT AT "WORKING LOAD"
28.JOINT LOAD
29.7 FX 2.03
30.4 FX 1.74
31.LOADING 2 DEAD (DL)
32.MEMBER LOAD
33.7 8 UNI GY -0.10
34.9 10 UNI GY -0.25
35.LOADING 3 LIVE (LL)
36.MEMBER LOAD
37.7 8 UNI GY -0.05
38.9 10 UNI GY -0.86
39.LOAD COMBINATION 4 DL+LL
40.2 1.0 3 1.0
41.LOAD COMBINATION 5 DL + 0.75 LL + 0.75 W
42.1 0.75 2 1.0 3 0.75
43.PERFORM ANALYSIS
```

P R O B L E M   S T A T I S T I C S

```
   NUMBER OF JOINTS/MEMBER+ELEMENTS/SUPPORTS =     9/   10/    3
   ORIGINAL/FINAL BAND-WIDTH =    3/   3
   TOTAL PRIMARY LOAD CASES  =    3, TOTAL DEGREES OF FREEDOM =    18
   SIZE OF STIFFNESS MATRIX =     216 DOUBLE PREC. WORDS
   REQRD/AVAIL. DISK SPACE = 12.02/ 2047.7 MB,  EXMEM = 1960.2 MB

++ Processing Element Stiffness Matrix.            17:49:28
++ Processing Global Stiffness Matrix.             17:49:28
++ Processing Triangular Factorization.            17:49:28
++ Calculating Joint Displacements.                17:49:28
++ Calculating Member  Forces.                     17:49:28


   44. PRINT MEMBER FORCES
```

```
MEMBER END FORCES      STRUCTURE TYPE = PLANE
-----------------
ALL UNITS ARE -- KIP   FEET
```

MEMBER	LOAD	JT	AXIAL	SHEAR-Y	SHEAR-Z	TORSION	MOM-Y	MOM-Z
1	1	1	-1.93	1.18	0.00	0.00	0.00	7.44
		4	1.93	-1.18	0.00	0.00	0.00	6.17
	2	1	1.54	-0.07	0.00	0.00	0.00	-0.26
		4	-1.54	0.07	0.00	0.00	0.00	-0.53
	3	1	4.03	-0.24	0.00	0.00	0.00	-0.93
		4	-4.03	0.24	0.00	0.00	0.00	-1.88
	4	1	5.56	-0.31	0.00	0.00	0.00	-1.20
		4	-5.56	0.31	0.00	0.00	0.00	-2.41
	5	1	3.11	0.64	0.00	0.00	0.00	4.62
		4	-3.11	-0.64	0.00	0.00	0.00	2.69
2	1	2	0.00	1.41	0.00	0.00	0.00	8.29
		5	0.00	-1.41	0.00	0.00	0.00	7.89
	2	2	4.13	0.00	0.00	0.00	0.00	0.00
		5	-4.13	0.00	0.00	0.00	0.00	0.00
	3	2	10.67	0.00	0.00	0.00	0.00	0.00
		5	-10.67	0.00	0.00	0.00	0.00	0.00
	4	2	14.81	0.00	0.00	0.00	0.00	0.00
		5	-14.81	0.00	0.00	0.00	0.00	0.00
	5	2	12.14	1.06	0.00	0.00	0.00	6.22
		5	-12.14	-1.06	0.00	0.00	0.00	5.92
3	1	3	1.93	1.18	0.00	0.00	0.00	7.42
		6	-1.93	-1.18	0.00	0.00	0.00	6.14
	2	3	1.54	0.07	0.00	0.00	0.00	0.26
		6	-1.54	-0.07	0.00	0.00	0.00	0.53
	3	3	4.03	0.24	0.00	0.00	0.00	0.93
		6	-4.03	-0.24	0.00	0.00	0.00	1.88
	4	3	5.56	0.31	0.00	0.00	0.00	1.20
		6	-5.56	-0.31	0.00	0.00	0.00	2.41
	5	3	6.01	1.14	0.00	0.00	0.00	6.53
		6	-6.01	-1.14	0.00	0.00	0.00	6.55
4	1	4	-0.55	0.56	0.00	0.00	0.00	2.69
		7	0.55	-0.56	0.00	0.00	0.00	3.50
	2	4	0.45	-0.11	0.00	0.00	0.00	-0.67
		7	-0.45	0.11	0.00	0.00	0.00	-0.52
	3·	4	0.33	-0.26	0.00	0.00	0.00	-1.95
		7	-0.33	0.26	0.00	0.00	0.00	-0.95
	4	4	0.78	-0.37	0.00	0.00	0.00	-2.62
		7	-0.78	0.37	0.00	0.00	0.00	-1.47
	5	4	0.28	0.12	0.00	0.00	0.00	-0.11
		7	-0.28	-0.12	0.00	0.00	0.00	1.39
5	1	5	0.00	0.90	0.00	0.00	0.00	4.85
		8	0.00	-0.90	0.00	0.00	0.00	5.09

MEMBER END FORCES   STRUCTURE TYPE = PLANE

ALL UNITS ARE -- KIP FEET

MEMBER	LOAD	JT	AXIAL	SHEAR-Y	SHEAR-Z	TORSION	MOM-Y	MOM-Z
	2	5	1.17	0.00	0.00	0.00	0.00	0.00
		8	-1.17	0.00	0.00	0.00	0.00	0.00
	3	5	0.37	0.00	0.00	0.00	0.00	0.00
		8	-0.37	0.00	0.00	0.00	0.00	0.00
	4	5	1.54	0.00	0.00	0.00	0.00	0.00
		8	-1.54	0.00	0.00	0.00	0.00	0.00
	5	5	1.44	0.68	0.00	0.00	0.00	3.64
		8	-1.44	-0.68	0.00	0.00	0.00	3.82
6	1	6	0.55	0.56	0.00	0.00	0.00	2.70
		9	-0.55	-0.56	0.00	0.00	0.00	3.50
	2	6	0.45	0.11	0.00	0.00	0.00	0.67
		9	-0.45	-0.11	0.00	0.00	0.00	0.52
	3	6	0.33	0.26	0.00	0.00	0.00	1.95
		9	-0.33	-0.26	0.00	0.00	0.00	0.95
	4	6	0.78	0.37	0.00	0.00	0.00	2.62
		9	-0.78	-0.37	0.00	0.00	0.00	1.47
	5	6	1.11	0.73	0.00	0.00	0.00	4.16
		9	-1.11	-0.73	0.00	0.00	0.00	3.85
7	1	7	1.47	-0.55	0.00	0.00	0.00	-3.30
		8	-1.47	0.55	0.00	0.00	0.00	-2.35
	2	7	0.11	0.45	0.00	0.00	0.00	0.36
		8	-0.11	0.58	0.00	0.00	0.00	-1.07
	3	7	0.26	0.33	0.00	0.00	0.00	0.83
		8	-0.26	0.18	0.00	0.00	0.00	-0.08
	4	7	0.37	0.78	0.00	0.00	0.00	1.19
		8	-0.37	0.77	0.00	0.00	0.00	-1.15
	5	7	1.41	0.28	0.00	0.00	0.00	-1.49
		8	-1.41	1.13	0.00	0.00	0.00	-2.89
8	1	8	0.56	-0.55	0.00	0.00	0.00	-2.35
		9	-0.56	0.55	0.00	0.00	0.00	-3.30
	2	8	0.11	0.58	0.00	0.00	0.00	1.07
		9	-0.11	0.45	0.00	0.00	0.00	-0.36
	3	8	0.26	0.18	0.00	0.00	0.00	0.08
		9	-0.26	0.33	0.00	0.00	0.00	-0.83
	4	8	0.37	0.77	0.00	0.00	0.00	1.15
		9	-0.37	0.78	0.00	0.00	0.00	-1.19
	5	8	0.73	0.31	0.00	0.00	0.00	-0.64
		9	-0.73	1.11	0.00	0.00	0.00	-3.46
9	1	4	1.12	-1.39	0.00	0.00	0.00	-8.37
		5	-1.12	1.39	0.00	0.00	0.00	-5.89
	2	4	-0.04	1.09	0.00	0.00	0.00	0.82
		5	0.04	1.48	0.00	0.00	0.00	-2.84
	3	4	-0.02	3.70	0.00	0.00	0.00	2.52
		5	0.02	5.15	0.00	0.00	0.00	-9.99

```
MEMBER END FORCES     STRUCTURE TYPE = PLANE
-----------------
ALL UNITS ARE -- KIP  FEET

MEMBER  LOAD  JT    AXIAL   SHEAR-Y  SHEAR-Z   TORSION    MOM-Y     MOM-Z

         4    4    -0.06     4.79     0.00      0.00      0.00       3.33
              5     0.06     6.64     0.00      0.00      0.00     -12.83
         5    4     0.79     2.83     0.00      0.00      0.00      -3.57
              5    -0.79     6.39     0.00      0.00      0.00     -14.75

  10     1    5     0.62    -1.38     0.00      0.00      0.00      -5.88
              6    -0.62     1.38     0.00      0.00      0.00      -8.35
         2    5    -0.04     1.48     0.00      0.00      0.00       2.84
              6     0.04     1.09     0.00      0.00      0.00      -0.82
         3    5    -0.02     5.15     0.00      0.00      0.00       9.99
              6     0.02     3.70     0.00      0.00      0.00      -2.52
         4    5    -0.06     6.64     0.00      0.00      0.00      12.83
              6     0.06     4.79     0.00      0.00      0.00      -3.33
         5    5     0.41     4.31     0.00      0.00      0.00       5.93
              6    -0.41     4.90     0.00      0.00      0.00      -8.97
```

************** END OF LATEST ANALYSIS RESULT **************

45. PRINT JOINT DISPLACEMENT

JOINT DISPLACEMENT (INCH RADIANS)   STRUCTURE TYPE = PLANE

JOINT	LOAD	X-TRANS	Y-TRANS	Z-TRANS	X-ROTAN	Y-ROTAN	Z-ROTAN
1	1	0.00000	0.00000	0.00000	0.00000	0.00000	0.00000
	2	0.00000	0.00000	0.00000	0.00000	0.00000	0.00000
	3	0.00000	0.00000	0.00000	0.00000	0.00000	0.00000
	4	0.00000	0.00000	0.00000	0.00000	0.00000	0.00000
	5	0.00000	0.00000	0.00000	0.00000	0.00000	0.00000
2	1	0.00000	0.00000	0.00000	0.00000	0.00000	0.00000
	2	0.00000	0.00000	0.00000	0.00000	0.00000	0.00000
	3	0.00000	0.00000	0.00000	0.00000	0.00000	0.00000
	4	0.00000	0.00000	0.00000	0.00000	0.00000	0.00000
	5	0.00000	0.00000	0.00000	0.00000	0.00000	0.00000
3	1	0.00000	0.00000	0.00000	0.00000	0.00000	0.00000
	2	0.00000	0.00000	0.00000	0.00000	0.00000	0.00000
	3	0.00000	0.00000	0.00000	0.00000	0.00000	0.00000
	4	0.00000	0.00000	0.00000	0.00000	0.00000	0.00000
	5	0.00000	0.00000	0.00000	0.00000	0.00000	0.00000
4	1	0.38823	0.00140	0.00000	0.00000	0.00000	-0.00123
	2	-0.00003	-0.00111	0.00000	0.00000	0.00000	-0.00026
	3	-0.00001	-0.00291	0.00000	0.00000	0.00000	-0.00091
	4	-0.00004	-0.00402	0.00000	0.00000	0.00000	-0.00116
	5	0.29113	-0.00225	0.00000	0.00000	0.00000	-0.00186
5	1	0.38745	0.00000	0.00000	0.00000	0.00000	-0.00039
	2	0.00000	-0.00299	0.00000	0.00000	0.00000	0.00000
	3	0.00000	-0.00772	0.00000	0.00000	0.00000	0.00000
	4	0.00000	-0.01071	0.00000	0.00000	0.00000	0.00000
	5	0.29059	-0.00878	0.00000	0.00000	0.00000	-0.00029
6	1	0.38703	-0.00140	0.00000	0.00000	0.00000	-0.00123
	2	0.00003	-0.00111	0.00000	0.00000	0.00000	0.00026
	3	0.00001	-0.00291	0.00000	0.00000	0.00000	0.00091
	4	0.00004	-0.00402	0.00000	0.00000	0.00000	0.00116
	5	0.29031	-0.00434	0.00000	0.00000	0.00000	0.00002
7	1	0.62817	0.00178	0.00000	0.00000	0.00000	-0.00049
	2	0.00007	-0.00142	0.00000	0.00000	0.00000	-0.00011
	3	0.00018	-0.00314	0.00000	0.00000	0.00000	0.00001
	4	0.00026	-0.00456	0.00000	0.00000	0.00000	-0.00010
	5	0.47134	-0.00244	0.00000	0.00000	0.00000	-0.00048
8	1	0.62715	0.00000	0.00000	0.00000	0.00000	-0.00017
	2	0.00000	-0.00380	0.00000	0.00000	0.00000	0.00000
	3	0.00000	-0.00797	0.00000	0.00000	0.00000	0.00000
	4	0.00000	-0.01177	0.00000	0.00000	0.00000	0.00000
	5	0.47036	-0.00978	0.00000	0.00000	0.00000	-0.00013
9	1	0.62676	-0.00178	0.00000	0.00000	0.00000	-0.00049
	2	-0.00007	-0.00142	0.00000	0.00000	0.00000	0.00011
	3	-0.00018	-0.00314	0.00000	0.00000	0.00000	-0.00001
	4	-0.00026	-0.00456	0.00000	0.00000	0.00000	0.00010
	5	0.46986	-0.00511	0.00000	0.00000	0.00000	-0.00026

REFERENCES

1. Breyer, D. E., et al, 1999. *Design of Wood Structures,* 4th ed. New York: McGraw-Hill.
2. Steinbrugge, K. V. 1970. *Earthquake Engineering.* Englewood Cliffs, NJ: Prentice-Hall.
3. Waite, T. 2000. *Steel-Frame House Construction.* NAHB Research Center, Upper Marlboro, MD.

CHAPTER 9

MATRICES IN ENGINEERING

9.1 USE OF MATRICES IN ENGINEERING

A coupled mass–spring system that corresponds to a system of linear equations is illustrated in Figure 9.1. Often it is convenient to express the equation system in matrix form,

$$10x - 4y + 2z = 0$$
$$6x \qquad + 8z = 0$$

and the coefficients of x, y, z grouped in compact matrix form as

$$\mathbf{A} = \begin{bmatrix} 10 & -4 & 2 \\ 6 & 0 & 8 \end{bmatrix}$$

where the coefficients become the entries of the so-called coefficient matrix $\mathbf{A}$. The entire equation will be represented as

$$\begin{bmatrix} 10 & -4 & 2 \\ 6 & 0 & 8 \end{bmatrix} \begin{bmatrix} x \\ y \\ z \end{bmatrix} = \begin{bmatrix} 0 \end{bmatrix}$$

where x, y, z form a column matrix often called a vector matrix. The flexibility coefficient matrix expresses the load–deflection relationship of the cantilever structure shown in Figure 9.2.

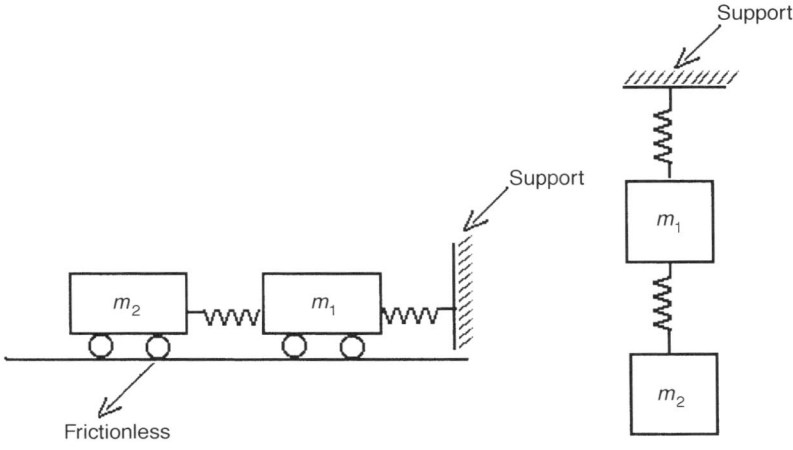

Figure 9.1

$$\frac{\mathbf{L}^2}{\mathbf{EI}} \left[\frac{1}{18} + \frac{4}{18} \right] \left[\begin{matrix} \mathbf{Q}_1 \\ \mathbf{Q}_2 \end{matrix} \right] = \left[\begin{matrix} \theta_1 \\ \theta_2 \end{matrix} \right]$$

where θ_1 is the rotation due to $\mathbf{Q}_1$ and θ_2 is the rotation caused by $\mathbf{Q}_2$ at the cantilever end **A.** The shorthand notation of the matrix equation is

$$\left(\frac{\mathbf{L}^2}{\mathbf{EI}} \right) \mathbf{AB} = \mathbf{C}$$

where **A** is the coefficient matrix, **B** is the force vector, and **C** is the elastic deformation response matrix, also a vector column. Note that both vectors **B** and **C** have direction and in engineering application, line of action, which is why they are considered vector matrices, in contrast with the coefficient matrix, which consists of scalar entries. Matrices are represented by uppercase letters and denoted between square brackets. A matrix that has *n* rows and *m*

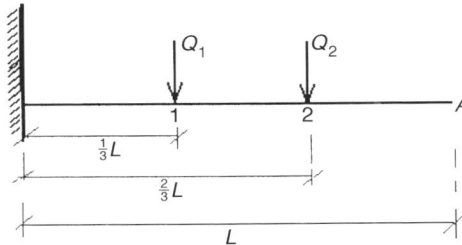

Figure 9.2 Cantilever structure expressed by the flexibility coefficient matrix.

columns is called an $n \times m$ (pronounced n by m) matrix. The $n \times m$ is called the *size* of a matrix. The location—*entry*—is referred to as a_{ij}, which represents the entry in the ith row and jth column.

Equality of Matrices

Two matrices **A** and **B** are equal (**A**=**B**) if and only if they have the same size and *all* of their corresponding entries are equal:

$$\mathbf{A} = [a_{jk}] \qquad \text{and} \qquad \mathbf{B} = [b_{jk}]$$

9.2 MATRIX ADDITION AND MULTIPLICATION

Matrix operations must follow well-defined rules.

Matrix Addition

Two matrices $\mathbf{A} = [a_{jk}]$ and $\mathbf{B} = [b_{jk}]$ can be added only if they have the same size $n \times m$. The addition is done by adding the corresponding entries:

$$\mathbf{A} + \mathbf{B} = [\alpha_{j\kappa} + b_{j\kappa}] = \mathbf{C} = [c_{jk}]$$

Matrix Multiplication

Two matrices **A** and **B** can be multiplied as $\mathbf{C} = \mathbf{AB}$ if and only if the number of columns of **A** is equal to the number of rows of **B**. Stated otherwise, if **A** is an $n \times m$ matrix and **B** is an $r \times p$ matrix, the operation is defined if and only if $r = m$. The product **C** will become an $n \times p$ matrix. The operation is done by multiplying each entry of row i of matrix **A** by the corresponding entry of column k of matrix **B** and adding all products together to obtain the corresponding entries:

$$\mathbf{C} = [c_{ik}]$$

Matrix **A** in $\mathbf{AB} = \mathbf{C}$ product is called the premultiplier to **B**. The order of multiplication cannot be changed, that is, $\mathbf{AB} \neq \mathbf{BA}$. In other words, matrix multiplication is *not commutative*. For instance,

$$\begin{bmatrix} 3 & 1 \\ -4 & 0 \end{bmatrix}\begin{bmatrix} 1 & 2 \\ 2 & 5 \end{bmatrix} = \begin{bmatrix} 3.1 + 1.2 & 3.2 + 1.5 \\ (-4).1 + 0.2 & (-4).2 + 0.5 \end{bmatrix} = \begin{bmatrix} 5 & 11 \\ -4 & -8 \end{bmatrix}$$

whereas

$$\begin{bmatrix} 1 & 2 \\ 2 & 5 \end{bmatrix} \begin{bmatrix} 3 & 1 \\ -4 & 0 \end{bmatrix} = \begin{bmatrix} 1.3 + 2.(-4) & 1.1 + 2.0 \\ 2.3 + 5.(-4) & 2.1 + 5.0 \end{bmatrix} = \begin{bmatrix} -5 & 3 \\ -14 & 7 \end{bmatrix}$$

Scalar Multiplication

The product of an $n \times m$ matrix and a scalar number c is achieved by multiplying each entry of $\mathbf{A} = [a_{jk}]$ by c:

$$c\mathbf{A} = \mathbf{A}c = [ca_{jk}] = [a_{jk}c]$$

Examples

1. Add matrices $\mathbf{A}$ and $\mathbf{B}$:

$$\mathbf{A} = \begin{bmatrix} 3 & 2 & 1 \\ 2 & 4 & 2 \\ 3 & 1 & 2 \end{bmatrix} \qquad \mathbf{B} = \begin{bmatrix} -2 & 2 & -3 \\ 2 & 1 & -6 \\ -1 & -2 & 0 \end{bmatrix}$$

Then

$$\mathbf{A} + \mathbf{B} = \begin{bmatrix} 3-2 & 2+2 & 1-3 \\ 2+2 & 4+1 & 2-6 \\ 3-1 & 1-2 & 2+0 \end{bmatrix} = \begin{bmatrix} 1 & 4 & -2 \\ 4 & 5 & -4 \\ 2 & -1 & 2 \end{bmatrix}$$

2. Multiply $\mathbf{A}$ and $\mathbf{B}$:

$$\mathbf{AB} = \begin{bmatrix} 3 & 1 \\ 2 & 2 \\ 3 & 2 \end{bmatrix} \begin{bmatrix} -5 & 2 \\ 2 & -2 \end{bmatrix}$$

$$= \begin{bmatrix} 3(-5) + 1.2 & 3.2 + 1(-2) \\ 2(-5) + 2.2 & 2.2 + 2(-2) \\ 3(-5) + 2.2 & 3.2 + 2(-2) \end{bmatrix} = \begin{bmatrix} 13 & 4 \\ -6 & 0 \\ -11 & 2 \end{bmatrix} = \mathbf{C}$$

Note that $\mathbf{A} = n \times m = 3 \times 2$ matrix and $\mathbf{B} = r \times p = 2 \times 2$. With $r = m$ the product yielded a $\mathbf{C} = n \times p = 3 \times 2$ matrix product of column and row vectors. If the 3×1 matrix

$$\mathbf{A} = \begin{bmatrix} 2 \\ 4 \\ 8 \end{bmatrix}$$

is multiplied by the 1×3 matrix

$$\mathbf{B} = [3 \quad 6 \quad 1]$$

expect that the product $\mathbf{C}$ will be a 3×3 matrix ($n = 3$, $p = 3$):

$$\mathbf{AB} = \begin{bmatrix} 2.3 & 2.6 & 2.1 \\ 4.3 & 4.6 & 4.1 \\ 8.3 & 8.6 & 8.1 \end{bmatrix} = \begin{bmatrix} 6 & 12 & 2 \\ 12 & 24 & 4 \\ 24 & 48 & 8 \end{bmatrix} = \mathbf{C}$$

9.3 MATRIX FORMS

A matrix with a number of rows that are equal to its number of columns, $u = m$, is called a *square matrix*.

A special case of the square matrix is the *symmetric matrix* in which entries above and below are symmetrical about the main diagonal.

A *skew-symmetric matrix* is like the symmetric matrix except the signs of the entries are reversed (minus instead of plus sign, or vice versa).

A *diagonal matrix* is a square matrix $\mathbf{A} = [a_{jk}]$ whose entries above and below the main diagonal are all zero.

A diagonal matrix ($\mathbf{S}$) whose entries (c) on the main diagonal are all equal is called a *scalar matrix*. The matrix commutes with any $n \times n$ matrix and the multiplication has the same effect as the multiplication by a scalar:

$$\mathbf{AS} = \mathbf{SA} = c\mathbf{A}$$

A special case of the scalar matrix is the *unit matrix* ($\mathbf{I}$), whose entries on the main diagonal are 1:

$$\mathbf{AI} = \mathbf{IA} = \mathbf{A}$$

Examples

$$\mathbf{D} = \begin{bmatrix} 4 & 0 & 0 \\ 0 & 0 & 0 \\ 0 & 0 & -10 \end{bmatrix} \qquad \mathbf{S} = \begin{bmatrix} c & 0 & 0 \\ 0 & c & 0 \\ 0 & 0 & c \end{bmatrix} \qquad \mathbf{I} = \begin{bmatrix} 1 & 0 & 0 \\ 0 & 1 & 0 \\ 0 & 0 & 1 \end{bmatrix}$$

$\qquad\qquad$ diagonal matrix $\qquad\qquad$ scalar matrix $\qquad\qquad$ unit matrix

9.4 TRANSPOSITION

So far we have covered most aspects of matrix algebraic operations except matrix division. To do a matrix division we have to go through a number of operations, one of them being *matrix transposition*. Transposition means changing all rows into columns by keeping the same order; that is, transpose the first row into the first column, then the nth row into the nth column. If $\mathbf{A}$ is the original matrix, its transpose is $\mathbf{A}^{\mathrm{T}}$. Assume you have formed the ratio $\mathbf{B}/\mathbf{A}$, where $\mathbf{B}$ and $\mathbf{A}$ are $n \times n$ square matrices and $\mathbf{A}$ is the divisor and $\mathbf{B}$ the dividend. The ratio a/b in plain algebra can also be expressed as $a \times b^{-1}$, where b^{-1} is the inverse of b. In matrix algebra the division is also carried out by postmultiplying the dividend matrix $\mathbf{B}$ by the inverse of the divisor matrix $\mathbf{A}$. The inverse of a matrix $\mathbf{A}$ is denoted $\mathbf{A}^{-1}$. Unlike other matrix operations, forming the inverse of a matrix is quite elaborate. The process involves creating a matrix $\mathbf{M}$ of *minors* of all entries (a_{jk}) then creating a cofactor matrix $\mathbf{C}$ of $\mathbf{A}$, transpose the cofactor matrix and divide by the determinant D of the original matrix $\mathbf{A}$.

9.5 MINOR AND COFACTOR MATRICES

The minor of the a_{jk} entry of a matrix is formed by *replacing* the entry by the *determinant* of a submatrix. This is obtained by eliminating the row j and column k intercepting a_{jk}. This will be illustrated by a 3×3 square matrix:

$$\begin{bmatrix} a_{11} & a_{12} & a_{13} \\ a_{21} & a_{22} & a_{23} \\ a_{31} & a_{32} & a_{33} \end{bmatrix}$$

To form the minor of a_{11}, the first row and the first column are crossed out leaving a 2×2 submatrix with a determinant that can be easily evaluated as

$$\begin{bmatrix} a_{22} & a_{23} \\ a_{32} & a_{33} \end{bmatrix} \qquad a_{22}a_{33} - a_{23}a_{32} = m_{11}$$

m_{11} being the minor of the a_{11} entry. Similarly, the minor of a_{12} is

$$m_{12} = a_{21}a_{33} - a_{23}a_{31}$$

Once all the minors are evaluated, the matrix $\mathbf{M}$ is formed:

$$\mathbf{M} = \begin{bmatrix} m_{11} & m_{12} & m_{13} \\ m_{21} & m_{22} & m_{23} \\ m_{31} & m_{32} & m_{33} \end{bmatrix}$$

The cofactor matrix will be formed by giving the appropriate signs to the minors (n_{jk}). The rule is

$$\begin{bmatrix} + & - & + & - & + \\ - & + & - & + & - \\ + & - & + & - & + \end{bmatrix}$$

Thus the cofactor matrix is

$$\mathbf{C} = \begin{bmatrix} c_{11} & c_{12} & c_{13} \\ c_{21} & c_{22} & c_{23} \\ c_{31} & c_{32} & c_{33} \end{bmatrix} = \begin{bmatrix} m_{11} & m_{12} & m_{13} \\ m_{21} & m_{22} & m_{23} \\ m_{31} & m_{32} & m_{33} \end{bmatrix}$$

The determinant of a matrix $\mathbf{A}$ is obtained as follows:

1. Form the cofactor matrix as described.
2. Transpose the cofactor matrix, called the *adjoint matrix* Adj[$\mathbf{A}$]:

$$\text{Adj}[\mathbf{A}] = \begin{bmatrix} c_{11} & c_{21} & c_{31} \\ c_{12} & c_{22} & c_{32} \\ c_{13} & c_{23} & c_{33} \end{bmatrix}$$

3. Divide Adj[$\mathbf{A}$] by the determinant of $\mathbf{A}$.

9.6 DETERMINANT OF A MATRIX

The computation of the second-order determinant (second order = 2×2 matrix) was already shown as

$$\mathbf{A} = \begin{bmatrix} a_{11} & a_{12} \\ a_{21} & a_{22} \end{bmatrix} \qquad D = a_{11}a_{22} - a_{12}a_{21}$$

The determinant of a 3×3 (third-order matrix)

$$\mathbf{A} = \begin{bmatrix} a_{11} & a_{12} & a_{13} \\ a_{21} & a_{22} & a_{23} \\ a_{31} & a_{32} & a_{33} \end{bmatrix}$$

can be expanded by row or column. For instance, by the first row, according to the Laplace expansion,

$$D = a_{11}(a_{22}a_{33} - a_{23}a_{32}) - a_{12}(a_{21}a_{33} - a_{23}a_{31}) + a_{13}(a_{21}a_{32} - a_{22}a_{31})$$

9.7 INVERSE OF A MATRIX

Problem 9.1   Find the inverse of

$$\mathbf{A} = \begin{bmatrix} 3 & 2 & 1 \\ 2 & 4 & 2 \\ 3 & 1 & 2 \end{bmatrix}$$

Solution   The cofactors of **A** are

$$\begin{array}{lll} \mathbf{A}_{11} = & 6 & \mathbf{A}_{12} = & 2 & \mathbf{A}_{13} = & -10 \\ \mathbf{A}_{21} = & -3 & \mathbf{A}_{22} = & 3 & \mathbf{A}_{23} = & 3 \\ \mathbf{A}_{31} = & 0 & \mathbf{A}_{32} = & -4 & \mathbf{A}_{33} = & 8 \end{array}$$

The adjoint of [**A**] is given as

$$\begin{bmatrix} 6 & -3 & 0 \\ 2 & 3 & -4 \\ -10 & 3 & 8 \end{bmatrix}$$

$$\det[\mathbf{A}] = 3(6) - 2(-2) + 1(-10) = 18 + 4 - 10 = 12$$

The inverse of **A** is

$$\mathbf{A}^{-1} = \begin{bmatrix} \frac{1}{2} & -\frac{1}{4} & 0 \\ \frac{1}{6} & \frac{1}{4} & -\frac{1}{3} \\ -\frac{5}{6} & \frac{1}{4} & \frac{2}{3} \end{bmatrix}$$

If one wants to check:

$$\mathbf{AA}^{-1} = \begin{bmatrix} 3 & 2 & 1 \\ 2 & 4 & 2 \\ 3 & 1 & 2 \end{bmatrix} \begin{bmatrix} \frac{1}{2} & -\frac{1}{4} & 0 \\ \frac{1}{6} & \frac{1}{4} & -\frac{1}{3} \\ -\frac{5}{6} & \frac{1}{4} & \frac{2}{3} \end{bmatrix}$$

First column
$$\begin{bmatrix} 1\frac{1}{2} & +\frac{2}{6} & -\frac{5}{6} = 1 \\ 1 & +\frac{4}{6} & -\frac{10}{6} = 0 \\ 1\frac{1}{2} & +\frac{1}{6} & -\frac{10}{6} = 0 \end{bmatrix}$$

Second column
$$\begin{bmatrix} -\frac{3}{4} & +\frac{2}{4} & +\frac{1}{4} = 0 \\ -\frac{2}{4} & +\frac{4}{4} & +\frac{2}{4} = 1 \\ -\frac{3}{4} & +\frac{1}{4} & +\frac{2}{4} = 0 \end{bmatrix}$$

Therefore

$$\mathbf{AA}^{-1} = \begin{bmatrix} 1 & 0 & 0 \\ 0 & 1 & 0 \\ 0 & 0 & 1 \end{bmatrix}$$

This gives **I**, the *unit matrix*.

9.8 LINEAR SYSTEMS OF EQUATIONS

Consider the problem of a two-degree (two vibrating masses, $\mathbf{M_1}$ and $\mathbf{M_2}$) coupled together by a spring system of $\mathbf{K_1}$ and $\mathbf{K_2}$ spring constants. We realize that the equations of motion—no damping considered for simplification—are coupled together in a system of linear equations as

$$\ddot{y}_1 - \frac{\mathbf{K_2}}{\mathbf{M_1}}(y_2 - y_1) + \frac{\mathbf{K_1}}{\mathbf{M_1}} + \frac{\mathbf{K_1}}{\mathbf{M_1}} y_1 = 0$$

$$\ddot{y}_2 + \frac{\mathbf{K_2}}{\mathbf{M_2}}(y_2 - y_1) = 0$$

where $\ddot{y}_1$ and $\ddot{y}_2$ are accelerations and y_1 and y_2 are displacements of $\mathbf{M_1}$ and $\mathbf{M_2}$ masses, respectively.

The basic pattern of a large number of methods originates from the *Gaussian and Gauss–Jordan elimination methods*. Consider an example of three equations:

$$2x_1 + 4x_2 + 5x_3 = 4$$

$$-4x_1 + 12x_2 + 4x_3 = 12$$

$$6x_1 + 2x_2 + 10x_3 = 2$$

or in matrix form

$$
\begin{bmatrix}
a_{11} & + & a_{12} & + & a_{13} \\
a_{21} & + & a_{22} & + & a_{23} \\
a_{31} & + & a_{32} & + & a_{33} \\
\vdots & \vdots & \vdots & \vdots & \vdots \\
a_{n_1} & + & a_{n_2} & + & a_{n_3}
\end{bmatrix}
\begin{bmatrix}
x_1 \\ x_2 \\ x_3 \\ \vdots \\ x_n
\end{bmatrix}
=
\begin{bmatrix}
b_1 \\ b_2 \\ b_3 \\ \vdots \\ b_n
\end{bmatrix}
$$

The aim of the method is to separate x_n from the rest of the unknowns, so we have

$$
c_n a_{n_3} = b'_n \qquad \text{or} \qquad a_{n_3} = \frac{b'_n}{c_n}
$$

where b'_n is the last entry of the vector **[B]** of given quantities and the prime notation indicates a transformed value after carrying out row operations.

We achieve the separation by eliminating the coefficients a_{21} and a_{31} from the second and third rows and eliminating a_{32} in the following step. The tool is an elementary row operation that allows multiplying any row by a scalar number, adding any row to another row, and interchanging rows without altering the mathematical matrices of the original equations.

We can proceed by multiplying the first row, $\mathbf{R}_1$, by 2 and adding it to the second row, $\mathbf{R}_2$, and then multiplying the first row by -3 and adding the product to the third row. After the first step the system of equations becomes

$$
2x_1 + 4x_2 + 5x_3 = 4
$$

$$
20x_2 + 14x_3 = 20
$$

$$
-10x_2 - 5x_3 = -10
$$

or in matrix form the augmented coefficient matrix

$$
\begin{bmatrix}
2 & 4 & 5 & 4 \\
20 & 14 & 20 \\
-2 & -5 & -10
\end{bmatrix}
$$

The next step is to eliminate a_{32} by multiplying the second row, $\mathbf{R}_2$, by $\frac{1}{2}$ and adding it to the third row, $\mathbf{R}_3$, resulting in

$$
2x_1 + 4x_2 + 5x_3 = 4
$$

$$
20x_2 + 14x_3 = 20
$$

$$
2x_3 = 0
$$

expressed in augmented matrix form augmented by the [**b**] column vector on the right-hand side of the system of equations in the coefficient matrix (*note that identical row operations were used in the matrix operations*):

$$\begin{bmatrix} 2 & 4 & 5 & 4 \\ & 20 & 14 & 20 \\ & & 2 & 0 \end{bmatrix}$$

which results in $x_3 = 0$, and by back substituting

$$20x_2 = 20$$

$$x_2 = 1$$

and

$$x_1 = \frac{[-4(20) + 4]}{2} = -38$$

$$2x_1 = 4 - 4(1) \qquad x_1 = 0$$

At this point, instead of back substituting we will rearrange the diagonal matrix so that we have only zeros, even above the main diagonal. By adding $-2\frac{1}{2}$ times row 3 to row 1, we obtain

$$\begin{bmatrix} 2 & 4 & 0 & 4 \\ & 20 & 14 & 20 \\ & & 2 & 0 \end{bmatrix}$$

and by adding -7 times row 3 to row 2, the next form of the matrix will be

$$\begin{bmatrix} 2 & 4 & 0 & 4 \\ & 20 & 0 & 20 \\ & & 2 & 0 \end{bmatrix}$$

The next step is to multiply row 2 by $-\frac{1}{5}$ and add it to row 1,

$$\begin{bmatrix} 2 & 0 & 0 & 0 \\ & 20 & 0 & 20 \\ & & 2 & 0 \end{bmatrix}$$

As a final step, each row is divided by its appropriate entry on the main diagonal to obtain

$$\begin{bmatrix} 1 & 0 & 0 & 0 \\ & 1 & 0 & 1 \\ & & 2 & 0 \end{bmatrix}$$

meaning

$$1x_1 + 0x_2 + 0x_3 = 0$$

$$1x_2 + 0x_3 = 1$$

$$2x_3 = 0$$

giving

$$x_1 = 0 \qquad x_2 = 1 \qquad x_3 = 0$$

9.9 ELEMENTARY ROW OPERATIONS

The operations described above are meant to demonstrate the Gaussian elimination method. For large systems of equations the procedure would require a rather large number of operations and computer memory. An advanced version is more suitable for computer programming, requiring less computation time and storage. The method can be characterized as follows:

Cycle 1 Eliminate all a_{i1} entries of the first column ($m = 1$) by multiplying the first row by a_{i1}/a_{11} and adding the product with its appropriate sign (negative or positive) to each ith row starting with $i = 2$ and ending with the nth row of the $n \times m$ matrix.

Cycle 2 Eliminate all a_{i2} entries in the second column by multiplying the first row ($\mathbf{R}_1$) by a_{i2}/a_{12} and adding the first row thus obtained to each ith row with appropriate sign (negative or positive) as required.

Cycle 3 Repeat the same procedure for the rest of the columns, ending the elimination with $m = 1$ to obtain a diagonal matrix with *all zeros below* the main diagonal.

Cycle 4 Rearrange the augmented matrix so as to have all *zeros below and above* the main diagonal of the coefficient matrix.

Cycle 5 Divide each row with the value of the appropriate entry on the main diagonal pertaining to the ith row. The result will be a unit matrix to the left of the finalized [**b**] column matrix. Print the results $x_1 = \ldots, x_2 = \ldots, x_u = \ldots.$

Although this was not meant to be a complete algorithm, the method described is systematic, well suited for computer programming of large systems. Naturally, a more detailed algorithm needs to be written.

Using the previous examples the following augmented matrices emerge:

Start

$$\begin{bmatrix} 2 & 4 & 5 & : & 4 \\ -4 & 12 & 4 & : & 12 \\ -6 & 2 & 10 & : & 2 \end{bmatrix}$$

End of cycle 1

$$\begin{bmatrix} 2 & 4 & 5 & 4 \\ 0 & 20 & 14 & 20 \\ 0 & -10 & -5 & -10 \end{bmatrix}$$

Cycle 2

$$\begin{bmatrix} 2 & 4 & 5 & 4 \\ 0 & 20 & 14 & 20 \\ -0 & 0 & 2 & 0 \end{bmatrix}$$

Cycle 4

$$\begin{bmatrix} 2 & 0 & 0 & 0 \\ 0 & 20 & 0 & 20 \\ 0 & 0 & -2 & 0 \end{bmatrix}$$

Cycle 5

$$\begin{bmatrix} 1 & 0 & 0 & 0 \\ 0 & 1 & 0 & 1 \\ 0 & 0 & 1 & 0 \end{bmatrix} \quad \begin{matrix} x_1 = 0 \\ x_2 = 1 \\ x_3 = 3 \end{matrix}$$

9.10 SUMMARY OF MATRIX OPERATIONS

Matrix operation can be summarized as follows:

1. *Addition*: $\mathbf{A} + \mathbf{B} = \mathbf{B} + \mathbf{A}$. The commutative law applies.
2. *Multiplication* by a scalar: $c\mathbf{A} = [ca_{jk}]$, obtained by multiplying each entry in $\mathbf{A}$ by c.
3. *Product of two matrices*: $\mathbf{AB} \neq \mathbf{BA}$ is not commutative, except if both $\mathbf{A}$ and $\mathbf{B}$ are diagonal matrices.
4. $\mathbf{A}(\mathbf{B} + \mathbf{C}) = \mathbf{AB} + \mathbf{AC}.$ The distributive law applies.
5. $\mathbf{A}(\mathbf{BC}) = (\mathbf{AB})\mathbf{C}.$ The association law applies and can be written as $\mathbf{ABC}.$
6. $(\mathbf{A} + \mathbf{B})^T = \mathbf{A}^T + \mathbf{B}^T$ but $(\mathbf{AB})^T = \mathbf{B}^T \mathbf{A}^T$. When the parentheses are removed, the product of the transposed matrices is in reverse order.
7. $\mathbf{A}^T = \mathbf{A}$ for symmetric matrices but $\mathbf{A}^T = -\mathbf{A}$ for skew symmetric matrices, as

$$\begin{bmatrix} 0 & 2 & -4 \\ -2 & 0 & 3 \\ 4 & -3 & 0 \end{bmatrix}$$

8. $\mathbf{IA} = \mathbf{AI} = \mathbf{A}$

Definition: If a matrix $\mathbf{A}$ is invertible, that is, it has an inverse, it is called *nonsingular:*

$$\mathbf{B} = \mathbf{A}^{-1}$$

If $\mathbf{B}$ fails to exist (no inverse), $\mathbf{A}$ is said to be *singular*.

CHAPTER 10

DIFFERENTIAL EQUATIONS

10.1 BASIC CONCEPTS

A differential equation can be solved by integrating the expression

$$y' = 2x$$

that is,

$$y = \int 2x \, dx = x^2 + c$$

The constant c, an indispensable companion of the indefinite integral, is the *general solution* in differential equation terminology and indicates that we have an infinite number of values or solutions. Indeed there are infinite identical curves in the x–y plane depending on the value of c chosen.

In physics and engineering, a point is given by ordered pairs (triples, tuples, etc., as the case might be), such as ($x = 0$, $y = 2$), to define the location through which the curve must pass. The terms $x = 0$, $y = 2$ are the *initial value,* and the procedure is the *initial value problem.* The value of c is defined by some given data (e.g., experiments) and the solution is termed the *particular solution.* For the equation above with

$$y_{(0)} = 2$$

meaning the value of $y = 2$ at $x = 0$, we have

$$y_{(0)} = 2 = 0^2 + c \qquad \text{or} \qquad c = 2$$

and the particular solution is

$$y = x^2 + 2$$

Not all problems can be solved so easily, and in most cases more refined methods are needed.

10.2 FIRST-ORDER DIFFERENTIAL EQUATIONS

In this chapter we will deal only with ordinary differential equations. An *ordinary differential equation* is an equation that contains one or several derivatives of an unknown function. Moreover, that unknown function depends only on one independent variable (x or t), as compared to *partial differential equations,* with two or more variables. The task is to find the unknown function.

Following is a classification of differential equations according to their *order:* The order of a differential equation is defined by the highest derivative that appears in the equation. For example,

$$y' + 6y = 12x \tag{1}$$

$$y'' + 8y = 0 \tag{2}$$

$$y''' \cos x + 2x^2 y'' + y = 0 \tag{3}$$

Equation (1) is a *first-order* differential equation; Equations (2) and (3) are of *second* and *third orders,* respectively.

10.3 SEPARATION OF VARIABLES

We will now present a method for solving first-order differential equations. Consider the differential equation

$$\frac{dy}{dx} = \frac{4y}{x} \tag{4}$$

for $x > 0$ and $y > 0$. We note that the function is continuous in the first quadrant and assumes the existence of a unique continuous solution in that quadrant. By separating the variables, we have

$$\frac{dy}{y} = \frac{4\,dx}{x}$$

and obtain a family of solutions

$$\ln|y| = 4\ln|x| + c$$

Raising it to the exponential yields

$$e^{\ln y} = e^{(\ln x^2)+c}$$

As the exponential and the logarithm cancel each other,

$$y = e^c x^2$$

or letting e^c be a constant c_1, we can write

$$y = c_1 x^2 \tag{5}$$

with $c_1 > 0$, where (5) is the general solution of (4).

10.4 EXACT EQUATIONS

There are differential equations that contain two parts,

$$M(x_1 y)\,dx + N(x_1 y_1)\,dy = 0 \tag{6}$$

whose partial derivatives are equal,

$$\frac{\partial M}{\partial y} = \frac{\partial N}{\partial x} \qquad \text{as} \qquad \frac{\partial M}{\partial y} = \frac{\partial^2 u}{\partial y \partial x} \tag{7}$$

and

$$\frac{\partial N}{\partial x} = \frac{\partial^2 u}{\partial x\,\partial y}$$

where

$$u(x_1 y) = c \tag{8}$$

is the general solution as

$$\frac{du}{\partial x} = M \tag{9a}$$

$$\frac{\partial u}{\partial y} = N \tag{9b}$$

Problem 10.1   Solve the differential equation

$$\frac{dy}{dx} = \frac{-(2xy + y^2)}{x^2 + 2xy} \tag{10}$$

Solution   Rewritten in differential form,

$$(2xy + y^2)\, dx + (x^2 + 2xy)\, dy = 0$$

where

$$M = 2xy + y^2$$

and

$$N = x^2 + 2xy$$

Note that

$$u(x,y) = x^2y + y^2x = c$$

is the general solution.

PROOF

$$M = \frac{\partial u}{\partial x} = 2xy + y^2 \qquad \text{and} \qquad N = \frac{\partial u}{\partial y} = x^2 + 2yx$$

Testing for exactness by (7),

$$\frac{\partial M}{\partial y} = \frac{\partial^2 u}{\partial y\, \partial x} = \frac{\partial}{\partial x}(x^2 + 2yx) = 2x + 2y$$

and

$$\frac{\partial N}{\partial x} = \frac{\partial^2 u}{\partial x\,\partial y} = \frac{\partial}{\partial y}\,(2xy + y^2) = 2x + 2y$$

Because the two partial derivatives are equal, it follows, from the fundamental principle of the total or exact differential (calculus), that

$$u(x_1 y) = \int M\,dx + k(y)$$

where $k(y)$ plays the role of a *constant* and can be obtained from (9b). Thus

$$u \int M\,dx = x^2 y + y^2 + K(y)$$

Now using (9b),

$$\frac{\partial u}{\partial y} = x^2 + 2yx + \frac{dk}{dy} = N = x^2 + 2xy$$

Hence

$$\frac{dk}{dy} = 0 \qquad \text{and} \qquad k = \int 0\,dy = c \qquad \text{(a constant)}$$

Therefore

$$u(x_1 y) = x^2 y + y^2 x + c$$

setting $c_1 = -c$ or

$$x^2 y + y^2 x = c_1$$

because c_1 can take any values. That is, we have proved, following prescribed rigorous steps, that the differential equation (10) is indeed *exact*.

10.5 INTEGRATING FACTOR

A basic yet powerful tool to solve differential equations is to use an integrating factor, especially when separation of variables is not possible. Let a first-order linear differential equation be of the form

$$A(x) \frac{dy}{dx} + B(x)y = C(x) \tag{11}$$

where $A(x)$, $B(x)$, $C(x)$ are some functions of x. By dividing each member of the equation by $A(x)$, we obtain

$$\frac{dy}{dx} + P(x)y = Q(x) \tag{12}$$

The expression is said to be of *standard form.*

The procedure to solve the equation follows: $P(x)$ premultiplying y forms the base of the integrating factor. Then:

(a) Integrate $P(x)$:

$$\int P(x) \, dx$$

(b) Raise $\int P(x) \, dx$ to the exponential to obtain

$$e^{\int P(x)dx} = I(x) \tag{13}$$

   This is the integrating factor.

(c) Multiply both sides of the equation by the integrating factor to obtain

$$I(x) \frac{dy}{dx} + I(x)P(x)y = I(x)Q(x) \tag{14}$$

The idea is to choose $I(x)$ so that

$$D(x)[I(x)y] = I(x) \frac{dy}{dx} + I(x)P(x)y \tag{15}$$

However, by the definition of the derivative, on the left side we have

$$I(x) \frac{dy}{dx} + I'(x)y = I(x) \frac{dy}{dx} + I(x)P(x)y \tag{16}$$

This implies that

$$I'(x) = I(x)P(x) \tag{17}$$

$$\frac{I'(x)}{I(x)} = P(x) \tag{18}$$

Integrating both sides and remembering that $\int ☺'/☺ = \ln ☺$ yields

$$\ln |I(x)| = \int P(x)\,dx \tag{19}$$

Taking the exponential on both sides yields

$$e^{\ln |Ix|} = e^{\int P(x)dx}$$

Because the exponential and natural logarithm cancel each other, we have

$$I(x) = e^{\int P(x)dx} \tag{20}$$

This completes the derivation of the integrating factor.

10.6   SECOND-ORDER LINEAR EQUATIONS

An nth-order linear differential equation has the form

$$y^{(n)} + p_1(x)y^{(n-1)} + p_2(x)y^{(n-2)} + \cdots + p_{n-1}(x)y' + p_n(x)y = Q(x) \tag{21}$$

Two cases are noted:

1. When $Q(x) = 0$, the equation is *homogeneous.*
2. When $Q(x) \neq 0$, the equation is *nonhomogeneous.*

Only homogeneous differential equations are considered in this section. Homogeneous linear differential equations have the property that, for any two solutions y_1 and y_2 *and* any pair of constants c_1 and c_2, the function described by

$$y(x) = c_1 y_1(x) + c_2 y_2(x) \tag{22}$$

is *also* a solution for the homogeneous equation.

There is no standard elementary technique for solving *general* homogeneous *n*th-order linear differential equations. In this chapter we focus our attention on the special case when the functions $p_1, p_2, \ldots, p_n$ *are all constants.* Moreover, we simplify matters even further by considering only the case when $n = 2$, that is, second-order linear differential equations. Equations of motion in structural dynamics are in this important category. In summary, we will address linear differential equations of the form

$$y''(x) + py'(x) + qy(x) = 0 \tag{23}$$

where p and q are constants.

A special case for (23) is when the $py'(x)$ term is missing:

$$y'(x) + qy(x) = 0 \quad \text{or} \quad y'(x) = -qy(x) \tag{24}$$

which has the solution

$$y(x) = ce^{-qx} \tag{25}$$

associated with problems of exponential growth and decay. The solution for equation (23) is of the form

$$y(x) = e^{mx} \tag{26}$$

with properly chosen *constant m*. The first and second derivatives of (26) are given as

$$y'(x) = me^{mx} \tag{27}$$

$$y''(x) = m^2 e^{mx} \tag{28}$$

Substituting (27) and (28) into Equation (23) yields

$$0 = y'' + py' + qy = m^2 e^{mx} + pme^{mx} + qe^{mx} = (m^2 + pm + q)e^{mx}$$

As $e^{mx} \neq 0$, the equation is satisfied if and only if

$$m^2 + pm + q = 0 \tag{29}$$

Equation (29) is termed the *characteristic equation* of (23) and has solutions

$$m_1 = -\frac{p}{2} + \frac{\sqrt{p^2 - 4q}}{2} \qquad \text{and} \qquad m_2 = -\frac{p}{2} - \frac{\sqrt{p^2 - 4q}}{2}$$

The solutions assume three different forms depending on the value of $p^2 - 4q$:

Case 1 When $p^2 - 4q > 0$, m_1 and m_2 are distinct real numbers, yielding

$$y_1(x) = e^{m_1 x} \qquad \text{and} \qquad y_2(x) = e^{m_2 x}$$

and the general solution is

$$y(x) = c_1 e^{m_1 x} + c_2 e^{m_2 x} \tag{30}$$

for any constants c_1 and c_2.

Case 2 When $p^2 - 4q = 0$, the characteristic equation has only one root and the solution is

$$y_1(x) = e^{m_1 x} = e^{-px/2}$$

Yet an additional solution exists of the form $xe^{-px/2}$ to be added to form the *general equation*

$$y(x) = c_1 e^{-px/2} + c_2 x e^{-px/2} \tag{31}$$

Case 3 When $p^2 - 4q < 0$, the roots m_1 and m_2 are distinct but *complex numbers.* Let

$$r = -\tfrac{1}{2}p \qquad \text{and} \qquad s = \tfrac{1}{2}\sqrt{4q - p^2}$$

Then

$$m_1 = -\tfrac{1}{2}p + \tfrac{1}{2}\sqrt{4q - p^2}\, i = r + is \qquad m_2 = r - is$$

The general solution can be expressed as

$$y(x) = c_1 e^{m_1 x} + c_2 e^{m_2 x} = c_1 e^{(r+is)x} + c_2 e^{(r-is)x}$$

leading to

$$y(x) = c_1 e^{rx} e^{sxi} + c_2 e^{rx} e^{-sxi} = e^{rx}(c_1 e^{sxi} + c_2 e^{-sxi})$$

and using Euler's equation

$$e^{ix} = \cos x + i \sin x$$

The general solution takes the form

$$y(x) = e^{rx}\{c_1(\cos sx + i \sin sx) + c_2[\cos(-sx) + i \sin(-sx)]\}$$

or, since

$$\cos(-sx) = \cos sx \qquad \text{and} \qquad \sin(-sx) = -\sin sx$$

we have

$$y(x) = e^{rx}(k_1 \cos sx + k_2 \sin sx) \tag{32}$$

with

$$k_1 = c_1 + c_2 \qquad \text{and} \qquad k_2 = c_1 - c_2$$

Problem 10.2 (Case 1)   Solve the differential equation

$$y'' - 8y' - y = 0 \tag{33}$$

with

$$y(x) = e^{mx} \qquad y'(x) = me^{mx} \qquad y''(x) = m^2 e^{mx}$$

$$y'' - 8y' - 4y = (m^2 - 8m - 4)e^{mx} = 0 \tag{34}$$

$$m_1 = \tfrac{8}{2} + \tfrac{1}{2}\sqrt{64 + 16} = 4 + 5 = 9 \qquad m_2 = 4 - 5 = -1$$

The general solution is

$$y(x) = c_1 e^{9x} + c_2 e^{-x} \tag{35}$$

The explanation for the above procedure follows: The solution of the differential equation is expected to be

$$[y] = [\mathbf{C}]e^{\omega t} \tag{36}$$

where $[\mathbf{y}]$ is a vector matrix and $[\mathbf{C}]$ a column matrix that has yet to be determined. Then

$$[\ddot{y}] = \omega^2[C]e^{\omega t} \tag{37}$$

and by letting $\lambda = \omega^2$,

$$[A][C]e^{\omega t} = \lambda[C]e^{\omega t} \tag{38}$$

and canceling $e^{\omega t}$ on both sides of the equation yields

$$[A - \lambda I][C] = 0 \tag{39}$$

where I is the *identity matrix*.

10.7 HOMOGENEOUS DIFFERENTIAL EQUATIONS

When the differential equation is reduced to the form of Equation (39), it is termed a *homogeneous differential equation.* Equation (39) has a nontrivial solution

$$[C] \neq [0]$$

if and only if its coefficient determinant is zero.

10.8 CHARACTERISTIC EQUATION

$$\det(A - \lambda I) = 0 \tag{40}$$

Equation (40) is said to be the characteristic equation of A. The column vector $[C]$ is termed the *eigenvector* of the system. It is evaluated by substituting first the value of λ_1 into Equation (39) and solving it for $[C]_1$. This will give the first eigenvector. To find the second eigenvector, λ_2 is substituted in Equation (39) followed by solving for the eigenvector $[C]_2$. *One vital feature of the eigenvectors is to provide the natural mode shapes of the vibration pertaining to the ω_1, ω_2, . . . , ω_n frequencies, respectively.*

CHAPTER 11

NUMERICAL METHODS AND
ENGINEERING APPLICATIONS

11.1 INTRODUCTION TO DYNAMIC ANALYSIS

First we will look at *structural vibrations*. Perhaps one of the most important applications of second-order differential equations is related to mechanical vibrations. The simplest structure to analyze is the single mass–single spring assembly, called a *single-degree system*. Initially for simplicity we will assume no damping and will focus on basic concepts.

A weight W is suspended by a coil spring from a firm support above and is in static equilibrium. (Figures 11.1 and 11.2) At this point a sudden impulse [e.g., downward force, velocity $y(0)$] at time $t = 0$ on the suspended mass $m = W/g$ causes the system to oscillate. As no additional external force is being applied, we wish to observe and describe mathematically the movement of the mass from a reference point A that corresponds to zero amplitude or to the starting point of the system at rest. If $F_{(t)}$ is identically zero, that is, if there is *no* forcing function, the resulting motion, *unaffected by any external force*, is called *free vibration*.

Call the time dependent displacement $y = y(t)$, the independent variable t for time, k the spring constant, and $m = W/g$, where $g = 32.2$ ft/sec^2 or 980 cm/sec^2 is the acceleration of gravity.

At any time t the system is in dynamic equilibrium; that is, the sum of all forces at any time is *zero*. This is called the *D'Alembert principle of equilibrium*.

Since there are no external forces acting on the mass once it starts moving, the only forces acting on the bouncing mass m, are the mass acceleration force, according to Newton's second law, and the opposing spring force ky.

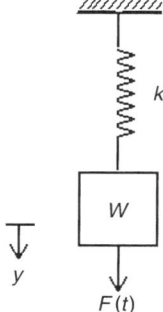

Figure 11.1 One-degree spring–mass system without damping, dynamic equilibrium.

11.2 EQUATION OF MOTION

The acceleration is given as

$$\ddot{y} = \frac{d^2 y}{dt^2}$$

and the resulting differential equation of motion is

$$m\ddot{y} + ky = 0 \tag{1}$$

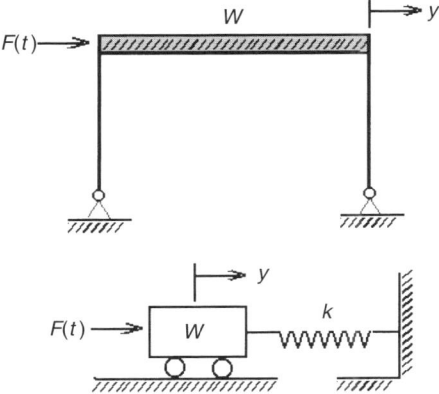

Figure 11.2 Moment frame structure idealized as spring–mass system.

or

$$\dot{y} + \frac{k}{m} y = 0 \tag{2}$$

The solution of the differential equation is

$$y(t) = C_1 \sin \sqrt{\frac{k}{M}} t + C_2 \cos \sqrt{\frac{k}{M}} \tag{3}$$

and letting $\sqrt{k/M} = \omega$,

$$y(t) = C_1 \sin \omega t + C_2 \cos \omega t \tag{4}$$

The displacement-versus-time diagram (Figure 11.3) is a cosine curve and, because there is no damping, it is repetitive. The natural period of the system, T, governs the repetition. It is governed only by its basic parameters or characteristics *mass M* and *spring coefficient k*. The natural period is given as

$$T = \frac{2\pi}{\omega} = 2\pi \sqrt{\frac{M}{k}} \tag{5}$$

expressed in seconds. Equally vital is the *natural frequency*, defined as the inverse of the natural period:

$$f = \frac{1}{T} = \frac{1}{2\pi} \sqrt{\frac{k}{M}} \tag{6}$$

normally expressed in cycles per second (cps).

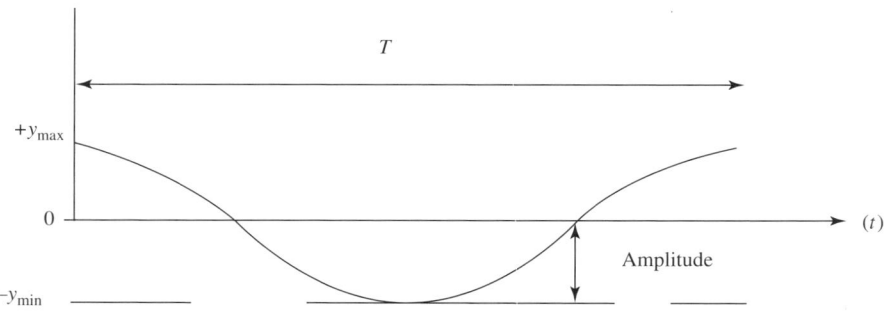

Figure 11.3 Displacement-versus-time diagram.

The equation of motion (4) indicates free vibration with no damping. It is interesting to note that textbooks seldom present the derivation of the solution of such differential equations; instead, only the end results are quoted without providing the reader with the benefit of a proof.

Derivation of Differential Equation of Motion

The free-body diagram shows all forces acting on the oscillating mass m of the one-degree mass–spring system (Figure 11.4).

Equations (1) and (2) represent the D'Alembert principle of dynamic equilibrium:

$$m\ddot{y} + ky = 0$$

or, after dividing by m,

$$\ddot{y} + \frac{k}{m}y = 0$$

The solution of these equations must be in the form

$$y = e^{rt} \tag{7}$$

where t is time and r needs to be found. Substituting (7) into (2) and differentiating twice $\ddot{y} = r^2 e^{rt}$, Equation (2) will take the form

$$r^2 e^{rt} + \frac{k}{m} e^{rt} = 0 \tag{8}$$

Since the equation is *homogeneous,* that is, the right-hand side equals zero, (8) reduces to

$$r^2 + \frac{k}{m} = 0 \tag{9}$$

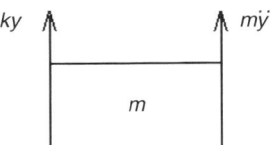

Figure 11.4 Dynamic equilibrium, without damping, of a single spring–mass system.

The solution of (9) yields

$$r^2 = -\frac{k}{m} \quad \text{or} \quad r = \sqrt{-\frac{k}{m}} = i \pm \sqrt{\frac{k}{m}} \tag{10}$$

which results in complex roots involving the *imaginary number i:*

$$r_1 = i\sqrt{\frac{k}{m}} \quad \text{and} \quad r_2 = -i\sqrt{\frac{k}{m}}$$

where i is a troublesome feature and our hope is that it will drop out in the derivation process. This hope will be fulfilled by *Euler's solution*. The Taylor series of e^x is given as

$$e^x = 1 + x + \frac{x^2}{2!} + \frac{x^3}{3!} + \frac{x^4}{4!} + \frac{x^5}{5!} + \cdots + \frac{x^n}{n!}$$

$$\sin x = \qquad x \qquad\qquad \frac{-x^3}{3!} \qquad \frac{+x^5}{5!} \qquad \frac{-x^7}{7!} \cdots \tag{11}$$

$$\cos x = 1 \qquad\qquad \frac{-x^2}{2!} \qquad \frac{+x^4}{4!} \qquad \frac{-x^6}{6!} \tag{12}$$

Therefore an expression like $e^{i\theta}$ can be put together by *combining* sin and cos functions:

$$e^{i\theta} = 1 + i\theta + \frac{(i\theta)^2}{2!} + \frac{(i\theta)^3}{3!} + \frac{(i\theta)^4}{4!} + \frac{(i\theta)^5}{5!} + \frac{(i\theta)^6}{6!} + \cdots$$

$$= 1 + i\theta - \frac{\theta^2}{2!} - \frac{i\theta^3}{3!} - \frac{\theta^4}{4!} + \frac{i\theta^5}{5!} - \frac{\theta^6}{6!} - \frac{i\theta^7}{7!} + \frac{\theta^8}{8!} + \cdots$$

$$= \underbrace{\left(1 \quad - \frac{\theta^2}{2!} \quad + \frac{\theta^4}{4!} - \frac{\theta^6}{6!} + \frac{\theta^8}{8!}\right)}_{\cos \theta} + i \underbrace{\left(\theta - \frac{\theta^3}{3!} + \frac{\theta^5}{5!} - \frac{\theta^7}{7!} \cdots\right)}_{i \sin \theta} \tag{13}$$

Hence *Euler's form of the single spring–mass differential equation* is

$$\boxed{e^{i\theta} = \cos \theta + i \sin \theta} \tag{14}$$

Remember that

$$i^2 = \sqrt{-1}\sqrt{-1} = -1 \qquad \text{and} \qquad i^4 = \sqrt{-1}\sqrt{-1}\sqrt{-1}\sqrt{-1} = 1$$

$$e^{i\pi} = -1$$

$$r = i\left(\pm\sqrt{\frac{k}{m}}\right) = \pm i\sqrt{\frac{4}{2}} = \pm i\sqrt{2} \tag{15}$$

Therefore the solution is

$$y = c_1 e^{(\sqrt{k/m})it} + c_2 e^{-(\sqrt{k/m})it}$$

which will be expressed as

$$y = c_1\left(\cos\sqrt{\frac{k}{m}}\,t + i\sin\sqrt{\frac{k}{m}}\,t\right) + c_2\left[\cos\left(-\sqrt{\frac{k}{m}}\,t\right) + i\sin\left(-\sqrt{\frac{k}{m}}\,t\right)\right]$$

$$= c_1\cos\sqrt{\frac{k}{m}}\,t + c_1 i\sin\sqrt{\frac{k}{m}}\,t + c_2\cos\sqrt{\frac{k}{m}}\,t - c_2 i\sin\sqrt{\frac{k}{m}}\,t \tag{16}$$

Factoring yields $\cos(\sqrt{k/m})t$ and $i\sin(\sqrt{k/m})t$ and collecting terms gives

$$(c_1 + c_2)\cos\sqrt{\frac{k}{m}}\,t + [i(c_1 - c_2)]\sin\sqrt{\frac{k}{m}}\,t = C_1\cos\sqrt{\frac{k}{m}}\,t + iC_2\sin\sqrt{\frac{k}{m}}\,t \tag{17}$$

Consider only the cases in which $c_1 + c_2$ and $i(c_1 - c_2)$ are real. Such combinations of c_1 and c_2 exist.

EXAMPLE

Let

$$c_1 = C + Li$$

$$c_2 = C - Li$$

$$c_1 - c_2 = 0 + 2Li \tag{18}$$

where L is a real number and the term $2Li$ is complex. Assume

$$c_1 + c_2 = \text{real number} \qquad \text{and} \qquad i(c_1 + c_2) = \text{multiple of } i$$

Therefore

$$[i(c_1 - c_2)] = i(2Li) = 2L$$

The reader will note that, because of the complementary nature of the complex conjugates $a + bi$, $a - bi$, i drops out of the equation and

$$y = C_1 \cos\sqrt{\omega}t + C_2 \sin\sqrt{\omega}t \tag{19}$$

is the *equation of motion*.

The derivation proves that y, the time-dependent position of the mass in free oscillation, can be expressed by the combination of sin and cos functions that involve ω and two arbitrary constants C_1 and C_2. The values of the constants are determined by the initial value condition of the problem.

Problem 11.1   Given $m = 2$ lb $- s^2/ft$, $k = 4$ lb/ft, calculate

(a) the natural circular frequency ω,
(b) the natural period T, and
(c) the natural frequency f

for the one-degree spring–mass system.

Solution

(a) $\sqrt{k/m} = \sqrt{\frac{4}{2}} = \sqrt{2}$

$$r = i(\pm \sqrt{k/m}) = \pm i\sqrt{2}$$

$$y = c_1 e^{\sqrt{2}it} + c_2 e^{-\sqrt{2}it}$$

$$y = c_1[\cos\sqrt{2}\,t + i\sin\sqrt{2}\,t] + c_2[\cos(-\sqrt{2}\,t) + i\sin(-\sqrt{2}\,t)]$$

$$= c_1 \cos\sqrt{2}\,t + c_1 i \sin\sqrt{2}\,t + c_2 \cos\sqrt{2}\,t - c_2 i \sin\sqrt{2}\,t$$

which, as shown above, reduces to

$$y = C_1 \cos\sqrt{2}\,t + C_2 \sin\sqrt{2}\,t$$

where $\sqrt{2} = \omega$ is the natural circular frequency
(b) The natural period is given as

$$T = 2\pi\sqrt{\frac{m}{k}} = 2\pi\sqrt{2} = 8.88 \text{ cps}$$

(c) The natural frequency is the inverse of the natural period,

$$f = \frac{1}{8.88} = 0.1125 \text{ s}$$

11.3 DAMPING: DAMPED FREE VIBRATION

EXAMPLE

The equation of a one-degree system is

$$m\ddot{y} + c\dot{y} + ky = 0$$

where $\dot{y}$ is the velocity of the excited mass and c is a constant of proportionality. The $c\dot{y}$ term is named the *viscous damping force* and gives the simplest mathematical treatment of the problem. As before, we assume a solution of the form

$$y = e^{st}$$

where s is a constant. Substituting into the differential equation and noting that $\dot{y} = se^{st}$ and $\ddot{y} = s^2 e^{st}$ give

$$(ms^2 + cs + k)e^{st} = 0$$

for all values of t as $e^{st} \neq 0$. Thus

$$s^2 + \left(\frac{c}{m}\right)s + \left(\frac{k}{m}\right) = 0$$

This equation is known as the *characteristic equation,* a vital tool in solving differential equations. The equation has two roots:

$$s_1 = -\frac{c}{2m} + \sqrt{\left(\frac{c}{2m}\right)^2 - \frac{k}{m}} \qquad s_2 = -\frac{c}{2m} - \sqrt{\left(\frac{c}{2m}\right)^2 - \frac{k}{m}}$$

Thus we have two answers that we will add to obtain the general solution:

$$y = Ae^{s_1 t} + Be^{s_2 t}$$

$$y = \exp\left[-\left(\frac{c}{2m}\right)t\right]\left(A \exp\left[\sqrt{\left(\frac{c}{2m}\right)^2 - \frac{k}{m}}\, t\right]\right.$$

$$\left. + B \exp\left[-\sqrt{\left(\frac{c}{2m}\right)^2 - \frac{k}{m}}\, t\right]\right)$$

We examine the three possible cases under the radical:

1. If $(c/2m)^2$ is larger than k/m, the exponents in the equation above are *real* numbers and no oscillation can take place.
2. If the damping term $(c/2m)^2$ is less than $(k/m)(c/2m)^2 < k/m$, the exponent becomes an imaginary number: $\pm\, i \sqrt{k/m - (c/2m)^2}\,t$ as

$$e^{\pm i \sqrt{\frac{k}{m} - \left(\frac{c}{2m}\right)^2}\, t} = \cos\sqrt{\frac{k}{m} - \left(\frac{c}{2m}\right)^2}\, t \pm i \sin\sqrt{\frac{k}{m} - \left(\frac{c}{2m}\right)^2}\, t$$

3. Assume $(c/2m)^2 = k/m$ is the limiting case between oscillatory and nonoscillatory motion as it reduces the radical to zero. Therefore the value of c, which removes all vibration, is called *critical damping c_c*:

$$\left(\frac{c_c}{2m}\right)^2 = \frac{k}{m} = \omega^2$$

where ω^2 is the circular frequency and $c_c = 2\sqrt{km} = 2m\omega$.

In this case the *overdamped system* merely *creeps back* to its neutral position. In practice it is convenient to express the value of the actual damping—inherent to the structure—in terms of the critical damping by the nondimensional damping ratio

$$\zeta = \frac{c}{c_c}$$

Hence we talk of critical damping of 1%, 2%, and so on.

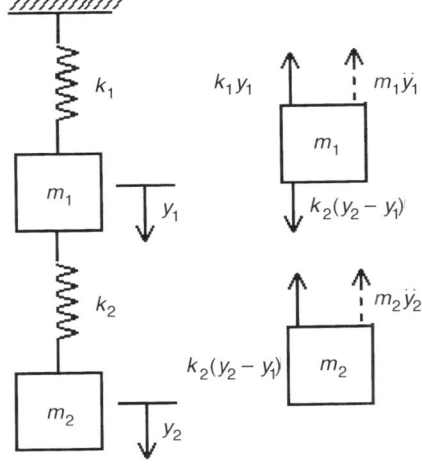

Figure 11.5

11.4 FREE VIBRATIONS: TWO-DEGREE SYSTEMS

EXAMPLE

It is evident that the two systems in Figure 11.5 are coupled: The individual masses cannot move independently. The equation of motion is

$$m_1\ddot{y}_1 + k_1 y_1 - k_2 (y_2 - y_1) = 0$$

$$m_2\ddot{y}_2 + k_2 (y_2 - y_1) = 0$$

We have a system of equations that is better expressed in matrix form. The tool to solve such system is the solution of the eigenvalue, which is discussed next.

11.5 EIGENVALUES AND EIGENVECTORS

We proceed by writing down the coefficient matrix, subtracting λ from the *diagonal* elements, and setting the determinant to zero:

$$\begin{pmatrix} a_{11} & a_{12} \\ a_{21} & a_{22} \end{pmatrix} \quad \det \begin{vmatrix} a_{11} - \lambda & a_{12} \\ a_{21} & a_{22} - \lambda \end{vmatrix} = 0$$

$$(a_{11} - \lambda)(a_{22} - \lambda) - a_{12}a_{22} = 0$$

Problem 11.2   Determine the *eigenvalues* $\lambda_n = \omega_n^2$ and the eigenvectors V_n associated with the system of equations

$$x' = -x + 6y \qquad y' = x - 2y$$

Solution   This system of equations can be expressed in matrix form as

$$\begin{bmatrix} -1 & 6 \\ 1 & -2 \end{bmatrix}$$

(a) $\det \begin{vmatrix} -1 - \lambda & 6 \\ 1 & -2 - \lambda \end{vmatrix} = (\lambda + 2)(\lambda + 1) - 6 = \lambda^2 + 3\lambda - 4 = 0$,
where $\lambda_1 = 1$ and $\lambda_2 = -4$. The λ's are the eigenvalues of the system.

(b) With $\lambda_1 = 1$,

$$\begin{bmatrix} -1 - (+1) & 6 \\ 1 & -3 \end{bmatrix} \begin{bmatrix} K_1 \\ K_2 \end{bmatrix} = 0 \qquad \begin{array}{l} -2K_1 + 6K_2 = 0 \\ -K_1 + 3K_2 = 0 \\ K_1 - 3K_2 = 0 \end{array} \qquad K_1 = 3K_2$$

the *first* eigenvector is

$$K_1 = \begin{bmatrix} 3 \\ 1 \end{bmatrix}$$

With $\lambda_2 = -4$,

$$\begin{bmatrix} -1 - (-4) & 6 \\ 1 & -2 - (-4) \end{bmatrix} \begin{bmatrix} K_1 \\ K_2 \end{bmatrix} = 0$$

$$3K_1 + 6K_2 = 0$$

and simplifying gives

$$K_1 + 2K_2 = 0$$

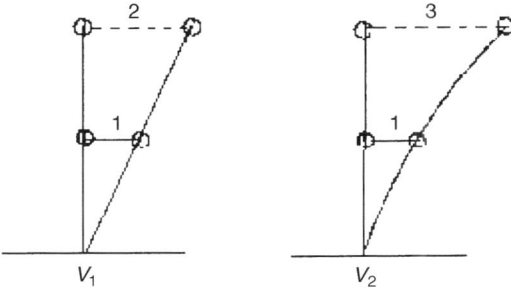

Figure 11.6

This yields the *second* eigenvector,

$$K_2 = \begin{bmatrix} 2 \\ 4 \end{bmatrix} = \begin{bmatrix} 2 \\ 1 \end{bmatrix}$$

Graphically, the eigenvectors correspond to the mode shapes in Figure 11.6.

Problem 11.3 Two-Degree System For the two-degree coupled system shown in Figure 11.7 and *using the eigenvalue matrix solution,* determine

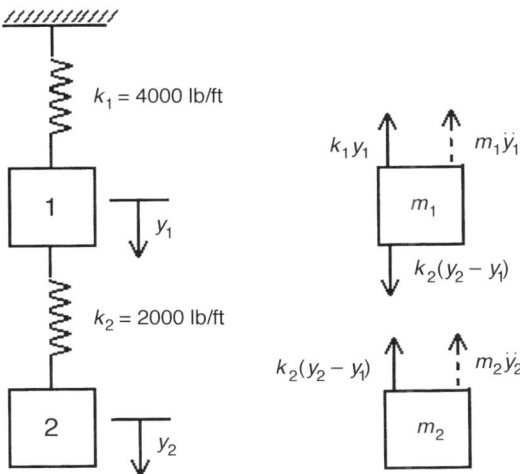

Figure 11.7 Two-degree coupled system.

(a) the two natural frequencies and
(b) the two natural periods of the system.

$$\text{Weight:} \qquad W_1 = 64.4 \text{ lb} \qquad W_2 = 32.2 \text{ lb force}$$

$$m_1 = \frac{64.4}{g = 32.2} = 2 \text{ lb s}^2/\text{ft expressed in mass}$$

$$m_2 = \frac{32.2}{32.2} = 1 \text{ lb s}^2/\text{ft expressed in mass}$$

$$K_1 = 333.3 \text{ lb/in.} = 4000 \text{ lb/ft}$$

$$K_2 = 166.6 \text{ lb/in.} = 2000 \text{ lb/ft}$$

Solution

$$2\ddot{y}_1 + 4000y_1 - 2000(y_2 - y_1) = 0$$

$$\ddot{y}_1 + 2000y_1 - 1000(y_2 - y_1) = 0 \qquad \text{equation of free vibration} \quad (20)$$

$$1\ddot{y}_2 + \qquad\qquad 1000(y_2 - y_1) = 0$$

In matrix form

$$\ddot{y}_1 = -3000y_1 + 1000y_2 \qquad \ddot{y}_2 = 1000y_1 - 1000y_2$$

$$\begin{bmatrix} \ddot{y}_1 \\ \ddot{y}_2 \end{bmatrix} = \begin{bmatrix} -3000 & +1000 \\ 1000 & -1000 \end{bmatrix} \begin{bmatrix} y_1 \\ y_2 \end{bmatrix} \tag{21}$$

$$\det \begin{vmatrix} -3000 - \lambda & 1000 \\ 1000 & -1000 - \lambda \end{vmatrix} = 0 \tag{22}$$

$$(-3 \times 10^3 - \lambda)(-1 \times 10^3 - \lambda) - 1.0 \times 10^6 = 0$$

$$3 \times 10^6 + 3 \times 10^3\lambda + 10^3\lambda + \lambda^2 - 10^6 = 0$$

$$\lambda^2 + 4 \times 10^3\lambda + 2 \times 10^6 = 0 \tag{23}$$

$$\lambda_{1,2} = \frac{-4 \times 10^3 \pm \sqrt{16 \times 10^6 - 8 \times 10^6}}{2} = \frac{(-4000 \pm 2828)}{2}$$

$$\lambda_1 = -585.8 \qquad \text{and} \qquad \lambda_2 = 3414$$

yielding

$$\omega_1 = \sqrt{585.8} = 24.2 \text{ rad/s} \quad \text{and} \quad \omega_2 = 58.2 \text{ rad/s}$$

The *natural periods* are

$$T_1 = \frac{2\pi}{\omega_1} = 0.26 \text{ s} \qquad T_2 = 0.108 \text{ s}$$

The *natural frequencies* are

$$f_1 = \frac{24.2}{2\pi} = 3.85 \text{ cps} \qquad f_2 = \frac{58.2}{2\pi} = 9.26 \text{ cps}$$

$$\lambda_n = \omega^2 n \qquad \omega_n = \text{natural circular frequency} = 2\pi f \text{ rad/s}$$

$$f_n = \frac{\omega_n}{2\pi} = \text{natural frequency in cps or Hz}$$

$$T_n = \frac{1}{f} = \text{natural period}$$

The system just analyzed has two ω's, two f's, and two T's.

11.6 MODELING ACTUAL STRUCTURES

EXAMPLE

The moment-resisting frame in Figure 11.8 can also be represented by a *spring–mass–dashpot* (damping) one-degree system. A horizontal force, representing mass times acceleration caused by the earthquake, is applied at the top of the frame. Three basic assumptions make the problem real:

(a) The weight of columns and walls is negligible.
(b) The mass of the floor is concentrated along the beam.
(c) The girder is sufficiently rigid to prevent significant rotation.

The weight of the floor—the portion of the floor capable of producing seismic body forces—is given as

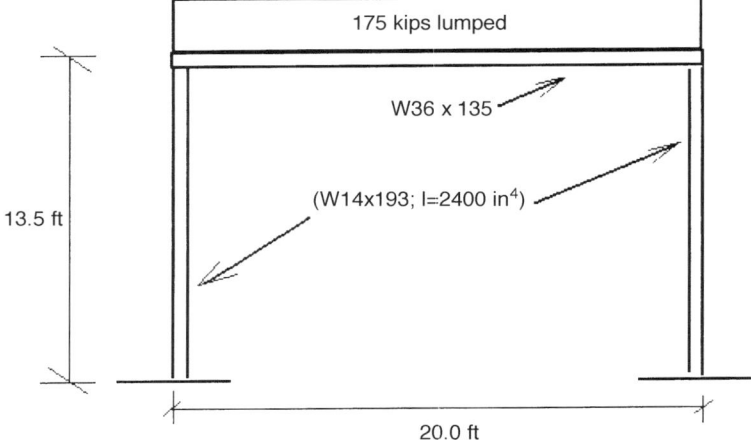

Figure 11.8   Dynamic model of steel moment frame, one-degree system.

$$W = 175{,}000 \text{ lb} \qquad m = \frac{175}{32.2} = 5.40 \text{ kip-s}^2/\text{ft}$$

The spring constant representing the combined stiffness of the two legs is the force that causes unit deflection at the top:

$$k = 12\, E\, \frac{2I}{n^3} = 12 \times 29 \times 10^6 \times 2 \times \frac{2400}{(13.5)^3 \times 144}$$

$$= 4.715 \times 10^6 \text{ lb/ft}$$

or 4715 kips/ft. Then the natural period of the structure is

$$T = 2\pi \sqrt{\frac{m}{k}} = 2\pi \sqrt{\frac{5.40}{4715}} = 0.2126 \text{ s}$$

and the natural frequency is

$$\frac{I}{T} = 4.70 \text{ cps or } 4.70 \text{ Hz}$$

This is vital information. Through his research in the Santa Monica area, the author established that *the ground in Santa Monica started vibrating horizontally with 4.0 Hz at the arrival of the s waves, 4.0 s after the start of the Northridge earthquake.*

This structure would have gone into resonance and broken up between 4 and 8 s during the earthquake.

11.7 THREE-DEGREE SYSTEMS

Problem 11.4   The diagram in Figure 11.9 shows a three-degree system. No damping has been assigned to the structural system. Develop the equation of motion without external driving function $F_{(t)}$.

Solution

$$m_1\ddot{y}_1 + k_1y_1 - k_2\,(y_2 - y_1) = 0 \qquad \text{(first equation)}$$

$$m_2\ddot{y}_2 + k_2\,(y_2 - y_1) - k_3\,(y_3 - y_2) = 0 \qquad \text{(second equation)}$$

$$m_3\ddot{y}_3 + k_3\,(y_3 - y_2) = 0 \qquad \text{(third equation)}$$

Problem 11.5   Determine (a) The eigenvalues and (b) the eigenvectors for the system shown in Problem 11.4.

Solution   To have a matrix representation and a matrix solution of the system, it is advisable to complete the following steps:

1. Isolate $\ddot{y}_1$, $\ddot{y}_2$ . . . , $\ddot{y}_n$ at each row by dividing by m_1, m_2, . . . , m_n each at the appropriate nth equation so that $\ddot{y}_1$, $\ddot{y}_2$, . . . , $\ddot{y}_n$ will remain on the right-hand side of the appropriate nth equation. The rest of the terms of such an equation will be rearranged.

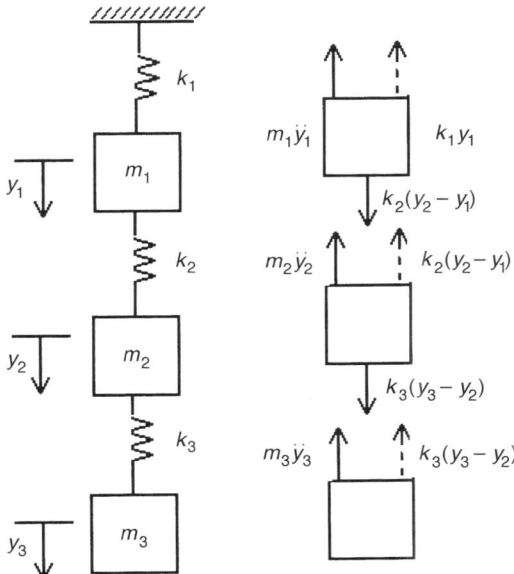

Figure 11.9   A three-degree system without damping.

2. Reorganize the system of equations in matrix form, such as

$$\begin{bmatrix} \ddot{y}_1 \\ \ddot{y}_2 \\ \ddot{y}_3 \end{bmatrix} = \underbrace{\begin{pmatrix} a_{11} & a_{12} & a_{13} \\ a_{21} & a_{22} & a_{23} \\ a_{31} & a_{32} & a_{33} \end{pmatrix}}_{\text{coefficient matrix}} \begin{bmatrix} y_1 \\ y_2 \\ y_3 \end{bmatrix}$$

The left side will consist of the $n \times n$ coefficient matrix $\mathbf{A}$ multiplied by the $n \times 1$ vector matrix of y_n.

3. Subtract from the coefficient matrix $\mathbf{A}_{n \times n}$ the identity matrix $\mathbf{I}$ multiplied by $\lambda_n = \omega_n^2$, $(\mathbf{A} - \lambda \mathbf{I})(y) = 0$, and set it to zero:

$$\begin{pmatrix} a_{11} & a_{12} & a_{13} \\ a_{21} & a_{22} & a_{23} \\ a_{31} & a_{32} & a_{33} \end{pmatrix} - \begin{pmatrix} \lambda & 0 & 0 \\ 0 & \lambda & 0 \\ 0 & 0 & \lambda \end{pmatrix} \begin{bmatrix} y_1 \\ y_2 \\ y_3 \end{bmatrix} = 0$$

4. Set the determinant of the resulting coefficient matrix to zero and solve for λ_n's:

$$\det \begin{vmatrix} a_{11} - \lambda & a_{12} & a_{13} \\ a_{21} & a_{22} - \lambda & a_{23} \\ a_{31} & a_{32} & a_{33} - \lambda \end{vmatrix} = 0$$

To fulfill the equation, the determinant must be zero.

Note. $\sqrt{\lambda} = \omega$ is the *natural circular frequency in radians*. The natural circular frequency is related to the *natural frequency f* by the expression

$$\omega = 2\pi f \quad \text{or} \quad f = \frac{\omega}{2\pi}$$

The natural period T is the *inverse* of the natural frequency:

$$T = \frac{I}{f}$$

An n-degree system has n natural period T_n, n natural frequencies f_n, and n natural circular frequencies ω_n.

Problem 11.6   Find (a) the eigenvalues and (b) the eigenvectors of the two-degree system

$$m_1 = m_2 = 1 \quad k_1 = 3 \quad k_2 = 2$$

Solution Set up the system equations:

$$\ddot{y}_1 = -5y_1 + 2y_2 \qquad \ddot{y}_2 = 2y_1 - 2y_2$$

$$\begin{bmatrix} \ddot{y}_1 \\ \ddot{y}_2 \end{bmatrix} = \overset{A}{\begin{bmatrix} -5 & 2 \\ 2 & -2 \end{bmatrix}} \begin{bmatrix} y_1 \\ y_2 \end{bmatrix} = \begin{bmatrix} \ddot{y}_1 \\ \ddot{y}_2 \end{bmatrix}$$

We know that vibration has the following solution:

$$y = xe^{\omega t} \qquad \ddot{y} = \omega^2 e^{\omega t}$$
$$= Ae^{\theta t}$$

where ω is the natural circular frequency. Letting $\lambda = \omega^2$ and dividing both sides by $e^{\omega t}$,

$$Ax = \lambda x$$

or, in matrix form $(A - \lambda I)x = 0$,

$$\underset{Ax}{\left[\begin{bmatrix} -5 & 2 \\ 2 & -2 \end{bmatrix}} - \begin{bmatrix} \lambda & 0 \\ 0 & \lambda \end{bmatrix}x\right] = \underset{(A - \lambda I)}{\begin{bmatrix} -5-\lambda & 2 \\ 2 & -2-\lambda \end{bmatrix}}\begin{bmatrix} x_1 \\ x_2 \end{bmatrix} = 0\underset{I}{\begin{bmatrix} 1 & 0 \\ 0 & 1 \end{bmatrix}} = \underset{\substack{\text{diagonal} \\ \text{matrix}}}{\begin{bmatrix} \lambda & 0 \\ 0 & \lambda \end{bmatrix}}$$

where $(A - \lambda I)x = 0$ is the *identity matrix* that must be used to carry out matrix subtraction.

$$\begin{vmatrix} -5-\lambda & 2 \\ 2 & -2-\lambda \end{vmatrix} = 0$$

$$\lambda^2 + 7\lambda + 6 = 0 \qquad (\lambda + 1)(\lambda + 6) = 0$$

This results in the two roots

$$\lambda_1 = -1 \qquad \lambda_2 = -6$$

These are eigenvalues* and we claim they represent the *natural* frequencies of the system. Plugging in the λ values into the equation above yields

---

Eigen means "individual property"; that is, it cannot be separated from a particular system, such as the natural frequency and the time-dependent mode of vibration of the *deflected shape*. (See Figures 11.10–11.12.)

$$\lambda_1: \begin{bmatrix} -5 - (-1) & 2 \\ 2 & -2 - (-1) \end{bmatrix} \begin{bmatrix} x_1 \\ x_2 \end{bmatrix} = 0$$

We obtain

$$-4 + 2x_2 = 0 \qquad \begin{matrix} x_2 = 2 \\ x_1 = 1 \end{matrix} \times \begin{bmatrix} 1 \\ 2 \end{bmatrix}$$

Similarly, with λ_2

$$x_1 = 2 \qquad x_2 = -1$$

$$x^{\lambda_1} = \begin{bmatrix} 1 \\ 2 \end{bmatrix} \qquad x^{\lambda_2} = \begin{bmatrix} 2 \\ -1 \end{bmatrix}$$

The same will happen if you imagine a pole swaying laterally (Figure 11.10):

Eigenvectors: Deflected Mode Shapes   First mode, deflected shape (Figure 11.11)
Second mode, deflected shape (Figure 11.12)

Problem 11.7   We will attempt to solve a three-degree system for its three eigenvalues and three sets of eigenvectors. This will lead us to the natural frequencies and mode shapes of vibration. A brief review of the steps to solve it for the determinant of a 3×3 matrix is in order. Given

$\mathbf{A}_{3 \times 3}$ coefficient matrix

$$\begin{pmatrix} a_{11} & a_{12} & a_{13} \\ a_{21} & a_{22} & a_{23} \\ a_{31} & a_{32} & a_{33} \end{pmatrix}$$

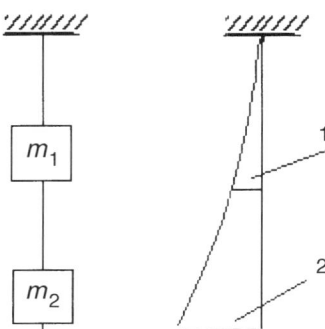

Figure 11.10

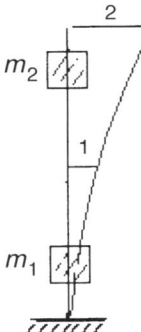

Figure 11.11

A

Calculate the determinant of the **[A]** coefficient matrix.

Solution Remember that the system we used for the 2 × 2 matrix can be extended to establish the determinant of a 3 × 3 matrix:

$$\det\begin{pmatrix} a_{11} & a_{12} & a_{13} \\ a_{21} & a_{22} & a_{23} \\ a_{31} & a_{32} & a_{33} \end{pmatrix} = \begin{pmatrix} a_{11} & a_{12} & a_{13} & a_{11} & a_{12} \\ a_{21} & a_{22} & a_{23} & a_{21} & a_{22} \\ a_{31} & a_{32} & a_{33} & a_{31} & a_{32} \end{pmatrix}$$

that is,

$$\begin{pmatrix} a_{11} & a_{12} & a_{13} & a_{11} & a_{12} \\ a_{21} & a_{22} & a_{23} & a_{21} & a_{22} \\ a_{31} & a_{32} & a_{33} & a_{31} & a_{32} \end{pmatrix}$$

The steps are as follows:

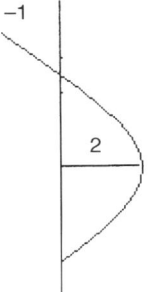

Figure 11.12

(a) Add two more columns by adding columns 1 and 2 to the original matrix.
(b) Cross-multiply the elements along the diagonal, left to right, without changing sign.
(c) Cross-multiply the elements along the right-to-left diagonals and change the sign for each cross-product along the line.

By adding up

$$(a_{11} \quad a_{22} \quad a_{33}) + (a_{12} \quad a_{23} \quad a_{31}) + (a_{13} \quad a_{21} \quad a_{32}) - (a_{13} \quad a_{22} \quad a_{31})$$

$$- (a_{11} \quad a_{23} \quad a_{32}) - (a_{12} \quad a_{21} \quad a_{33}) = \det[\mathbf{A}]$$

11.8 EXISTENCE AND UNIQUENESS THEORY: WRONSKIAN

The eigenvectors must be linearly independent. It is advisable to group the eigenvalues column by column, in one block of the coefficient matrix. Thus, if $X_1, X_2, \ldots, X_n$ are solution eigenvectors of a homogeneous system, then the vectors are linearly independent if (and only if) the Wronskian determinant is *unequal to zero* to guarantee a unique solution:

$$W(X_1, X_2, \ldots, X_n) = |X_1, X_2 \ldots, X_n| \neq 0$$

Should the Wronskian* determinant be zero, it would be an indication that the original set of differential equations is not independent. For instance, one row of the coefficient matrix must be dependent on another row.

Problem 11.8   Prove that the eigenvectors of the two-degree system in Problem 11.6 are linearly independent.

Solution

$$W(X_1, X_2) = \begin{vmatrix} 1 & 2 \\ 2 & -1 \end{vmatrix} = -1 - (2 \times 2) = -5 \neq 0$$

Therefore the eigenvectors are linearly independent.

11.9 DRIVING FUNCTION (F_t): SEISMIC GROUND MOTION AS F_t

So far we have concentrated on differential equations and described mechanical vibrations without the driving function $F(t)$. The tools for those equations

---

*Wronski (1778–1853) was a Polish-born mathematician who changed his name from I. M. Höne to Wronski.

proved powerful to determine the characteristics of a structure, natural fre-quency, and mode shape of vibrations. Can we apply the same system to the uneven nature of ground motion? Ground motion involves displacement and acceleration reflected in the jagged trace of graphical representations with sharp peaks and sudden reversals that defy the closed-form differential equa-tions.

Will the differential equations that require the driving functions of a well-defined, smooth curve of a mathematical formulation be applicable to the structural deformations, stresses, and strain on the structure at any given time during an earthquake? The reader will realize that such an approach will not be feasible. In most cases it would be virtually impossible to reveal the re-lationship between the ground shaking and the structural response that would give a picture of deformations, stresses, and strains at any given time during the earthquake.

To answer these questions, the author explored a new avenue. The work was verified by field measurements of structural damage caused by the 1994 Northridge/Santa Monica earthquake. Proceeding carefully, first with hand calculations, then with digital computer analysis, he applied the lumped-impulse method of numerical integration to tackle the noncontinuous, jagged function of earthquake-induced ground motion as driving function input data.

The procedure was applied to an existing multistory steel frame damaged in the Northridge/Santa Monica seismic event of 1994. *What emerged from the procedure was a time-dependent data of structural response, lateral de-formations, and floor accelerations that offered an insight of what actually happened to the just built steel moment frame structure during the earthquake. The method is described in Chapter 12.*

CHAPTER 12

METHODS AND TOOLS TO UNRAVEL
SECRETS OF EARTHQUAKES

12.1 ELEMENTS OF AN EARTHQUAKE

After a seismic event it is a challenge to find out what happened during the earthquake. The author's research on steel frame buildings subjected to earthquake forces confirmed the existence of extremely high frequency vibrations of 2–3 Hz coupled with exceptionally large reversals of lateral floor displacements exceeding 9 in.

Perhaps one of the first steps is to follow the path of the earthquake while its traces are still undisturbed by human interference. The author visited areas in the vicinity of the epicenter of the Northridge earthquake and noted power poles snapped off like match sticks (Figure 12.1); concrete curbs cracked, broken, or shifted to a different location; chimneys and fences collapsed; buildings ripped in half; and newly built portions of stone floors split by deep, widening cracks.

Combining visual observation with organized engineering measurements and records can give us an insight into the physics and mechanics involved in the structural response, movement, and resulting damage during an earthquake. Answering questions such as what happened, how did the ground behave, how did the structure respond during those crucial seconds or milliseconds takes facts, instrumentation, knowledge of physics, soil and structural dynamics, structural analysis, knowledge of the building materials used, and mathematics.

At the time of the 1971 San Fernando earthquake the author was doing research on earthquakes, studying virtually all available information related to earthquake engineering, including soil dynamics, structural dynamics, properties of soil, and brittle fracture of metals. The most significant information

336

Figure 12.1 Overall damage on Balboa Boulevard near the epicenter of the North-ridge earthquake.

that came through the university civil engineering department was that the main building of the Olive View Hospital in the San Fernando Valley and its separated elevator tower moved several feet.

Another interesting piece of information was that the accelerogram readings at the Pacoima Dam reached 100% g in the lateral direction and nearly 100% g vertical. All of these indicated that the UBC design provisions of about 13% g lateral and 0% g vertical were woefully inadequate. When the author brought up these facts at a SEAOC meeting at the Los Angeles Convention Center, the reaction was that upgrading the design forces would increase construction costs. No further consideration was given to the measured large accelerations and significant vertical component and no code upgrading took place in the following period of quieter seismic activity. It took 23 years and a destructive earthquake such as Northridge for the vertical-component design to be incorporated into UBC 1997.

In the case of the San Fernando Valley Olive View Hospital severely damaged by the 1971 earthquake the consensus until then had been that towers having different dynamic response, natural period, and frequency must be separated from each other. Such design preference was influenced by SEAOC. However, such separation, especially if it involves a tall, slender structure adjacent to another building, can be of disastrous consequences. Because of the relatively strong swaying movement the Olive View Hospital slender el-

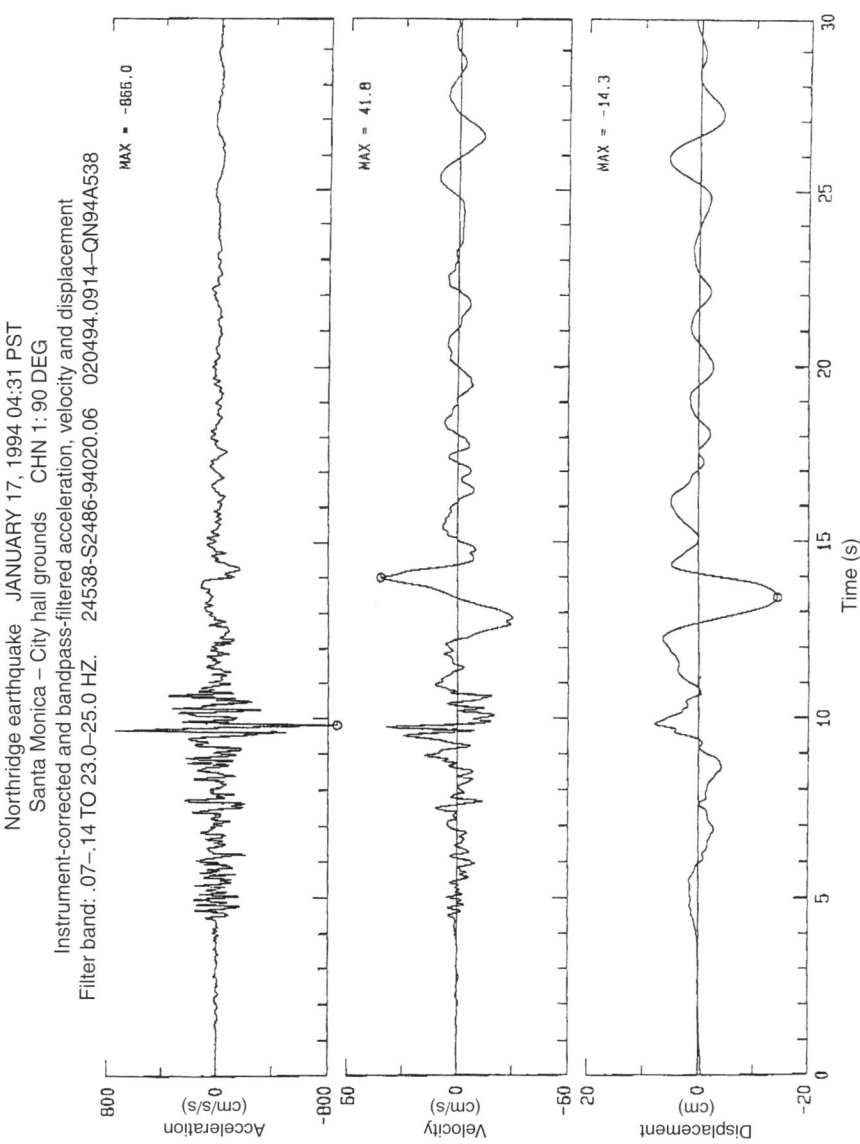

Figure 12.2 Northridge earthquake. Seismograph readings at the Santa Monica City Hall showing the east–west lateral acceleration of the ground.

evator tower knocked down part of the main building. The 2–3-in. *seismic separation joint* considered essential for a *good* design did not work out, mainly because the elevator tower started rocking on its foundation under seismic excitation and acted as a sledgehammer on the adjacent building. It would have been better to build the two structures together rather than creating a seismic separation joint.

The Northridge earthquake confirmed the author's prediction: *Both horizontal and vertical ground acceleration components* recorded at the Cedar Hills Nursery in Tarzana, California, were close to 200% *g*. Similar high-acceleration values were recorded at several other locations in downtown Los Angeles and adjacent areas, including the University of Southern California Medical Center.

12.2 VERTICAL-ACCELERATION COMPONENT

The extremely high vertical acceleration of the Northridge earthquake surprised even those who were studying earthquakes. The impact of vertical acceleration on structures was equally surprising. The roof structure, composed of sheet metal and lightweight concrete, of the Great Western bank building in Northridge caved in due to the bouncing vibrations of air conditioning and other equipment, ending up on the floor below. Although banned in Europe, this roof construction, known as Robinson roof decking, had been considered reliable until the Northridge event.

As stated earlier, accurate electronic data of site-specific seismograph readings are essential to retrace the response history of a structure during an earthquake. The next step is what to do with the data. From a geological point of view and as result of the extensive California statewide network, the data have been recorded and are available to the public. From a civil/structural engineering point of view, the work is just about to begin. The first question is how to obtain actual time-dependent deformations, internal forces, and stresses from the raw geological data. Some experts on the subject would say *"no problem, use spectral analysis."* Yet spectral analysis is based on a *single-degree, single-mass, single-dashpot system,* while the structures to be analyzed are complex and consist of multiple-mass, multiple-spring and multiple-damping system.

12.3 NEW METHOD OF DYNAMIC ANALYSIS

Following the 1994 Northridge earthquake and upon the building owner's request, the author investigated the damage suffered by a four-story steel moment frame just completed in the City of Santa Monica that had been designed and detailed by the UBC seismic regulations in force (1991). The damage was severe and extensive. All beam-to-column moment joints were damaged, beam flange-to-column flange welds cracked, column flanges were torn from their web, and, more significantly, some column webs were torn, the tearing going through the column flanges.

The extensive damage was clear indication that the actual internal forces and stresses caused by the earthquake were at least *one magnitude larger than the UBC-recommended design values.* What do we mean by *one magnitude larger?* Let's take, for instance, a value of 20 obtained by rational analysis. If the actual stresses were 200 instead of 20, the design assumption would have been wrong by *one whole magnitude larger than assumed* as compared to the actual forces. The implications are serious for a practicing structural engineer. A small percent deviation, say 10%, from the actual forces acting on a structure normally would not be catastrophic; yet when the earthquake forces are *one magnitude larger* than the assumed design forces, the result might be the collapse of the structure.

Anyone who wants to determine the time-dependent deformation of a structure such as the moment frame that occurs at some specific time during an earthquake will be faced with the dilemma of how to do the analysis. A closed-form analysis can only be possible if the graph of the driving function (ground acceleration, displacement, etc.) represents a well-behaved function which is smooth with continuous curves and no breaks of discontinuity or abrupt changes. From a mathematical point of view, the trace of an earthquake is far from a well-behaved function. It has sharp points that represent sudden, abrupt changes in ground acceleration and displacement. We can thus conclude that a closed-form differential equation will not yield accurate results.

12.4 BACKGROUND OF RESEARCH

Following a thorough review of the performance and mechanism of virtually all compression testing machines built and used since the turn of the twentieth century, it became evident to the writer that conventional testing machines do not allow free lateral expansion or contraction—triaxial test—of the specimen at the machine–platen–specimen interface.[1,2]

It was not difficult to imagine the lateral restraint that would occur if a steel specimen were welded to the machine platen. *Yet exactly the same phenomenon occurs at the beam–column joint interface of a welded moment frame.* The flanges of the beam are solidly welded to massive column flanges, which act like the machine platen. The laterally and almost infinitely stiff column prevents free lateral contraction of the beam, which is vital to initiate yielding. Contrary to expectations, no yielding would take place at the highest stress gradient, the beam–column interface.[3]

The author overcame the problem by numerical integration using the lumped-impulse method (Figure 12.3). He succeeded in writing an algorithm capable of accepting the ground acceleration–displacement input as the driv-

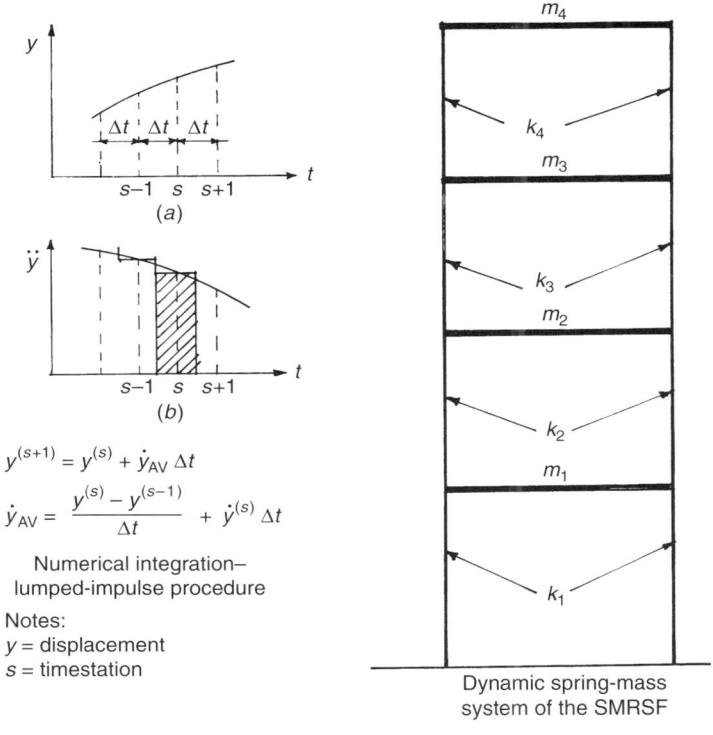

$$y^{(s+1)} = y^{(s)} + \dot{y}_{AV}\,\Delta t$$

$$\dot{y}_{AV} = \frac{y^{(s)} - y^{(s-1)}}{\Delta t} + \ddot{y}^{(s)}\,\Delta t$$

Numerical integration–
lumped-impulse procedure

Notes:
y = displacement
s = timestation

Dynamic spring-mass
system of the SMRSF

Notes:
k = spring constant of given story
m = floor mass

Figure 12.3 Numerical integration of the lumped-impulse procedure.

ing function created by the impact of the unpredictable, violent, randomly changing earthquake. The system keeps track of each earthquake-generated impulse received by the foundation at each fraction of time.

The peak lateral east–west ground acceleration exceeded 0.9 g and the vertical acceleration 0.41 g as measured by a seismograph located close to the actual four-story project analyzed further in this chapter. When combined these values give us over a 100% resulting ground acceleration associated with a magnitude larger than the UBC 1994 design force. There is no doubt that such large disparity between the code-prescribed seismic and actual driving force must have caused the extensive damage to the beam-to-column joints verified by the author's analysis.

Another major factor, not reflected in a purely static analysis, was the sensitivity of a framed structure to undergo self-inflicted oscillations that created a large internal reaction of forces and stresses.[3]

12.5 ANALYSIS OF ACTUAL STRUCTURE

The subject structure analyzed was a four-story SMRSF in the City of Santa Monica damaged by the 6.7-Richter-magnitude Northridge earthquake of January 1994. Figure 12.4 shows the layout of the frames. Three main intriguing aspects motivated the research:

1. The damaged structure had been designed according to the latest seismic provisions of the leading and—from a seismic design point of view—most effective building code (UBC 1991) in force at the time of the earthquake.
2. Although structurally complete when the earthquake hit, a significant portion of the mass of the building was still missing, including architectural features such as a heavy stone-clad facade and partitions. It was most surprising that, having perhaps the best chance to escape the event unscathed, the structure was severely damaged by the earthquake.
3. The type of structural damage was brittle fracture.

The structural configuration consists of a four-story main *tower* and a three-story lower wing attached to the tower by a passageway. The moment-

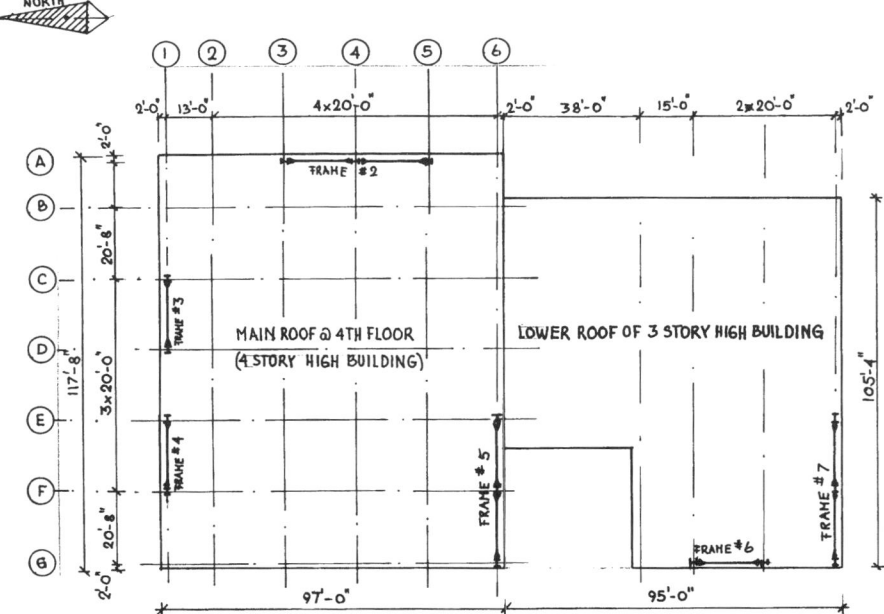

Figure 12.4 Layout of SMRSF frames. Frames carrying gravity loads only not shown for simplicity.

resisting steel structure sits atop cast-in-place walls of a two-story concrete basement that constitutes a stiff box below ground level. The analysis concentrated on the four-story tower portion of the building. Masses of tributary floor areas were used in the model to generate earthquake-related inertial forces at floor level. The results of the analysis of frame 3 are discussed here (Figure 12.5). The grade of steel was A36 with $F_y = 36$ ksi and $F_u = 58$–80 ksi.

The four-story tower and the three-story wing were considered to act independently in this analysis. It was assumed that the tower would resist its own generated body seismic forces and the wing would resist its own seismic forces without transmitting any substantial effect to the tower. No attempt was made to account for torsion that might have been caused by the nonsymmetrical structural arrangement of the three-story wing or for additional sway components due to the rocking motion of the basement. No doubt the torsion and rocking of the substructure or basement would have increased lateral deformations and stresses in the frame, causing further structural impairment; however, the purpose of the analysis was to determine a lower bound for

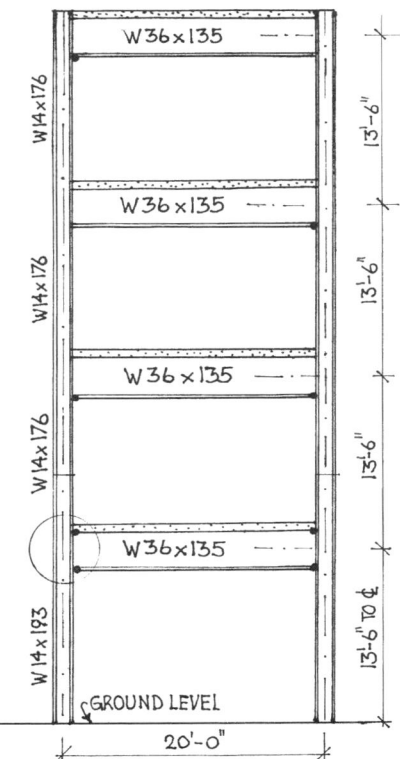

Figure 12.5 Frame 3, the subject of the analysis. (●) Locations of brittle fracture.

deformations, forces, and stresses. If these were found critical without added adverse effects, the earthquake-induced stresses, contrary to the 1991 or 1994 UBC lateral coefficient method predictions, would be destructive enough to cause permanent structural damage.[4]

12.6 RESULTS AND FINDINGS

Figure 12.6 shows the dynamic structural response of first and second floors between 8 and 10 s of seismic impact from the seismograph-recorded ground acceleration. The trace of the seismograph indicates a sudden peak of 0.9 *g* east–west ground acceleration at 9.8 s. The north–south component of the lateral acceleration, also significant, reached a peak of 0.414 *g* at the same time. The two vector components produced an acceleration vector of 0.98 *g*, something to bear in mind when designing buildings in seismic zones.

The ground acceleration data used directly as input were obtained from the January 17, 1994, readings of the Strong Motion Seismograph at the Santa Monica City Hall in the vicinity of the subject structure. The information was released by the California Department of Conservation, Division of Mines and Geology.

There was a marked dynamic structural response to ground motion even before the peak was reached. The displacements are measured in feet from the unstressed, undeformed positions of the structure before the earthquake. The results of the computer analysis indicated a dramatic buildup of lateral deformations between 4 and 14 s—the *effective time period*—of the quake, when most of the significant structural deformations took place.

A sudden increase in structural displacement can be noted in the graph between 9 and 10 s, when the structure went on a wild ride. At least 16 first-floor lateral displacements $(\Delta_1 - \Delta_0)$, $(\Delta_2 - \Delta_1)$ of 6–7 in. and six peak displacements of 8–10 in. occurred during this effective time period. These displacements correspond to 12–18 times the allowable *story drift* specified by most codes based on a *static* lateral coefficient method. All other floor displacements were equally high, all corresponding to significant column deformations and internal stresses affecting the beam-to-column joint.

According to the author's research, the yield strength of the base metal of the structure was exceeded several times during the earthquake, sending it straight into strain hardening. Contrary to expected ductile behavior, almost all joints fractured in the actual structure showing no significant sign of beam yielding or *plastic hinge* formation. What made the MRSF take an alternate course for structural response, choosing a brittle fracture mechanism instead of plastic hinge forming, is analyzed here.

12.7 NATURE AND CAUSES OF JOINT FAILURE

As a *moment-resisting frame* the steel moment frame resists seismic forces by bending—nature's way of countering lateral forces by large deflections.

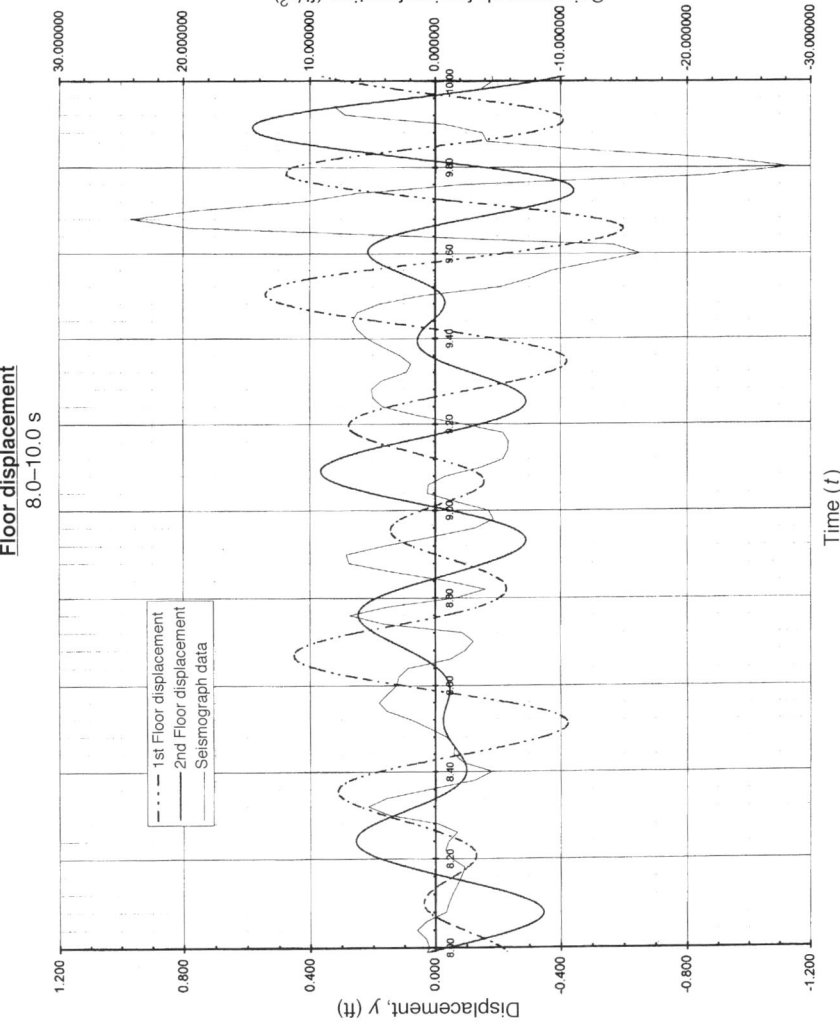

Figure 12.6 Dynamic structural response of first and second floors between 8 and 10 s of seismic impact.

Significant inertial forces generated by the dynamic response of the relatively large mass of the concrete floors, architectural cladding, and other features bend the structure back and forward with increasing amplitude.

When all the flexural strains of the infinitesimal Δ_x long elements are *integrated* during the excessive sway caused by the earthquake, they add up to considerable structural deformations and large stresses in localized areas. Such critical stress concentrations occur at the beam–column joint where large beam flange reactions are evoked by the internal moments caused by H_n and H_{n-1} story shears. Because of their magnitude, there is every sign that these stresses cannot be managed. The moment-resisting frame is confronted with the virtually impossible task of resisting, by bending, exceptionally large lateral forces.[4]

Originating in Europe and used on the East Coast of the United States for essentially static conditions in regions where earthquakes are virtually unknown, is the steel moment frame suited to resist drift produced by earthquake-generated dynamic forces? Paradoxically, in Europe this type of

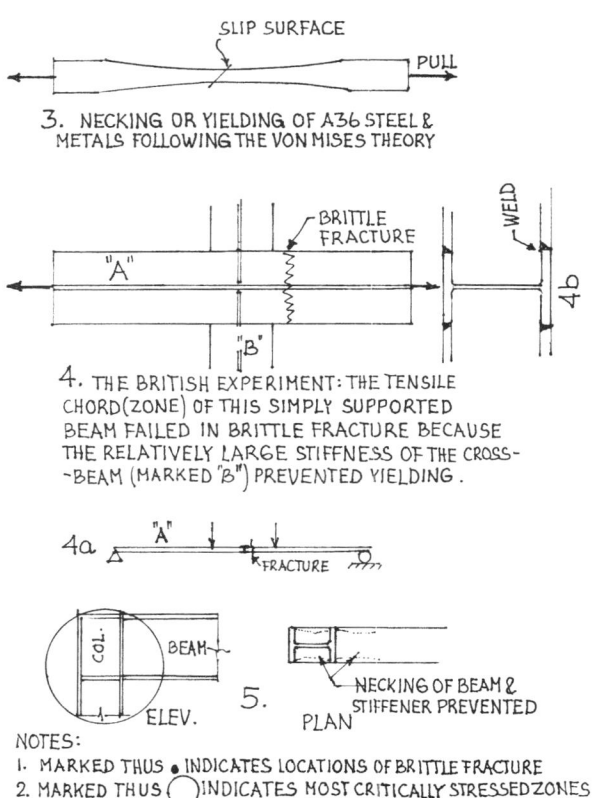

Figure 12.7 The British experiment.

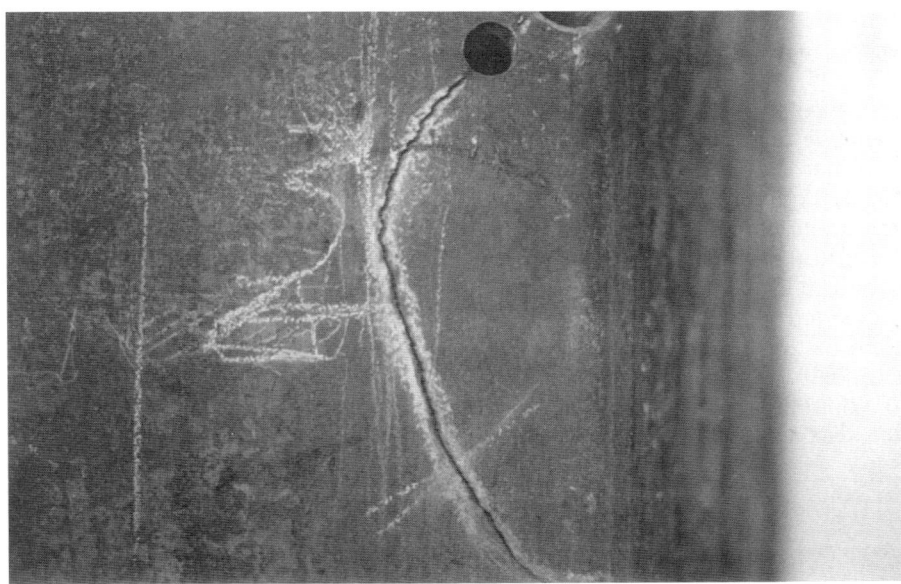

Figure 12.8   *Above:* Column web fracture within the connection panel zone in the subject structure during the Northridge earthquake. *Below:* The author during inspection of one of the first retrofittings of cracked welded steel moment-frame structure connections following the 1994 Northridge earthquake.

frame is known as a *rigid frame* because the beam–column joint, unlike other frame members such as beams and columns, does not deform appreciably. *The fact is that the joint is rigid, not ductile.* It performs well under *static* loads but exhibits inherent weakness under reversal of stresses caused by *dynamic* response to earthquakes.

What happens to the most critically stressed segment of the beam welded to the column? Contrary to prevalent engineering predictions, the moment-resisting frame beam does not yield before the frame columns are damaged by brittle fracture. The answer to the beam–column joint behavior may be found in a series of British tests[5] done in the 1970s on wide-flange, simply supported steel beams loaded to the *limit state,* the onset of a permanent deformation (Figure 12.7).

Identical beams were fabricated and divided into two sets: In set 1, where the bottom tensile flange of the load-carrying beam A was not welded to the tensile flange of cross-beam B, the beam yielded under the predicted load. In set 2, where the bottom tensile flanges of beams A and B were solidly welded together, the loaded beam A did not yield but failed instead in brittle fracture.

Why is welding of transverse structural components to tensile flanges so critical to the load-carrying characteristics of the member? The phenomenon of *necking* is essential to promote yielding of metal components subjected to tension. If necking takes place, a ductile behavior occurs under *static* tensile loading condition, satisfying the von Mises yield criterion. Conversely, if necking is prevented, brittle fracture occurs and all benefits of ductile structural behavior are frustrated.

During the British experiment the bottom flange of the cross beam, welded to the bottom flange of beam A, prevented lateral contraction due to the Poisson's ratio effect of the tensile zone of the load-carrying beam. According to the author's concept and findings, an identical phenomenon occurs at the beam–column joint of a MRSF. Massive W14, 1.5 column flanges of a full story-high length of virtually infinite lateral resistance prevent lateral contraction, necking, and yielding of the beam flanges and formation of a *ductile hinge,* leaving no alternative but brittle fracture of the overstressed joint. Figure 12.8 shows a column web fracture in the building analyzed.

REFERENCES

1. Erdey, C. K. 1979. "A New Load-Transmitting Medium to Measure Strength of Brittle Materials." *ASTM Journal of Testing and Evaluation,* Vol. 7, No. 6, pp. 317–325.
2. Erdey, C. K. 1980. "Finite Element Analysis and Tests with a New Load-Transmitting Medium to Measure Compression Strength of Brittle Materials." *Materials and Structures, RILEM,* Vol. 13, No. 74, pp. 83–90.
3. Erdey, C. K. 1997. "Earthquake-Caused Brittle Fracture in Steel Framed Buildings. *Structural Engineering Forum (SEF) journal,* Vol. 3, Issue 1, pp. 10–15.
4. Erdey, C. K. 1999. "Performance of Steel Moment Frames in an Earthquake." Paper presented at Eurosteel '99, Second European Conference on Steel Structures, Prague, May 26–28.
5. BCSA/CONSTRADO. 1978. Battersea College Seminar, England.

CHAPTER 13

RECENT AND FUTURE DEVELOPMENTS IN SEISMIC DESIGN

13.1 TESTS ON JOINTS

The University of California at San Diego,[1] University of Texas at Austin,[2] and others have conducted a limited number of quasi-static tests on joints. Two types of test methodologies were used:

1. Slowly and gradually applying static cyclic loading tests in compliance with the protocol of the SAC (Structural Engineers Association of California, Applied Technology Council, California Universities for Research in Earthquake Engineering) with a cycle period of approximately $T = 4$ min (approximate frequency $f = 0.00417$ Hz)
2. More accelerated cycle periods, $T = 1$ to 1.25 s (frequencies of 0.80–1.00 Hz)

None of the tests, however, are true representations of the high-frequency response vibration caused by an earthquake (2–3 Hz) detected by the author. Despite the mismatch between actual and laboratory simulated frequencies, the tests clearly indicate an early breakdown of the specimen under a relatively small joint rotation (2%) and story drift (3%). The author's analytical work shows that the joint rotation and story drift of the subject structure during the Northridge earthquake were at least 4% and 6%, respectively—*twice as critical as the laboratory tests*. Those progress reports are already an indication that the joint of the conventional frame would not yield at the beam–column interface, not even under static loading, let alone dynamic loading.

13.2 DOGBONE EXPERIMENT

Recent efforts indicate a trend to create a plastic response zone away from the troubled connection but sufficiently close to the column to avoid considerable drop in moment gradient by the use of a reduced beam section—a *dogbone*.[3] However, because of the high-frequency speed straining effect of the earthquake, the author has firmly concluded that the reduced dogbone section of the beam would not yield during an earthquake but rather would undergo rapid strain hardening that would make the most strained panel zone as brittle as glass.

13.3 JOINT STRAIN HARDENING: SPEED STRAINING

The author's evaluation of the results of the dynamic analysis reveals that an ultrahigh frequency vibration was imparted to the structure by the rapid high-frequency energy release of the ground during the earthquake. On average, 3 cps (3 Hz) lateral floor displacements occurred with associated large story drifts: $\Delta_n - \Delta_{d-1}$ at times exceeding 10 in. The lateral floor displacements were reversed in rapid succession of about 6 times/s, causing the columns to bend back and forth equally fast, that is, 6 times/s.

As a result of the high-frequency structural vibration added to the lateral deformations, large internal flange reaction forces are delivered to the column joint in a rapid succession of blows. The large reaction forces, as they reverse from tension to compression alternating at about 6 times/s, give rise to critically high alternating joint stresses of a magnitude fluctuating between F_y and F_u: $F_y < f_a \rightarrow F_u$.

During a period of 6–16 s more than 70 critical stress reversals occurred at the joint interface of the structure analyzed. On about 36 occasions the actual stress at beam flange–column interface exceeded the nominal yield strength F by 40%. On 30 occasions the intensity of actual stresses exceeded $1.5 \times F_y$ (54 ksi) and occasionally approached 80 ksi, the maximum recorded nominal strength of the base metal. On six occasions the actual induced stress would have exceeded $F_u = 80$ ksi by 30–40% had the structure not fractured at this point.

13.4 MECHANISM OF JOINT DEGRADATION

The extremely large internal forces carried by the beam flanges in the absence of yielding will destroy components of the assembly of the connected area— the weakest will fail in their path: The base metal of the column flange could be gouged (flange torn), torn by lamellar separation, or sheared off horizontally; the beam flange torn in tension; and the weld metal connecting beam-to-column fracture or the column web torn in brittle shear failure. Such failures, observed by the author, are also verified in a number of publications.[4]

How does degradation of the material occur? In the author's opinion, due to the high-frequency pulsating internal reactions and associated high stresses, the hysteresis loop of the base material and weld in the path of large stresses is flattened at the initial stages of the effective period (4–6 s in the analyzed case), leaving a strain-hardened material with virtually no fracture toughness. The large hysteresis loops reported in quasi-static tests, giving the impression of considerable energy dissipation, do not exist in an actual earthquake and thus they are misleading.

The author believes that the brittleness observed after an earthquake is not caused by the weld but is a consequence of the high-frequency speed straining generated by the earthquake. At such a high frequency of structural response, vibration speed straining occurs, resulting in a strain-hardened base and weld metal with properties entirely different from those not subjected to vibration. This explains why the earthquake-damaged A36 steel in the subject structure became brittle and unweldable. In summary, the joint of the steel moment frame will fail on two counts when subjected to simulated or actual seismic forces:

No necking, no ductility. Failure will occur even under static or quasi-static loading[5]: $f \leqq 1$ Hz if the actual stress f_a exceeds F_y and approaches F_u, that is, $F_y < f_a \rightarrow F_u$.

Speed straining accelerates the degradation and breakdown process. Speed straining coupled with high-frequency stress reversals of magnitude—the induced stress larger than the yield—changes the crystalline structure of the steel and causes extreme brittleness that affects not just localized areas but a considerably large zone around the entire panel joint.

13.5 CONCLUSIONS

We identified three basic problems with this type of structure in an earthquake:

(a) The rigid frame joint is incapable of undergoing plastic deformation and fractures.

(b) The parent metal of the joint region turns brittle by rapid strain hardening due to speed straining.

(c) Excessive story drift is an inherent feature of the sway frame or moment-resisting frame. Driven by its large floor mass the sway frame, unlike braced-frame or shear wall structures, responds to dynamic excitation by large lateral displacements of increasing magnitude. Large lateral movements of a structure during an earthquake are a hazard to the occupants as proven by the alarming internal damage to the 13-story steel-framed Santa Clara Civic Center Office Building caused by the Loma Prieta earthquake mentioned earlier in this book.

The undesirable vibration characteristic of the steel moment frame is aggravated by insufficient inherent damping: absence of built-in partitions on large open-floor areas and energy-dissipating external walls replaced by curtain walls, a trend increasingly popular for reasons of planning, economy, and efforts to reduce mass.

Due to basic conceptual design flaws, the conventional moment-resisting steel frame and its California variation SMRF (special moment-resisting frame) are prone to earthquake damage. Construction of this structure in California was temporarily suspended after the Northridge earthquake. It would be a mistake if the structure would be reinstated without attempting well-engineered improvement of the conventional design.

13.6 NEW TRENDS

New avenues should be explored, such as replacing the welded steel moment frame by a braced steel frame enhanced by added engineered damping with calibrated hydraulic piston dampers. A question is inevitable: How can we improve the performance of conventional MRFs and SMRFs during an earthquake? In the author's opinion, a vital principle is to control movement—lateral displacement—of the structure. *Seismic isolation* and *engineered damping* seem a good choice to achieve that goal.

There is a fairly extensive literature on engineered systems and projects accomplished. Robert D. Hanson[6] did a great deal of research on supplemental damping for improved seismic performance. He focused on *viscous damping, viscoelastic damping devices,* and *friction* and *yielding metallic damping* devices.

Examples of friction devices in large structures are the Canadian Space Agency and the Casino de Montréal in Montreal, Canada; the Ecole Polyvalente in Sorel, Canada; the Gorgas Hospital in Panama; and the UC Davis Water Tower in California. The system of metallic yield devices aims at providing supplemental damping/energy dissipation by movement and out-of-plane bending of series of engineered metal plates at strategic locations of the structure where movements are anticipated due to structural dynamic response to earthquakes. Examples of buildings equipped with this system are the Cardiology Building and the Social Security Building, both in Mexico City, Mexico, and the Wells Fargo Bank in San Francisco, California. Buildings provided with *viscous devices* are, among others, the Civic Center Building in San Francisco and the San Francisco Opera House in California; the

Hayward City Hall, Hayward, California; the Los Angeles City Hall, California; the San Bernardino County Medical Center, California; and the Light Towers, Rich Stadium in Buffalo, New York, equipped with viscous dampers (Taylor Devices). The *viscoelastic devices* aim at providing supplemental damping/energy dissipation, for instance, viscoelastic material such as neoprene sandwiched between steel plates that are forced to move with respect to one another by the dynamic structural response of the building during an earthquake. Examples of buildings equipped with this system are the Navy Supply Center, San Diego, California; the San Mateo County Hall of Justice, Redwood City, California; the Santa Clara County Building, Santa Clara, California; and the School Building in Phoenix, Arizona.

A number of practicing engineers and researchers believe that, by combining the benefits of supplemental damping/energy dissipation systems and well-developed systems such as the SCBF, significantly enhanced structural performance can be expected during earthquakes.

13.7 SEISMIC ISOLATION

As stated in reference 7 (p. 1):

> A seismic isolation system may be defined as a flexible or sliding interface positioned between a structure and its foundation, for the purpose of decoupling the horizontal motions of the ground from the horizontal motions of the structure, thereby reducing earthquake damage to the structure and its contents.

An in-depth description of the systems and design practices for seismic isolation systems is beyond the scope of this text. Only a brief summary is presented here.

In 1999 the AASHTO (American Association of State Highway and Transportation Officials) published a "Guide Specification for Seismic Isolation Design." The IBC provisions address the subject of seismically isolated structures in Chapter 16, "Structural Design." The 2000 IBC dedicates several pages to the topic, treating it in extensive detail in Section 1623, "Seismically Isolated Structures." The 2003 IBC mentions the subject in a few lines in Section 1623, "Seismically Isolated Structures," adopting the requirements of Section 9.13 of the ASCE 7:

> **1623.1 Design Requirements**. Every seismically isolated structure and every portion thereof shall be designed and constructed in accordance with the requirements of Section 9.13 of ASCE 7, except as modified in Section 1623.1.1 that refers to fire resistance requirements.

As discussed in Chapter 2 of this book, IBC 2006 has reduced significantly the sections on "Structural Design," with the last section 1613.6.2, referring briefly to "additional seismic-force-resisting systems for seismically isolated structures." Again, the requirements are adopted from ASCE 7 with an exception at the end of Section 17.5.4.2 of ASCE 7. The subject is discussed in further detail in Chapter 2, Section 2.4, IBC 2006.

Researchers on the subject of seismic isolated buildings state that proper application of this technology leads to better performing structures that will remain essentially elastic during large earthquakes.[8] There are about 1000 seismically isolated structures around the world. The number includes not only buildings but also bridges and tanks. Of these approximately 150 are in the United States. Literature on seismic isolation repeatedly quotes the University of Southern California Hospital in Los Angeles as a significant example of a seismically isolated, seven-story-plus-basement structure that survived the Northridge earthquake and remained operational. The technology of seismic isolation has recently made remarkable advancements since the concept was first put into practice with two buildings constructed on rollers: one in Mexico, the other in Sevastopol, Ukraine.

The first seismically isolated building with a rubber isolation system emerged in 1969 in Skopje, in former Yugoslavia. It is a three-story school building that rests on solid blocks of rubber without the inner horizontal steel-reinforcing plates as is done today.

The first bridge structure that utilized an isolation system, with added damping, was the Te Teko viaduct in New Zealand, built in 1988. The isolation system contains a *sandwich* of laminated steel and rubber bearing layers with a central lead core for energy dissipation. This type of isolation system, referred to as *lead–rubber bearing* (LRB), is now widely used. The first building supplied with LRB isolation was the William Clayton Building in Wellington, New Zealand, in 1981.

The first seismically isolated building in the United States was the Foothill Communities Law and Justice Center in Rancho Cucamonga, California, completed in 1985. It took some time until another isolated building was built in the United States. The reason for the reluctance was quite simple: Seismic isolated structures did not find their way into the building codes. Design professionals, on the other hand, were not able to show any appreciable savings to their clients by using this system.

Theoretically, in a perfectly functioning seismic isolation there will be no lateral seismic force acting on the isolated superstructure. Yet if the building codes and building officials persist in designing such superstructures to the same lateral forces pertaining to a fixed-base (nonisolated) structure, there will be no savings. This is because the superstructure will end up with equally heavy steel sections or massive reinforced-concrete sizes to counter relatively light lateral seismic design forces.

Common isolation systems are the *elastomeric* and *sliding* types. A third group, called *hybrid,* are elastomeric isolators combined with flat sliding type of isolators. Figures 13.1–13.3 show examples of seismic isolators designed to act as "spring" system.

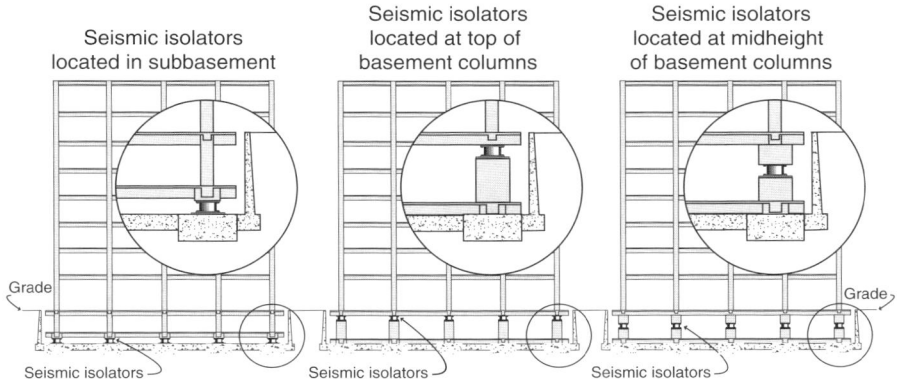

Figure 13.1 Seismic isolator placement alternatives. From left to right: in subbasement, at top of basement columns, and at midheight of basement columns. (Courtesy of Dynamic Isolation Systems, Inc.)

The Dynamic Isolation Systems (DIS) seismic (base) isolator consists of alternate layers of rubber and steel bonded together, with a cylinder of pure lead tightly inserted through a hole in the middle (Figure 13.4). The rubber layers allow the isolator to displace sideways, thus reducing the earthquake loads experienced by the building and its occupants. They are designed to also act as a spring to ensure that the structure returns to its original position once the shaking has stopped. Thick steel plates are bonded to the top and

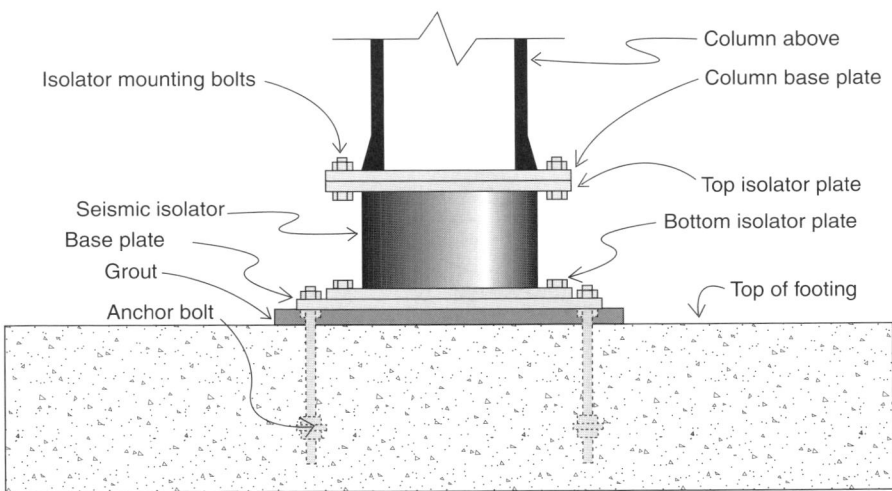

Figure 13.2 A seismic isolator assembly for new construction, alternate 1. (Courtesy of Dynamic Isolation Systems, Inc.)

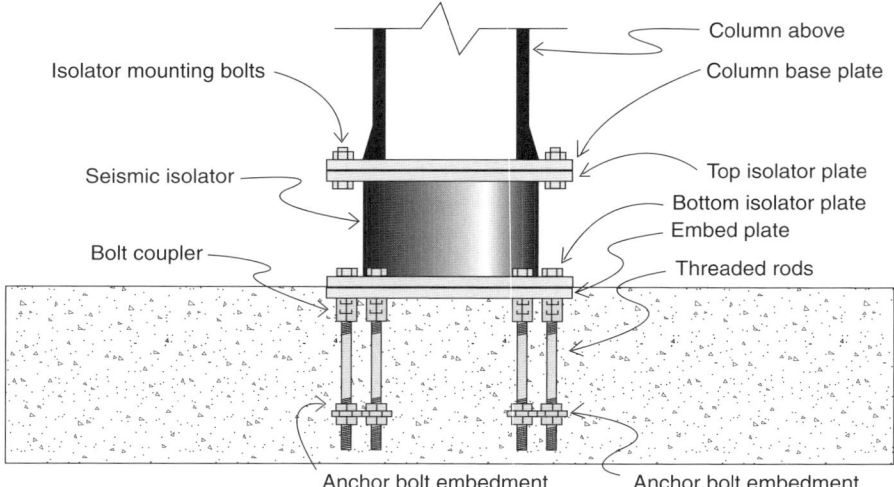

Figure 13.3 Seismic isolator assembly for new construction, alternate 2. Isolators can be placed on foundation footings, at the top of basement columns, or at column midheight. (Courtesy of Dynamic Isolation Systems, Inc.)

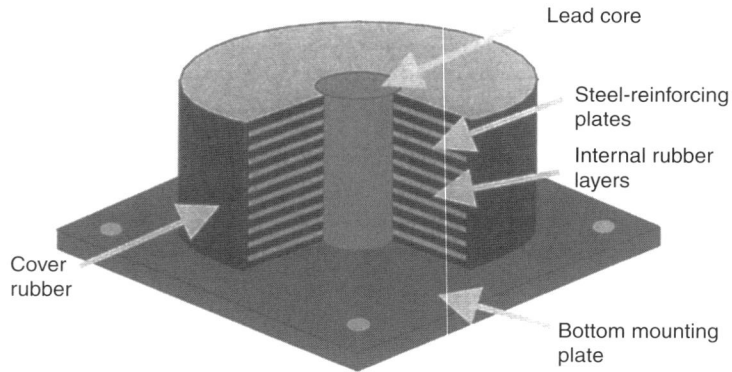

Figure 13.4 Cross section of a DIS Seismic Isolator™. Vulcanized rubber layers can move in any horizontal direction and are laminated between steel sheets to form a movable, flexible base. (Courtesy of Dynamic Isolation Systems, Inc.)

bottom surfaces to allow the isolator to be solidly bolted to the structure above and the foundation below. During earthquake events, the lead is pushed sideways by the rubber and steel layers absorbing a portion of the earthquake energy. This dampening effect helps to further reduce the earthquake forces and contributes to control the lateral displacement of the structure. Examples of buildings equipped with the DIS systems are the Rockwell International Corporation Headquarters Building 80 in Seal Beach, California; the Salt Lake City City Hall; the San Francisco City Hall; the University of Southern California Hospital in Los Angeles; the Oakland City Hall; and the Kerkhoff Hall, University of California at Los Angeles.

In general, the seismic isolation schemes described here are horizontal systems that act chiefly against lateral forces. California earthquakes have demonstrated that, contrary to common belief, there was a significant vertical component that occasionally exceeded 100% g. Stated otherwise, under the vertical-component force generated by an average California quake, the building is thrown up and down and becomes virtually weightless. Since it is not tied down to a vertical isolation system, the building can virtually fly off its base or overturn. Unless this problem is resolved, we do not have an optimum isolation system.

13.8 ENGINEERED DAMPING

While seismic isolation could be costly, especially for tall buildings of excessive mass and dimensions, *engineered damping* may be a practical and economical solution that can even be applied to existing structures to boost a deficient inherent damping. Engineered damping, whether applied to a new building or a retrofitted structure, basically consists of bracing that incorporates calibrated hydraulic piston dampers (Figure 13.5). The system reduces the damage caused by excessive sway in the moment frames (MRFs or SMRFs) and provides improved shock-absorbing capabilities to brace-framed buildings.

An example of the use of both isolators and dampers is the Arrowhead Regional Medical Center in Colton, California (Figure 13.6).

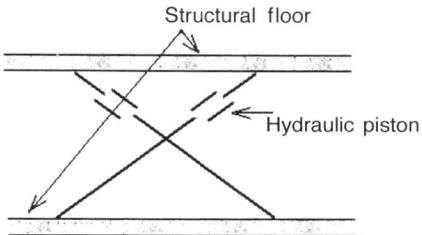

Figure 13.5 Bracing with calibrated hydraulic piston dampers.

Figure 13.6   Arrowhead Regional Medical Center. (Courtesy of KPFF Consulting Engineers.)

In the future, the number of conventional structures built in seismic zones will be gradually reduced and replaced by either seismic isolators or another engineered system. An entirely new field of technology has already emerged in California to retrofit existing buildings or constructing new ones with the above options, departing from traditional structures susceptible to earthquake damage.

REFERENCES

1. Uang, C. M., and Bondad, D. M. 1996. "Dynamic Testing of Full-Scale Steel Moment Connections." Paper presented at the Eleventh World Conference on Earthquake Engineering, Acapulco, Mexico.
2. Engelhardt, M. D., and Husain, A. S. 1993. "Cyclic-Loading Performance of Welded Flange-Bolted Connections." *Journal of Structural Engineering, ASCE,* Vol. 119, No. 12, pp. 3537–3550.
3. Iwankiw, N. R., and Carter, C. J. 1996. "The Dogbone: A New Idea to Chew On." *Modern Steel Construction,* AISC, April.
4. SAC 95-01. 1995. "Steel Moment Frame Connection." *Advisory,* No. 3.
5. Zekioglu, A., Mozzafarian, M., Le Chang, K., Uang, C. M., and Noel, S. 1996. "Development and Testing of Moment Connections for City of Hope National Medical Center." *AISC-Northridge Steel Update II.*
6. Hanson, R. D. 1997. "Supplemental Energy Dissipation for Improved Earthquake Resistance." University of Michigan, Ann Arbor, MI, assignment with FEMA.
7. ASCE, SEI. 2004. *Primer on Seismic Isolation.* Task Committee on Seismic Isolation, Reston, VA.
8. Naeim, F., and Kelly, J. M. 1999. *Design of Seismic Isolated Structures.* New York: Wiley.

ACRONYMS

AASHTO	American Association of State Highway and Transportation Officials
ACI	American Concrete Institute
AF&PA	American Forest and Paper Association
AISC	American Institute of Steel Construction
AISI	American Iron and Steel Institute
AITC	American Institute of Timber Construction
ANSI	American National Standards Institute
APA	The Engineered Wood Association
ASCE	American Society of Civil Engineers
ASD	Allowable stress design
ASTM	American Society for Testing and Materials
ATC	Applied Technology Council
AWS	American Welding Society
BOCA	Building Officials and Code Administrators International
BSSC	Building Seismic Safety Council
CABO	Council of American Building Officials
CALTRANS	California Department of Transportation
CBF	Concentrically braced frame
CEMCO	California Expanded Metal Products Co.
CRSI	Concrete Reinforcing Steel Institute
CSI	Construction Specifications Institute
FEMA	Federal Emergency Management Agency
GLB	Glued laminated beam

HSS	Hollow steel section
IBC	International Building Code
ICBO	International Conference of Building Officials
ICC	International Code Council
LGSEA	Light Gauge Steel Engineers Association
LRB	Lead–rubber bearing
LRFD	Load and resistance factor design
MRF	Moment-resisting frame
MSJC	Masonry Standards Joint Committee
NAHB	National Association of Home Builders
NASPEC (or NAS)	North American Specification for the Design of Cold-Formed Steel Structural Members (issued by AISI)
NASFA	North American Steel Framing Alliance
NBC	National Building Code
NDS	National Design Specification for Wood Construction
NEHRP	National Earthquake Hazard Reduction Program
NFPA	National Forest Products Association
NIBS	National Institute of Building Sciences
OCBF	Ordinary concentrically braced frame
OMF	Ordinary moment frame
OMRF	Ordinary moment-resisting frame
PCA	Portland Cement Association
SAC	A joint venture of SEAOC, ATC, and CUREe (California Universities for Research in Earthquake Engineering)
SBCCI	Southern Building Code Congress International
SCBF	Special concentrically braced frame
SEAOC	Structural Engineers Association of California
SEI	Structural Engineering Institute
SJI	Steel Joist Institute
SMF	Special moment frame
SMRF	Special moment-resisting frame
SMRSF	Special moment-resisting space frame
SUG	Seismic use group
TMS	The Masonry Society
UBC	Uniform Building Code
USD	Ulimate strength design
WSD	Working strength design

GLOSSARY

2 × 4, 2 × 6, 4 × 8 Nominal dimensions (thickness by width) of the cross sections of wood members (actual dimensions are about 0.5 in. less).

Absorption Amount of water that a body or unit absorbs when immersed in water for a certain length of time.

Adhesion bond The power of adhesion of grout or mortar to a masonry element or unit.

Admixture A material such as aggregate or cement added to concrete to alter its properties.

Aggregates An ingredient, namely granular, to produce hydraulic cement concrete or mortar. *See* Coarse aggregate and Fine aggregate.

Anchor bolts Steel rod threaded at one end used to secure structural members to concrete or masonry. Commonly formed in an L or J shape.

Angles In steel construction, rolled sections in an L shape.

Balanced condition When both steel and concrete yield simultaneously.

Balloon framing Residential building construction in which one-piece studs extend from the first floor line or sill to the roof plate. Joists for upper floors are nailed to the sides of studs and receive additional support from ledger boards.

Base plate Plate or steel slab upon which a column or scaffold section stands.

Base shear Total design lateral force V or shear at the base of the structure.

Beam A structural member that resists transverse loads. A *simple* beam is placed on supports at each end. A *cantilever* beam extends beyond the support or supports. A *continuous* beam is supported on more of two supports. A *fixed* beam has one or both ends restrained to prevent rotation.

Bearing capacity Maximum unit pressure any material will withstand before failure or deformation.

Bearing footing Foundation that carries vertical loads (weight) only.

Bearing wall Continuous vertical support for floors, other walls, roofs, or other structural loads. A concrete wall that supports more than 200 lbs per linear foot as superimposed load or a wall supporting its own weight for more than one story.

Bed joint In masonry construction, a layer of mortar upon which masonry units are laid.

Blocked plywood Plywood sheets attached to framing members located along the entire perimeter.

Blocking Piece of wood fastened between structural members to strengthen joint, provide structural support, or block air passage.

Box nails Fastener with a flat head and a shank not as thick as that of a common nail. Used in wood that splits easily.

Braced frames *Concentrically braced frame* (CBF)—when the members of a braced frame are subjected mainly to axial forces. *Eccentrically braced frame* (EBF)—steel braced frame designed in conformance with UBC 1997 Section 2213.10, IBC 2003 Section 1602.1. *Ordinary braced frame* (OBF)—steel-braced framed designed in accordance with the provisions of UBC 1997 Section 2213.8 or 2214.6. *Special concentrically braced frame* (SCBF)—a steel-braced frame designed in conformance with the provisions of UBC 1997 Section 2213.9, IBC 2003 Section 1602.1.

Brittle fracture Sudden splitting without (or minimal) previous elastic deformation.

California roof Type of roof with framing members installed above the structural roof diaphragm and framing. The upper framing is, therefore, an architectural component and does not participate in the structural resistance of the house.

Cap Name given to a masonry piece set on top of a masonry wall or pier. Metal caps are also used.

Cast in place Concrete cast at the job site for permanent placement.

Cast stone Stone manufactured of Portland cement concrete and used as fascia, veneer, or trim on buildings and other structures.

Ceiling joist Intermediate horizontal structural member used to support finished ceiling material.

Cement A binding ingredient used for concrete, mortar, and grout. *See* Portland cement.

Checking Development of shallow cracks at regular intervals on the surface of concrete, plaster, or paint.

Coarse aggregate Usually gravel from a river deposit or crushed stone.

Cold joint Joint or discontinuity in concrete resulting from delay in the placement of successive lifts.

Collar joint   A vertical joint between two wythes of masonry.

Collar tie   Horizontal member that ties rafters together above the wall plate.

Collector(s)   In a lateral-force-resisting system, collectors are elements used to transmit forces between floor diaphragms and other members.

Column sections   More suitable than I beams to be used as columns. The radius of gyration of an I beam about an axis is comparatively small.

Common nails   Nails with a smooth cylindrical shaft and a flat head. Used to join wood-framing members and in general carpentry work.

Compaction   Process of compressing, vibrating, or tamping to eliminate voids and reduce the volume of soil, subgrade material, or other porous material such as freshly placed concrete.

Composite column   A column that combines concrete and steel sections. Generally speaking, a concrete column reinforced longitudinally with steel shapes.

Composite special moment frame (C-SMF)   A moment frame of composite or reinforced-concrete column and either structural steel or composite beams.

Compressive stress   Results from a force that tends to shorten the member.

Concrete masonry unit (CMU)   Precast hollow or solid masonry unit made of Portland cement and fine aggregate with or without admixtures or pigments. Formed into modular or nonmodular dimensions to be laid with other similar units.

Concrete slab on grade   Concrete reinforced or unreinforced slab poured against the ground surface.

Concrete stem wall   In building construction, any wall made of reinforced concrete such as a basement wall.

Connectors   Fasteners.

Cooler nails   Nails with special size and head configuration for use in gypsum board application.

Corbel   Ledge formed when courses of masonry project out of the face of a wall.

Crawl space   Unfinished accessible area below the first floor of a structure. Commonly used for components such as ductwork and piping.

Creep   A plastic-type deformation that occurs with time in materials such as concrete when subjected to continuous stress.

Cripple wall   The wood-framed wall (usually between the footing and the first floor) not less than 14 in. and not more than 4 ft in height.

Curing   The method of maintaining adequate moisture and temperature conditions to develop the required strength and reduce shrinkage of concrete during its setting phase.

Curtain wall   Exterior wall supported by the frame of the building as opposed to being self-supported or load-bearing wall.

Damping   Loss of energy of a system when its vibratory motion dies out due to internal strains.

Dead load   A permanent load on a structure or member such as weight of materials: columns, floors, framing, partitions, roofs, and walls.

Deflection   Deformation or deviation of a member as a result of its own weight, an applied force, or other stress.

Design strength   Nominal strength multiplied by a strength reduction factor. *See* Strength, design.

Development length   Per ACI code, length of embedded reinforcement required to develop the design strength of reinforcement at a critical section.

DF.   *See* Douglas fir.

Diaphragm   A surface element such as floor slab, deck, or wall that resists forces in its own plane or transmits forces to resisting systems.

Doubly reinforced beam   When reinforcement is placed in the compression zone of a beam in addition to tensile reinforcement.

Douglas fir   (DF)   Curry-grained wood that resembles white pine used for plywood or laminated sheets.

Drag struts   Members used to transmit lateral loads from the diaphragms to the shear walls at points of irregularity of the diaphragms.

Dry joint   Joint or head without mortar.

Drywall   Interior surfacing material applied to framing members using dry construction. Fireproof gypsum core is encased with heavy paper on one side and liner paper on the other side.

Ductility   Capability of an element or structure to withstand a substantial amount of distortion without a significant loss of strength.

Dynamic forces   *Dynamic* forces and loads in a structure are generated by inertia forces in motion, as opposed to *static* forces, which are caused by its stationary weight. They are variable in nature and depend mostly on the individual dynamic characteristics of the structure in question. Dynamic forces are of special significance in relation to earthquake-generated motions.

Edge nailing   1. Nailing pattern along the perimeter of the diaphragm panels. 2. Hidden or toenailing of, for example, floor boards.

Embedment length   Per ACI regulations, length of embedded reinforcement provided beyond a critical section.

Expansion joint   A break or space in construction to allow for thermal expansion and contraction of the materials used in the structure.

Factored load   Load multiplied by appropriate load factors.

Fascia   1. Horizontal trim member at the lower end of roof rafters. 2. Complete assembly of exterior trim members at the lower end of an overhang.

Field nailing   Nailing pattern inside the diaphragm panels.

Fine aggregate   Sand without any substantial amount of silt, organic matter, or other materials.

Finish grade   Elevation of ground (walks, drives) and other improved surfaces after final grading operations.

Flexure (or bending) When a structural member, say a beam in horizontal position with ends supported on walls, is subjected to bending stresses under a load placed at any point on the beam.

Fly ash aggregate   Pulverized fuel ash (coal or coke) produced of lightweight aggregates by a specific process where pellets of the material are sintered at 1000–1200°C.

Footing   Commonly, a portion of a shallow bearing-type foundation, usually concrete poured directly into the excavation.

Furring   Finishing of the face of a masonry wall to make room for insulation, prevent moisture propagation, or provide a further finish to the wall.

Gable end   Three- or four-sided section of an end wall that extends from the top wall plate to the ridge.

Girder   Large horizontal structural member constructed of several steel, reinforced-concrete, or timber members that support loads at isolated points along its length.

GLB (glued-laminated beam)   Assembly of laminated lumber; the grain of all laminations is longitudinally parallel and bonded with adhesives.

Grout   From Swedish *groot* ("oatmeal"). Slump mixture of cement, aggregates, and water of rather fluid consistency poured into void spaces in masonry construction.

Header   1. Horizontal structural member over the top of a wall opening that distributes the load to either side of the opening. 2. Floor or ceiling joist or roof rafter that is perpendicular to the common joists or rafters. Distributes loads to other joists or rafters around openings. 3. Masonry member perpendicular to the face of the wall to tie the front and back portions of the wall together.

HD (hold-down)   See next.

Hold-downs   (bolts) Anchor bolts used to connect frame to foundation.

HSS (hollow structural section)   In steel design, the sections can be rectangular, square, or round.

I beam   In steel construction, American standard I beam, produced by most of the rolling mills.

Joist   Horizontal support member to which finish floor and ceiling materials are fastened. Common joist material includes wood, steel, and concrete.

Kern limit   Limit set for the eccentricity of compressive force while preventing tensile stress.

King stud   Vertical support installed along both sides of a framed opening for a window or door. Extends from the top to bottom plate.

Lateral-force-resisting system The portion of the structural system designed to resist design seismic forces.

Let-in bracing In wood house framing, the diagonal braces notched into studs.

Lightweight concrete Concrete made of aggregates that are the byproduct of industrial processes.

Limit state A condition reached by a structure when it can no longer fulfill its intended function due to brittle fracture, excessive deformation, collapse, or other factors that make it unusable for the purpose intended.

Live loads Generally speaking, all loads that are not part of a *dead load* and are therefore of *temporary* nature: people, vehicles, equipment, stored material, movable partitions, or natural phenomena such as winds, snow, and earthquakes.

Lumber grades Classification of lumber in regard to the strength of material.

Lumber species Classification of lumber in regard to the type of trees.

Masonry General term applied to brickwork, blockwork, and stonework.

Masonry cements Cements specifically used in mortar. *See* Mortar.

Modulus of elasticity Elasticity is the property of a body to resist permanent deformation. The modulus of elasticity involves the concepts of stress and strain and their relationship.

Modulus of rupture Upper bound of the measured tensile resistance of concrete.

Moisture content 1. Amount of moisture in a given air space. Expressed as grains of moisture per pound of dry air. 2. Amount of moisture in the cellular structure of a material such as aggregate. Expressed as a percentage of the dry weight of material.

Moment Tendency of a force to produce rotation about a certain given point or axis.

Moment of inertia The resultant of the *mass* of a body and the square of a length, named *radius of gyration*.

Monolithic A structural member built in one piece or assembly properly engineered to act as one solid element.

Mortar A mixture of plastic consistency of cementitious materials, fine aggregates, and water. It could be *fat* when it tends to stick to the trowel, *harsh* when difficult to spread for lack of sufficient plasticizer, or *lean* when it lacks sufficient cementitious ingredients and is also difficult to spread.

Mudsill Continuous timber placed on the ground that distributes a load and provides a level surface for scaffolding and shoring.

Nominal strength Strength of a member calculated per provisions and assumptions of strength design methods.

Nonbearing wall A wall that merely separates space into rooms but does not carry overhead partitions or floor joists except its own weight, as opposed to a load-bearing wall.

Open-web steel joists   Lightweight steel trusses normally used to support floor and roof panels *between* main supporting elements.

Ordinary moment-resisting frame (OMRF)   A moment-resisting frame not meeting special detailing requirements for ductile behavior.

OSB (oriented strand board)   A panel of compressed wood particles set in layers at right angles to one another, bonded with phenolic resin. Widely used in wood construction.

Overturning moment   An arithmetic sum resulting from the moments of all forces above the base of a structure multiplied by the heights above such a base.

Panel zone   In a beam-to-column connection, the area that transmits moment by means of a shear panel.

Parapet   Part of a wall that projects or extends above roof level.

Particleboard   Particles of wood or wood particles and fibers bonded together with synthetic resins or other bonding agents to produce a manufactured panel.

P-delta effect   A secondary effect (moment) caused by column axial loads and lateral deflections on a structural member.

Perimeter (edge) nailing   Pattern of nailing of a shear (horizontal or vertical) diaphragm along its perimeter (edge). More loaded and more responsible connection than field nailing.

Piers   Vertical support that provides bearing in the ground; it functions similarly to a column. The bottom of this support may be widened or bellied to enlarge the load-bearing area. Upper structural members are set on these supports.

Pilaster   A part of a wall that projects on one or both sides of the wall and functions as a vertical beam, a column, or even an architectural element.

Plastic zone   In general, the yielded zone of a member.

Plate   Horizontal framing member at top or bottom of a wood-framed wall. Studs bear on bottom plate. Joists and rafters rest on top plate.

Plate girder   A built-up section to compensate for the case when a rolled section does not meet the needed requirements. The plate or box girder is made up of plates and angles welded together to form an I section.

Platform framing   System using wood studs one story high finished with a platform consisting of the underflooring for the next story.

Poisson's ratio   Applies to an elastic body. Ratio of the transverse strain to the longitudinal strain when body is subjected to a longitudinal stress T.

Pony wall   *See* Cripple wall.

Portland cement   A product obtained by thoroughly mixing calcareous, silica, alumina, and iron-oxide-bearing materials, burning them, and grinding the resulting cinder. Depending on the ingredients and the method of producing it, cements can be *high-alumina, granulated blast furnace slag, pozzolanas, oil-well cements, magnesium oxychloride,* and *masonry ce-*

ments. The latter are used specifically for masonry construction and provide a more plastic mortar than ordinary Portland cement. *See* Masonry cements and Rapid-hardening Portland cement.

Posts   Columns or pillars; vertical support.

Precast concrete   As opposed to cast in place, precast concrete is cast at the plant or a location away from the construction site.

Purlin   A horizontal member resting usually on trusses and supporting the roof rafters.

Rafter   Sloped roof structural member that supports roof sheathing and roof loads.

Raised foundation   A foundation that raises the first floor of the structure above the grade, creating a crawl space. Usually consists of continuous-perimeter stem walls/footings and individual piers/footings inside.

Rapid-hardening Portland cement   Obtained by a development of manufacturing process, basically finer grinding and perfected method of mixing of the raw materials, for instance, higher lime contents.

Reactions   In general, upward forces at a beam support to keep it in balance against downward forces (or loads).

Read-out   Framing at the opening.

Rebar   Steel reinforcement bars of different size and shape commonly used to strengthen (reinforced) masonry.

Required strength   Strength of a member or cross section required to resist factored loads or related internal moments and forces.

Resistance factor (strength reduction factor)   Relates to deviations of the actual strength from the nominal strength and resulting possible failures.

Retaining wall   A concrete or masonry wall designed and built to resist lateral forces or pressure from or displacement of soil or other materials.

Ridge   1. Highest point of a sloping roof; roof peak. 2. Highest horizontal member of a sloping roof to which rafters are fastened.

Seismic coefficient   A function of soil type and seismic zone.

Service load   Load specified by general building codes (without load factors). It is also considered nominal load, expected to be supported by the structure under normal usage.

Shapes (or sections)   In steel construction, the products of the rolling mills.

Shearing force   An internal force that works tangential to the section of a member on which it acts.

Sheathing   A protective material that covers prestressing steel mainly to prevent bonding with the surrounding concrete or protects reinforcing bars against corrosion. Includes fiberboard, gypsum board, plywood, polystyrene, and lumber.

Slenderness ratio   Related to the radius of gyration in all structural steel column formulas: l/r, l being the effective length of the column and r the minimum radius of gyration, both values in inches.

Soffit   Underside of beams, roof overhangs, lintels, or reveals.

Spalling   Flaking off of the surface of concrete.

Spandrel   Portion of a wall between the head of a window and the windowsill above.

Special moment-resisting frame (SMRF)   Per UBC 1997, a moment-resisting frame detailed to provide ductile behavior and comply with the requirements given in Chapter 19 or 22.

Splice   Between two structural elements, splice is the connection that joins both members at their ends to produce a single, longer element.

Stirrups   Vertical steel bars added to a concrete beam to increase shear resistance.

Stone concrete   Normal-weight concrete. *See* Lightweight concrete.

Story drift   Lateral displacement of one floor with respect to the floor above or below.

Straps   Strip of metal or other material used to provide additional tensile reinforcement to a joint.

Strength, design   Per ACI code, nominal strength multiplied by strength reduction factor ϕ.

Strength reduction factor   *See* Resistance factor.

Stress   Intensity of force measured per unit area.

Stucco   Exterior finish material composed of Portland cement, lime, sand, and water.

Studs   1. Wood or metal vertical framing member in a wall. 2. Threaded fastener that is threaded on both ends and has no head. 3. Bolt anchored to another member at one end.

Subgrade   Compacted soil used to support a concrete slab or other structure.

Tees (structural tees)   Steel sections made by splitting W, M, and S shapes so they become WT, MT, and ST shapes.

Tensile stress   Stress that results from a force that tends to lengthen the member.

Treated lumber   Lumber coated or saturated with a stain or chemical to retard fire, insect damage, or decay resulting from exposure to weather.

Trimmer   A beam or joist into which a header is framed.

Truss   Structural member constructed of components commonly placed in a triangular arrangement.

T section   A T beam that can be handled like rectangular cross sections provided the depth of the rectangular stress block does not exceed the thickness of the T flange.

Unbraced length   Distance between the braced points of a structural member.

Vertical shear   Symbol: V. Refers to the propensity of one part of an element to move vertically with respect to an adjacent part.

Von Mises yield criterion Known also as maximum strain-energy-of-distortion theory. Based on the assumption that inelastic action starts at any point in a body under stresses when the strain energy of distortion, per unit volume, equals the distortion absorbed in a simple tensile bar stressed to its elastic limit under *uniaxial* stress.

Wall anchor A bearing plate type of anchor used in masonry construction to support a beam on a masonry wall.

Wide-flange section As a rule, wide-flange steel sections have greater flange widths than standard I beams.

Wire mesh Heavy-gauge wires joined in a grid to reinforce and increase tensile strength of concrete.

Working stress design Method based on using actual or *working* loads.

Wythe In masonry construction, the continuous vertical section of a wall, one masonry unit thick.

Young's modulus In 1807, while studying the behavior of elastic bodies subjected to external forces, Young discovered a constant. Assuming that T is the *stress* in the cross section of a thin rod and e the extension or *strain* associated with the stress, then $T = Ee$, where E is Young's modulus in tension.

REFERENCES

1. Ambrose, J. 1991. *Simplified Design of Masonry Structures.* Wiley, New York.
2. Lea, F. M. 1971. *The Chemistry of Cement and Concrete.* Chemical Publishing, New York.
3. Parker, H. 1967. *Simplified Design of Structural Steel,* 3rd ed. Wiley, New York.
4. IBC 2003, International Code Council, Country Club Hills, IL.

COMPUTER ANALYSIS

A. SMRF PROJECT PART I

```
***********************************************************
*                                                         *
*               S T A A D - III                           *
*               Revision 23.0                             *
*               Proprietary Program of                    *
*               Research Engineers, Inc.                  *
*               Date=    SEP 19, 1999                     *
*               Time=    8:50:37                          *
*               Build No.    1009.01.01                   *
*                                                         *
*******i                                           ********
```

```
 1. STAAD PLANE ICBO FRAME
 2. *LRFD DESIGN,SEISMIC Z4,70MPH WIND,EXPS.C,--4 STORY 6SMRF SYST.
 3. UNI KIP FEET
 4. JOINT COORDINATES
 5. 1 0. 0.; 2 20. 0.; 3 0. 13.5; 4 20. 13.5; 5 0. 27.; 6 20. 27.
 6. 7 0. 40.5; 8 20. 40.5; 9 0. 54.; 10 20.0 54.
 7. MEMBER INCIDENCE
 8. 1 1 3; 2 2 4; 3 3 4; 4 4 6; 5 5 6; 6 3 5; 7 6 8; 8 7 8; 9 5 7
 9. 10 8 10; 11 9 10; 12 7 9
10. MEMBER PROPERTY AMERICAN
11. 1 2 4 6 TABLE ST W14X257
12. 7 9 10 12 TABLE ST W14X145
13. 3 TABLE ST W36X160
14. 5 TABLE ST W24X68
15. 8 TABLE ST W24X68
16. 11 TABLE ST W14X48
17. UNIT INCHES
18. CONSTANTS
19. E 26000.0 ALL
20. MEMBER OFFSET
21. 3 5 8 11 START 7.0 0.0
22. 3 5 8 11 END  -7.0 0.0
23. PRINT MEMBER INFORMATION
```

*LRFD DESIGN,SEISMIC Z4,70MPH WIND,EXPS.C,--

MEMBER INFORMATION

MEMBER	START JOINT	END JOINT	LENGTH (INCH)	BETA (DEG)	RELEASES
1	1	3	162.000	0.00	
2	2	4	162.000	0.00	
3	3	4	226.000	0.00	
4	4	6	162.000	0.00	
5	5	6	226.000	0.00	
6	3	5	162.000	0.00	
7	6	8	162.000	0.00	
8	7	8	226.000	0.00	
9	5	7	162.000	0.00	
10	8	10	162.000	0.00	
11	9	10	226.000	0.00	
12	7	9	162.000	0.00	

************ END OF DATA FROM INTERNAL STORAGE ************

24. SUPPORT
25. 1 2 FIXED
26. MEMBER RELEASE
27. 1 2 START MP 0.865
28. DRAW MEMBER JOINT PROPERTY SUPPORT

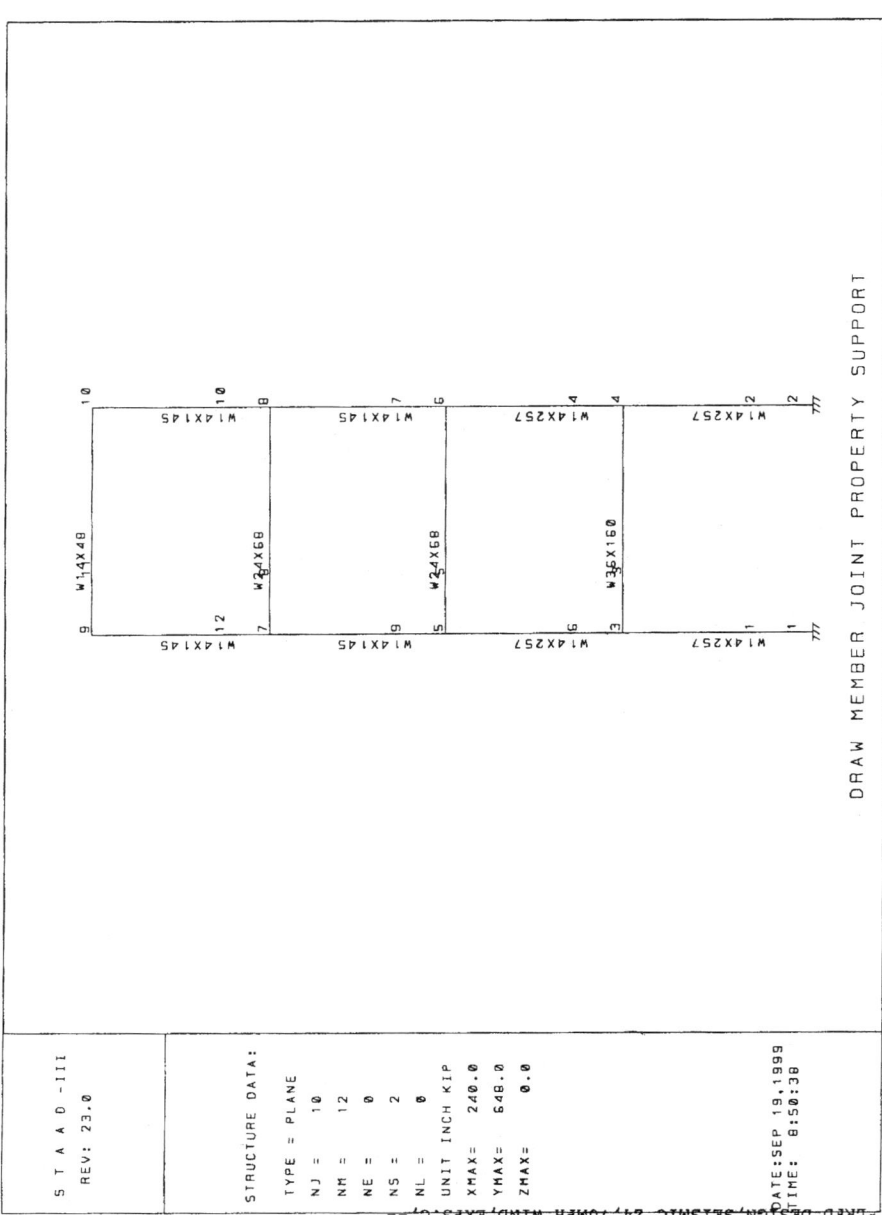

DRAW MEMBER JOINT PROPERTY SUPPORT

S T A A D - I I I
REV: 23.0

STRUCTURE DATA:

TYPE = PLANE

NJ = 10

NM = 12

NE = 0

NS = 2

NL = 0

UNIT INCH KIP

XMAX= 240.0

YMAX= 648.0

ZMAX= 0.0

DATE:SEP 19,1999
TIME: 8:50:38

*LRFD DESIGN,SEISMIC 24,70MPH WIND,EXPS.C,--

```
*LRFD DESIGN,SEISMIC Z4,70MPH WIND,EXPS.C,--
29. UNIT FT
30. LOADING 1 DEAD
31. MEMBER LOAD
32. 3 UNI GY -0.6
33. 5 UNI GY -0.6
34. 8 UNI GY -0.6
35. 11 UNI GY -0.55
36. LOADING 2 LIVE (L)
37. MEMBER LOAD
38. 3 UNI GY -0.4
39. 5 UNI GY -0.4
40. 8 UNI GY -0.4
41. LOADING 3 ROOF LIVE (LR)
42. MEMBER LOAD
43. 11 UNI GY -0.16
44. LOADING 4 EH FROM LEFT AT *"STRENGTH LEVEL"
45. JOINT LOAD
46. 3 FX 3.9
47. 4 FX 3.9
48. 5 FX 7.8
49. 6 FX 7.8
50. 7 FX 11.7
51. 8 FX 11.7
52. 9 FX 12.6
53. 10 FX 12.6
54. LOAD COMBINATION 5 144 PERCENT DL + 50 PERCENT L
55. 1 1.44 2 0.5
56. LOAD COMBINATION 6 195 PERCENT DL + 50 PERCENT L
57. 1 1.948 2 0.5
58. LOAD COMBINATION 7 595 PERCENT EQ FOR M. INELASTIC R. DISPLACEMENT
59. 4 5.95
60. LOAD COMBINATION 8 120 PERCENT DL + 160 PERCENT L + 50 PERCENT LR    (3-2)
61. 1 1.2 2 1.6 3 0.5
62. LOAD COMBINATION 9 120 PERCENT DL + 160 PERCENT LR + 50 PERCENT L    (3-3)
63. 1 1.2 3 1.6 2 0.5
64. LOAD COMBINATION 10 120 + EV= 144 PERC DL + 50 PERC L + 100 PERC EH  (3-5)
65. 1 1.44 2 0.5 4 1.0
66. LOAD COMBINATION 11 90 - EV= 68 PERCENT DL + 100 PERCENT EH          (3-6)
67. 1 0.68 4 1.0
68. LOAD COMBINATION 12 195 PERCENT DL + 50 PERCENT L + 340 PERCENT EH   (6-1)
69. 1 1.948 2 0.5 4 3.4
70. LOAD COMBINATION 13 15 PERCENT DL + 340 PERCENT EH                   (6-2)
71. 1 0.152 4 3.4
72. PERFORM ANALYSIS


    P R O B L E M   S T A T I S T I C S
    -----------------------------------

NUMBER OF JOINTS/MEMBER+ELEMENTS/SUPPORTS =    10/   12/    2
ORIGINAL/FINAL BAND-WIDTH =     2/    2
TOTAL PRIMARY LOAD CASES  =     4, TOTAL DEGREES OF FREEDOM =    24
SIZE OF STIFFNESS MATRIX =       216 DOUBLE PREC. WORDS
REQRD/AVAIL. DISK SPACE = 12.02/ 2047.7 MB,  EXMEM = 1963.5 MB
```

```
*LRFD DESIGN,SEISMIC Z4,70MPH WIND,EXPS.C,—  ++
Processing Element Stiffness Matrix. 8:50:38 ++
Processing Global Stiffness Matrix.  8:50:38 ++
Processing Triangular Factorization. 8:50:38 ++
Calculating Joint Displacements.     8:50:38 ++
Calculating Member Forces.           8:50:38
73. PRINT MEMBER FORCES
```

*LRFD DESIGN,SEISMIC Z4,70MPH WIND,EXPS.C,--

MEMBER END FORCES   STRUCTURE TYPE = PLANE

ALL UNITS ARE -- KIP FEET

MEMBER	LOAD	JT	AXIAL	SHEAR-Y	SHEAR-Z	TORSION	MOM-Y	MOM-Z
1	1	1	22.13	-0.39	0.00	0.00	0.00	-0.36
		3	-22.13	0.39	0.00	0.00	0.00	-4.86
	2	1	11.30	-0.26	0.00	0.00	0.00	-0.24
		3	-11.30	0.26	0.00	0.00	0.00	-3.28
	3	1	1.51	0.00	0.00	0.00	0.00	0.00
		3	-1.51	0.00	0.00	0.00	0.00	0.02
	4	1	-130.84	36.00	0.00	0.00	0.00	109.10
		3	130.84	-36.00	0.00	0.00	0.00	376.90
	5	1	37.52	-0.69	0.00	0.00	0.00	-0.64
		3	-37.52	0.69	0.00	0.00	0.00	-8.63
	6	1	48.76	-0.88	0.00	0.00	0.00	-0.83
		3	-48.76	0.88	0.00	0.00	0.00	-11.10
	7	1	-778.50	214.20	0.00	0.00	0.00	649.13
		3	778.50	-214.20	0.00	0.00	0.00	2242.57
	8	1	45.39	-0.88	0.00	0.00	0.00	-0.83
		3	-45.39	0.88	0.00	0.00	0.00	-11.07
	9	1	34.62	-0.59	0.00	0.00	0.00	-0.56
		3	-34.62	0.59	0.00	0.00	0.00	-7.44
	10	1	-93.32	35.31	0.00	0.00	0.00	108.45
		3	93.32	-35.31	0.00	0.00	0.00	368.27
	11	1	-115.79	35.74	0.00	0.00	0.00	108.85
		3	115.79	-35.74	0.00	0.00	0.00	373.60
	12	1	-396.10	121.52	0.00	0.00	0.00	370.10
		3	396.10	-121.52	0.00	0.00	0.00	1270.37
	13	1	-441.49	122.34	0.00	0.00	0.00	370.87
		3	441.49	-122.34	0.00	0.00	0.00	1280.73
2	1	2	22.13	0.39	0.00	0.00	0.00	0.36
		4	-22.13	-0.39	0.00	0.00	0.00	4.86
	2	2	11.30	0.26	0.00	0.00	0.00	0.24
		4	-11.30	-0.26	0.00	0.00	0.00	3.28
	3	2	1.51	0.00	0.00	0.00	0.00	0.00
		4	-1.51	0.00	0.00	0.00	0.00	-0.02
	4	2	130.84	36.00	0.00	0.00	0.00	109.10
		4	-130.84	-36.00	0.00	0.00	0.00	376.90
	5	2	37.52	0.69	0.00	0.00	0.00	0.64
		4	-37.52	-0.69	0.00	0.00	0.00	8.63
	6	2	48.76	0.88	0.00	0.00	0.00	0.83
		4	-48.76	-0.88	0.00	0.00	0.00	11.10
	7	2	778.50	214.20	0.00	0.00	0.00	649.13
		4	-778.50	-214.20	0.00	0.00	0.00	2242.57
	8	2	45.39	0.88	0.00	0.00	0.00	0.83
		4	-45.39	-0.88	0.00	0.00	0.00	11.07
	9	2	34.62	0.59	0.00	0.00	0.00	0.56
		4	-34.62	-0.59	0.00	0.00	0.00	7.44

*LRFD DESIGN,SEISMIC Z4,70MPH WIND,EXPS.C,--

MEMBER END FORCES   STRUCTURE TYPE = PLANE

ALL UNITS ARE -- KIP FEET

MEMBER	LOAD	JT	AXIAL	SHEAR-Y	SHEAR-Z	TORSION	MOM-Y	MOM-Z
	10	2	168.36	36.69	0.00	0.00	0.00	109.74
		4	-168.36	-36.69	0.00	0.00	0.00	385.54
	11	2	145.89	36.26	0.00	0.00	0.00	109.34
		4	-145.89	-36.26	0.00	0.00	0.00	380.21
	12	2	493.61	123.28	0.00	0.00	0.00	371.76
		4	-493.61	-123.28	0.00	0.00	0.00	1292.57
	13	2	448.22	122.46	0.00	0.00	0.00	370.98
		4	-448.22	-122.46	0.00	0.00	0.00	1282.21
3	1	3	-1.19	5.65	0.00	0.00	0.00	11.17
		4	1.19	5.65	0.00	0.00	0.00	-11.17
	2	3	-0.75	3.77	0.00	0.00	0.00	7.38
		4	0.75	3.77	0.00	0.00	0.00	-7.38
	3	3	-0.02	0.00	0.00	0.00	0.00	0.03
		4	0.02	0.00	0.00	0.00	0.00	-0.03
	4	3	0.00	-66.44	0.00	0.00	0.00	-625.66
		4	0.00	66.44	0.00	0.00	0.00	-625.66
	5	3	-2.09	10.02	0.00	0.00	0.00	19.77
		4	2.09	10.02	0.00	0.00	0.00	-19.77
	6	3	-2.70	12.89	0.00	0.00	0.00	25.44
		4	2.70	12.89	0.00	0.00	0.00	-25.44
	7	3	0.00	-395.33	0.00	0.00	0.00	-3722.70
		4	0.00	395.33	0.00	0.00	0.00	-3722.69
	8	3	-2.64	12.81	0.00	0.00	0.00	25.23
		4	2.64	12.81	0.00	0.00	0.00	-25.23
	9	3	-1.84	8.66	0.00	0.00	0.00	17.13
		4	1.84	8.66	0.00	0.00	0.00	-17.13
	10	3	-2.09	-56.42	0.00	0.00	0.00	-605.89
		4	2.09	76.46	0.00	0.00	0.00	-645.43
	11	3	-0.81	-62.60	0.00	0.00	0.00	-618.07
		4	0.81	70.28	0.00	0.00	0.00	-633.26
	12	3	-2.70	-213.01	0.00	0.00	0.00	-2101.81
		4	2.70	238.79	0.00	0.00	0.00	-2152.70
	13	3	-0.18	-225.04	0.00	0.00	0.00	-2125.56
		4	0.18	226.76	0.00	0.00	0.00	-2128.95
4	1	4	16.48	1.58	0.00	0.00	0.00	9.60
		6	-16.48	-1.58	0.00	0.00	0.00	11.71
	2	4	7.53	1.01	0.00	0.00	0.00	6.30
		6	-7.53	-1.01	0.00	0.00	0.00	7.38
	3	4	1.51	0.02	0.00	0.00	0.00	0.04
		6	-1.51	-0.02	0.00	0.00	0.00	0.18
	4	4	64.40	32.10	0.00	0.00	0.00	287.52
		6	-64.40	-32.10	0.00	0.00	0.00	145.83
	5	4	27.50	2.78	0.00	0.00	0.00	16.98
		6	-27.50	-2.78	0.00	0.00	0.00	20.55
	6	4	35.87	3.58	0.00	0.00	0.00	21.86
		6	-35.87	-3.58	0.00	0.00	0.00	26.50

*LRFD DESIGN,SEISMIC Z4,70MPH WIND,EXPS.C,--

MEMBER END FORCES   STRUCTURE TYPE = PLANE

ALL UNITS ARE -- KIP FEET

MEMBER	LOAD	JT	AXIAL	SHEAR-Y	SHEAR-Z	TORSION	MOM-Y	MOM-Z
	7	4	383.17	191.00	0.00	0.00	0.00	1710.73
		6	-383.17	-191.00	0.00	0.00	0.00	867.70
	8	4	32.58	3.52	0.00	0.00	0.00	21.63
		6	-32.58	-3.52	0.00	0.00	0.00	25.96
	9	4	25.95	2.43	0.00	0.00	0.00	14.75
		6	-25.95	-2.43	0.00	0.00	0.00	18.04
	10	4	91.89	34.88	0.00	0.00	0.00	304.50
		6	-91.89	-34.88	0.00	0.00	0.00	166.38
	11	4	75.60	33.17	0.00	0.00	0.00	294.05
		6	-75.60	-33.17	0.00	0.00	0.00	153.79
	12	4	254.82	112.72	0.00	0.00	0.00	999.42
		6	-254.82	-112.72	0.00	0.00	0.00	522.33
	13	4	221.46	109.38	0.00	0.00	0.00	979.02
		6	-221.46	-109.38	0.00	0.00	0.00	497.61
5	1	5	0.49	5.65	0.00	0.00	0.00	15.69
		6	-0.49	5.65	0.00	0.00	0.00	-15.69
	2	5	0.10	3.77	0.00	0.00	0.00	10.57
		6	-0.10	3.77	0.00	0.00	0.00	-10.57
	3	5	0.10	0.00	0.00	0.00	0.00	-0.05
		6	-0.10	0.00	0.00	0.00	0.00	0.05
	4	5	0.00	-30.24	0.00	0.00	0.00	-284.76
		6	0.00	30.24	0.00	0.00	0.00	-284.76
	5	5	0.76	10.02	0.00	0.00	0.00	27.88
		6	-0.76	10.02	0.00	0.00	0.00	-27.88
	6	5	1.01	12.89	0.00	0.00	0.00	35.85
		6	-1.01	12.89	0.00	0.00	0.00	-35.85
	7	5	0.00	-179.93	0.00	0.00	0.00	-1694.33
		6	0.00	179.93	0.00	0.00	0.00	-1694.33
	8	5	0.80	12.81	0.00	0.00	0.00	35.72
		6	-0.80	12.81	0.00	0.00	0.00	-35.72
	9	5	0.80	8.66	0.00	0.00	0.00	24.04
		6	-0.80	8.66	0.00	0.00	0.00	-24.04
	10	5	0.76	-20.22	0.00	0.00	0.00	-256.88
		6	-0.76	40.26	0.00	0.00	0.00	-312.64
	11	5	0.33	-26.40	0.00	0.00	0.00	-274.09
		6	-0.33	34.08	0.00	0.00	0.00	-295.43
	12	5	1.01	-89.93	0.00	0.00	0.00	-932.34
		6	-1.01	115.71	0.00	0.00	0.00	-1004.04
	13	5	0.07	-101.96	0.00	0.00	0.00	-965.80
		6	-0.07	103.68	0.00	0.00	0.00	-970.57
6	1	3	16.48	-1.58	0.00	0.00	0.00	-9.60
		5	-16.48	1.58	0.00	0.00	0.00	-11.71
	2	3	7.53	-1.01	0.00	0.00	0.00	-6.30
		5	-7.53	1.01	0.00	0.00	0.00	-7.38
	3	3	1.51	-0.02	0.00	0.00	0.00	-0.04
		5	-1.51	0.02	0.00	0.00	0.00	-0.18

*LRFD DESIGN,SEISMIC Z4,70MPH WIND,EXPS.C,--

MEMBER END FORCES   STRUCTURE TYPE = PLANE

ALL UNITS ARE -- KIP FEET

MEMBER	LOAD	JT	AXIAL	SHEAR-Y	SHEAR-Z	TORSION	MOM-Y	MOM-Z
	4	3	-64.40	32.10	0.00	0.00	0.00	287.52
		5	64.40	-32.10	0.00	0.00	0.00	145.83
	5	3	27.50	-2.78	0.00	0.00	0.00	-16.98
		5	-27.50	2.78	0.00	0.00	0.00	-20.55
	6	3	35.87	-3.58	0.00	0.00	0.00	-21.86
		5	-35.87	3.58	0.00	0.00	0.00	-26.50
	7	3	-383.17	191.00	0.00	0.00	0.00	1710.73
		5	383.17	-191.00	0.00	0.00	0.00	867.70
	8	3	32.58	-3.52	0.00	0.00	0.00	-21.63
		5	-32.58	3.52	0.00	0.00	0.00	-25.96
	9	3	25.95	-2.43	0.00	0.00	0.00	-14.75
		5	-25.95	2.43	0.00	0.00	0.00	-18.04
	10	3	-36.90	29.32	0.00	0.00	0.00	270.54
		5	36.90	-29.32	0.00	0.00	0.00	125.28
	11	3	-53.19	31.03	0.00	0.00	0.00	280.99
		5	53.19	-31.03	0.00	0.00	0.00	137.87
	12	3	-183.09	105.56	0,00	0.00	0.00	955.70
		5	183.09	-105.56	0.00	0.00	0.00	469.33
	13	3	-216.45	108.90	0.00	0.00	0.00	976.10
		5	216.45	-108.90	0.00	0.00	0.00	494.05
7	1	6	10.83	1.09	0.00	0.00	0.00	7.28
		8	-10.83	-1.09	0.00	0.00	0.00	7.42
	2	6	3.77	0.91	0.00	0.00	0.00	5.38
		8	-3.77	-0.91	0.00	0.00	0.00	6.92
	3	6	1.51	-0.08	0.00	0.00	0.00	-0.23
		8	-1.51	0.08	0.00	0.00	0.00	-0.86
	4	6	34.16	24.30	0.00	0.00	0.00	156.57
		8	-34.16	-24.30	0.00	0.00	0.00	171.48
	5	6	17.48	2.02	0.00	0.00	0.00	13.17
		8	-17.48	-2.02	0.00	0.00	0.00	14.15
	6	6	22.98	2.58	0.00	0.00	0.00	16.87
		8	-22.98	-2.58	0.00	0.00	0.00	17.92
	7	6	203.24	144.58	0.00	0.00	0.00	931.58
		8	-203.24	-144.58	0.00	0.00	0.00	1020.32
	8	6	19.78	2.72	0.00	0.00	0.00	17.23
		8	-19.78	-2.72	0.00	0.00	0.00	19.54
	9	6	17.29	1.63	0.00	0.00	0.00	11.05
		8	-17.29	-1.63	0.00	0.00	0.00	10.99
	10	6	51.64	26.32	0.00	0.00	0.00	169.74
		8	-51.64	-26.32	0.00	0.00	0.00	185.63
	11	6	41.52	25.04	0.00	0.00	0.00	161.52
		8	-41.52	-25.04	0:00	0.00	0.00	176.53
	12	6	139.12	85.20	0.00	0.00	0.00	549.20
		8	-139.12	-85.20	0.00	0.00	0.00	600.95
	13	6	117.78	82.79	0.00	0.00	0.00	533.44
		8	-117.78	-82.79	0.00	0.00	0.00	584.17

*LRFD DESIGN,SEISMIC Z4,70MPH WIND,EXPS.C,--

MEMBER END FORCES   STRUCTURE TYPE = PLANE

ALL UNITS ARE -- KIP FEET

MEMBER	LOAD	JT	AXIAL	SHEAR-Y	SHEAR-Z	TORSION	MOM-Y	MOM-Z
8	1	7	-1.05	5.65	0.00	0.00	0.00	15.58
		8	1.05	5.65	0.00	0.00	0.00	-15.58
	2	7	0.55	3.77	0.00	0.00	0.00	9.40
		8	-0.55	3.77	0.00	0.00	0.00	-9.40
	3	7	-0.54	0.00	0.00	0.00	0.00	0.43
		8	0.54	0.00	0.00	0.00	0.00	-0.43
	4	7	0.00	-26.23	0.00	0.00	0.00	-246.97
		8	0.00	26.23	0.00	0.00	0.00	-246.97
	5	7	-1.25	10.02	0.00	0.00	0.00	27.13
		8	1.25	10.02	0.00	0.00	0.00	-27.13
	6	7	-1.78	12.89	0.00	0.00	0.00	35.05
		8	1.78	12.89	0.00	0.00	0.00	-35.05
	7	7	0.00	-156.05	0.00	0.00	0.00	-1469.47
		8	0.00	156.05	0.00	0.00	0.00	-1469.47
	8	7	-0.66	12.81	0.00	0.00	0.00	33.95
		8	0.66	12.81	0.00	0.00	0.00	-33.95
	9	7	-1.86	8.66	0.00	0.00	0.00	24.08
		8	1.86	8.66	0.00	0.00	0.00	-24.08
	10	7	-1.25	-16.21	0.00	0.00	0.00	-219.84
		8	1.25	36.25	0.00	0.00	0.00	-274.10
	11	7	-0.72	-22.38	0.00	0.00	0.00	-236.38
		8	0.72	30.07	0.00	0.00	0.00	-257.56
	12	7	-1.78	-76.28	0.00	0.00	0.00	-804.65
		8	1.78	102.06	0.00	0.00	0.00	-874.75
	13	7	-0.16	-88.31	0.00	0.00	0.00	-837.33
		8	0.16	90.03	0.00	0.00	0.00	-842.07
9	1	5	10.83	-1.09	0.00	0.00	0.00	-7.28
		7	-10.83	1.09	0.00	0.00	0.00	-7.42
	2	5	3.77	-0.91	0.00	0.00	0.00	-5.38
		7	-3.77	0.91	0.00	0.00	0.00	-6.92
	3	5	1.51	0.08	0.00	0.00	0.00	0.23
		7	-1.51	-0.08	0.00	0.00	0.00	0.86
	4	5	-34.16	24.30	0.00	0.00	0.00	156.57
		7	34.16	-24.30	0.00	0.00	0.00	171.48
	5	5	17.48	-2.02	0.00	0.00	0.00	-13.17
		7	-17.48	2.02	0.00	0.00	0.00	-14.15
	6	5	22.98	-2.58	0.00	0.00	0.00	-16.87
		7	-22.98	2.58	0.00	0.00	0.00	-17.92
	7	5	-203.24	144.58	0.00	0.00	0.00	931.58
		7	203.24	-144.58	0.00	0.00	0.00	1020.32
	8	5	19.78	-2.72	0.00	0.00	0.00	-17.23
		7	-19.78	2.72	0.00	0.00	0.00	-19.54
	9	5	17.29	-1.63	0.00	0.00	0.00	-11.05
		7	-17.29	1.63	0.00	0.00	0.00	-10.99
	10	5	-16.68	22.28	0.00	0.00	0.00	143.40
		7	16.68	-22.28	0.00	0.00	0.00	157.34

*LRFD DESIGN,SEISMIC Z4,70MPH WIND,EXPS.C,--

MEMBER END FORCES   STRUCTURE TYPE = PLANE

ALL UNITS ARE -- KIP FEET

MEMBER	LOAD	JT	AXIAL	SHEAR-Y	SHEAR-Z	TORSION	MOM-Y	MOM-Z
	11	5	-26.79	23.56	0.00	0.00	0.00	151.62
		7	26.79	-23.56	0.00	0.00	0.00	166.43
	12	5	-93.16	80.04	0.00	0.00	0.00	515.46
		7	93.16	-80.04	0.00	0.00	0.00	565.12
	13	5	-114.49	82.45	0.00	0.00	0.00	531.23
		7	114.49	-82.45	0.00	0.00	0.00	581.91
10	1	8	5.18	2.14	0.00	0.00	0.00	11.45
		10	-5.18	-2.14	0.00	0.00	0.00	17.48
	2	8	0.00	0.37	0.00	0.00	0.00	4.68
		10	0.00	-0.37	0.00	0.00	0.00	0.26
	3	8	1.51	0.46	0.00	0.00	0.00	1.29
		10	-1.51	-0.46	0.00	0.00	0.00	4.97
	4	8	7.93	12.60	0.00	0.00	0.00	90.79
		10	-7.93	-12.60	0.00	0.00	0.00	79.31
	5	8	7.46	3.27	0.00	0.00	0.00	18.83
		10	-7.46	-3.27	0.00	0.00	0.00	25.30
	6	8	10.09	4.36	0.00	0.00	0.00	24.65
		10	-10.09	-4.36	0.00	0.00	0.00	34.18
	7	8	47.19	74.97	0.00	0.00	0.00	540.19
		10	-47.19	-74.97	0.00	0.00	0.00	471.91
	8	8	6.97	3.39	0.00	0.00	0.00	21.88
		10	-6.97	-3.39	0.00	0.00	0.00	23.87
	9	8	8.63	3.50	0.00	0.00	0.00	18.15
		10	-8.63	-3.50	0.00	0.00	0.00	29.06
	10	8	15.39	15.87	0.00	0.00	0.00	109.62
		10	-15.39	-15.87	0.00	0.00	0.00	104.61
	11	8	11.45	14.06	0.00	0.00	0.00	98.58
		10	-11.45	-14.06	0.00	0.00	0.00	91.20
	12	8	37.06	47.20	0.00	0.00	0.00	333.33
		10	-37.06	-47.20	0.00	0.00	0.00	303.84
	13	8	27.75	43.17	0.00	0.00	0.00	310.42
		10	-27.75	-43.17	0.00	0.00	0.00	272.32
11	1	9	2.14	5.18	0.00	0.00	0.00	14.46
		10	-2.14	5.18	0.00	0.00	0.00	-14.46
	2	9	0.37	0.00	0.00	0.00	0.00	0.26
		10	-0.37	0.00	0.00	0.00	0.00	-0.26
	3	9	0.46	1.51	0.00	0.00	0.00	4.09
		10	-0.46	1.51	0.00	0.00	0.00	-4.09
	4	9	0.00	-7.93	0.00	0.00	0.00	-74.69
		10	0.00	7.93	0.00	0.00	0.00	-74.69
	5	9	3.27	7.46	0.00	0.00	0.00	20.95
		10	-3.27	7.46	0.00	0.00	0.00	-20.95
	6	9	4.36	10.09	0.00	0.00	0.00	28.30
		10	-4.36	10.09	0.00	0.00	0.00	-28.30
	7	9	0.00	-47.19	0.00	0.00	0.00	-444.38
		10	0.00	47.19	0.00	0.00	0.00	-444.38

*LRFD DESIGN,SEISMIC 24,70MPH WIND,EXPS.C,--

MEMBER END FORCES   STRUCTURE TYPE = PLANE

ALL UNITS ARE -- KIP   FEET

MEMBER	LOAD	JT	AXIAL	SHEAR-Y	SHEAR-Z	TORSION	MOM-Y	MOM-Z
	8	9	3.39	6.97	0.00	0.00	0.00	19.81
		10	-3.39	6.97	0.00	0.00	0.00	-19.81
	9	9	3.50	8.63	0.00	0.00	0.00	24.03
		10	-3.50	8.63	0.00	0.00	0.00	-24.03
	10	9	3.27	-0.47	0.00	0.00	0.00	-53.73
		10	-3.27	15.39	0.00	0.00	0.00	-95.64
	11	9	1.46	-4.41	0.00	0.00	0.00	-64.85
		10	-1.46	11.45	0.00	0.00	0.00	-84.52
	12	9	4.36	-16.88	0.00	0.00	0.00	-225.64
		10	-4.36	37.06	0.00	0.00	0.00	-282.23
	13	9	0.33	-26.18	0.00	0.00	0.00	-251.73
		10	-0.33	27.75	0.00	0.00	0.00	-256.13
12	1	7	5.18	-2.14	0.00	0.00	0.00	-11.45
		9	-5.18	2.14	0.00	0.00	0.00	-17.48
	2	7	0.00	-0.37	0.00	0.00	0.00	-4.68
		9	0.00	0.37	0.00	0.00	0.00	-0.26
	3	7	1.51	-0.46	0.00	0.00	0.00	-1.29
		9	-1.51	0.46	0.00	0.00	0.00	-4.97
	4	7	-7.93	12.60	0.00	0.00	0.00	90.79
		9	7.93	-12.60	0.00	0.00	0.00	79.31
	5	7	7.46	-3.27	0.00	0.00	0.00	-18.83
		9	-7.46	3.27	0.00	0.00	0.00	-25.30
	6	7	10.09	-4.36	0.00	0.00	0.00	-24.65
		9	-10.09	4.36	0.00	0.00	0.00	-34.18
	7	7	-47.19	74.97	0.00	0.00	0.00	540.19
		9	47.19	-74.97	0.00	0.00	0.00	471.91
	8	7	6.97	-3.39	0.00	0.00	0.00	-21.88
		9	-6.97	3.39	0.00	0.00	0.00	-23.87
	9	7	8.63	-3.50	0.00	0.00	0.00	-18.15
		9	-8.63	3.50	0.00	0.00	0.00	-29.06
	10	7	-0.47	9.33	0.00	0.00	0.00	71.96
		9	0.47	-9.33	0.00	0.00	0.00	54.01
	11	7	-4.41	11.14	0.00	0.00	0.00	83.00
		9	4.41	-11.14	0.00	0.00	0.00	67.43
	12	7	-16.88	38.48	0.00	0.00	0.00	284.03
		9	16.88	-38.48	0.00	0.00	0.00	235.48
	13	7	-26.18	42.51	0.00	0.00	0.00	306.94
		9	26.18	-42.51	0.00	0.00	0.00	267.00

************* END OF LATEST ANALYSIS RESULT *************

74. PRINT JOINT DISPLACEMENT

*LRFD DESIGN,SEISMIC Z4,70MPH WIND,EXPS.C,--

JOINT DISPLACEMENT (INCH RADIANS)   STRUCTURE TYPE = PLANE

JOINT	LOAD	X-TRANS	Y-TRANS	Z-TRANS	X-ROTAN	Y-ROTAN	Z-ROTAN
1	1	0.00000	0.00000	0.00000	0.00000	0.00000	0.00000
	2	0.00000	0.00000	0.00000	0.00000	0.00000	0.00000
	3	0.00000	0.00000	0.00000	0.00000	0.00000	0.00000
	4	0.00000	0.00000	0.00000	0.00000	0.00000	0.00000
	5	0.00000	0.00000	0.00000	0.00000	0.00000	0.00000
	6	0.00000	0.00000	0.00000	0.00000	0.00000	0.00000
	7	0.00000	0.00000	0.00000	0.00000	0.00000	0.00000
	8	0.00000	0.00000	0.00000	0.00000	0.00000	0.00000
	9	0.00000	0.00000	0.00000	0.00000	0.00000	0.00000
	10	0.00000	0.00000	0.00000	0.00000	0.00000	0.00000
	11	0.00000	0.00000	0.00000	0.00000	0.00000	0.00000
	12	0.00000	0.00000	0.00000	0.00000	0.00000	0.00000
	13	0.00000	0.00000	0.00000	0.00000	0.00000	0.00000
2	1	0.00000	0.00000	0.00000	0.00000	0.00000	0.00000
	2	0.00000	0.00000	0.00000	0.00000	0.00000	0.00000
	3	0.00000	0.00000	0.00000	0.00000	0.00000	0.00000
	4	0.00000	0.00000	0.00000	0.00000	0.00000	0.00000
	5	0.00000	0.00000	0.00000	0.00000	0.00000	0.00000
	6	0.00000	0.00000	0.00000	0.00000	0.00000	0.00000
	7	0.00000	0.00000	0.00000	.0.00000	0.00000	0.00000
	8	0.00000	0.00000	0.00000	0.00000	ơ.00000	0.00000
	9	0.00000	0.00000	0.00000	0.00000	0.00000	0.00000
	10	0.00000	0.00000	0.00000	0.00000	0.00000	0.00000
	11	0.00000	0.00000	0.00000	0.00000	0.00000	0.00000
	12	0.00000	0.00000	0.00000	0.00000	0.00000	0.00000
	13	0.00000	0.00000	0.00000	0.00000	0.00000	0.00000
3	1	-0.00011	-0.00182	0.00000	0.00000	0.00000	-0.00004
	2	-0.00007	-0.00093	0.00000	0.00000	0.00000	-0.00002
	3	0.00000	-0.00012	0.00000	0.00000	0.00000	0.00000
	4	0.62416	0.01078	0.00000	0.00000	0.00000	-0.00135
	5	-0.00019	-0.00309	0.00000	0.00000	0.00000	-0.00006
	6	-0.00025	-0.00402	0.00000	0.00000	0.00000	-0.00008
	7	3.71374	0.06416	0.00000	0.00000	0.00000	-0.00801
	8	-0.00024	-0.00374	0.00000	0.00000	0.00000	-0.00008
	9	-0.00017	-0.00285	0.00000	0.00000	0.00000	-0.00005
	10	0.62396	0.00769	0.00000	0.00000	0.00000	-0.00141
	11	0.62408	0.00954	0.00000	0.00000	0.00000	-0.00137
	12	2.12189	0.03265	0.00000	0.00000	0.00000	-0.00466
	13	2.12212	0.03639	0.00000	0.00000	0.00000	-0.00458
4	1	0.00011	-0.00182	0.00000	0.00000	0.00000	0.00004
	2	0.00007	-0.00093	0.00000	0.00000	0.00000	0.00002
	3	0.00000	-0.00012	0.00000	0.00000	0.00000	0.00000
	4	0.62416	-0.01078	0.00000	0.00000	0.00000	-0.00135
	5	0.00019	-0.00309	0.00000	0.00000	0.00000	0.00006
	6	0.00025	-0.00402	0.00000	0.00000	0.00000	0.00008
	7	3.71374	-0.06416	0.00000	0.00000	0.00000	-0.00801
	8	0.00024	-0.00374	0.00000	0.00000	0.00000	0.00008
	9	0.00017	-0.00285	0.00000	0.00000	0.00000	0.00005

*LRFD DESIGN,SEISMIC Z4,70MPH WIND,EXPS.C,--

JOINT DISPLACEMENT (INCH RADIANS) STRUCTURE TYPE = PLANE

JOINT	LOAD	X-TRANS	Y-TRANS	Z-TRANS	X-ROTAN	Y-ROTAN	Z-ROTAN
	10	0.62435	-0.01388	0.00000	0.00000	0.00000	-0.00128
	11	0.62423	-0.01202	0.00000	0.00000	0.00000	-0.00132
	12	2.12239	-0.04068	0.00000	0.00000	0.00000	-0.00450
	13	2.12216	-0.03694	0.00000	0.00000	0.00000	-0.00457
5	1	0.00011	-0.00318	0.00000	0.00000	0.00000	-0.00006
	2	0.00002	-0.00155	0.00000	0.00000	0.00000	-0.00004
	3	0.00002	-0.00025	0.00000	0.00000	0.00000	0.00000
	4	1.11783	0.01609	0.00000	0.00000	0.00000	-0.00290
	5	0.00016	-0.00536	0.00000	0.00000	0.00000	-0.00010
	6	0.00022	-0.00697	0.00000	0.00000	0.00000	-0.00013
	7	6.65111	0.09574	0.00000	0.00000	0.00000	-0.01728
	8	0.00017	-0.00643	0.00000	0.00000	0.00000	-0.00013
	9	0.00017	-0.00499	0.00000	0.00000	0.00000	-0.00009
	10	1.11800	0.01073	0.00000	0.00000	0.00000	-0.00301
	11	1.11791	0.01393	0.00000	0.00000	0.00000	-0.00294
	12	3.80085	0.04774	0.00000	0.00000	0.00000	-0.01000
	13	3.80065	0.05423	0.00000	0.00000	0.00000	-0.00988
6	1	-0.00011	-0.00318	0.00000	0.00000	0.00000	0.00006
	2	-0.00002	-0.00155	0.00000	0.00000	0.00000	0.00004
	3	-0.00002	-0.00025	0.00000	0.00000	0.00000	0.00000
	4	1.11783	-0.01609	0.00000	0.00000	0.00000	-0.00290
	5	-0.00016	-0.00536	0.00000	0.00000	0.00000	0.00010
	6	-0.00022	-0.00697	0.00000	0.00000	0.00000	0.00013
	7	6.65111	-0.09574	0.00000	0.00000	0.00000	-0.01728
	8	-0.00017	-0.00643	0.00000	0.00000	0.00000	0.00013
	9	-0.00017	-0.00499	0.00000	0.00000	0.00000	0.00009
	10	1.11767	-0.02145	0.00000	0.00000	0.00000	-0.00280
	11	1.11776	-0.01825	0.00000	0.00000	0.00000	-0.00286
	12	3.80042	-0.06168	0.00000	0.00000	0.00000	-0.00974
	13	3.80062	-0.05519	0.00000	0.00000	0.00000	-0.00986
7	1	-0.00023	-0.00476	0.00000	0.00000	0.00000	-0.00006
	2	0.00012	-0.00210	0.00000	0.00000	0.00000	-0.00007
	3	-0.00012	-0.00047	0.00000	0.00000	0.00000	0.00001
	4	1.78563	0.02108	0.00000	0.00000	0.00000	-0.00258
	5	-0.00027	-0.00791	0.00000	0.00000	0.00000	-0.00012
	6	-0.00039	-0.01033	0.00000	0.00000	0.00000	-0.00015
	7	10.62447	0.12540	0.00000	0.00000	0.00000	-0.01534
	8	-0.00014	-0.00931	0.00000	0.00000	0.00000	-0.00018
	9	-0.00040	-0.00751	0.00000	0.00000	0.00000	-0.00009
	10	1.78536	0.01317	0.00000	0.00000	0.00000	-0.00270
	11	1.78547	0.01784	0.00000	0.00000	0.00000	-0.00262
	12	6.07074	0.06133	0.00000	0.00000	0.00000	-0.00892
	13	6.07109	0.07093	0.00000	0.00000	0.00000	-0.00877
8	1	0.00023	-0.00476	0.00000	0.00000	0.00000	0.00006
	2	-0.00012	-0.00210	0.00000	0.00000	0.00000	0.00007
	3	0.00012	-0.00047	0.00000	0.00000	0.00000	-0.00001
	4	1.78563	-0.02108	0.00000	0.00000	0.00000	-0.00258
	5	0.00027	-0.00791	0.00000	0.00000	0.00000	0.00012

*LRFD DESIGN,SEISMIC 24,70MPH WIND,EXPS.C,--

JOINT DISPLACEMENT (INCH RADIANS)   STRUCTURE TYPE = PLANE

JOINT	LOAD	X-TRANS	Y-TRANS	Z-TRANS	X-ROTAN	Y-ROTAN	Z-ROTAN
	6	0.00039	-0.01033	0.00000	0.00000	0.00000	0.00015
	7	10.62447	-0.12540	0.00000	0.00000	0.00000	-0.01534
	8	0.00014	-0.00931	0.00000	0.00000	0.00000	0.00018
	9	0.00040	-0.00751	0.00000	0.00000	0.00000	0.00009
	10	1.78590	-0.02898	0.00000	0.00000	0.00000	-0.00245
	11	1.78578	-0.02431	0.00000	0.00000	0.00000	-0.00254
	12	6.07151	-0.08198	0.00000	0.00000	0.00000	-0.00861
	13	6.07116	-0.07238	0.00000	0.00000	0.00000	-0.00876
9	1	0.00066	-0.00552	0.00000	0.00000	0.00000	-0.00019
	2	0.00011	-0.00210	0.00000	0.00000	0.00000	0.00003
	3	0.00014	-0.00069	0.00000	0.00000	0.00000	-0.00007
	4	2.33959	0.02223	0.00000	0.00000	0.00000	-0.00283
	5	0.00101	-0.00900	0.00000	0.00000	0.00000	-0.00026
	6	0.00134	-0.01180	0.00000	0.00000	0.00000	-0.00036
	7	13.92055	0.13228	0.00000	0.00000	0.00000	-0.01683
	8	0.00104	-0.01033	0.00000	0.00000	0.00000	-0.00022
	9	0.00108	-0.00877	0.00000	0.00000	0.00000	-0.00033
	10	2.34060	0.01324	0.00000	0.00000	0.00000	-0.00309
	11	2.34004	0.01848	0.00000	0.00000	0.00000	-0.00296
	12	7.95594	0.06379	0.00000	0.00000	0.00000	-0.00998
	13	7.95470	0.07475	0.00000	0.00000	0.00000	-0.00965
10	1	-0.00066	-0.00552	0.00000	0.00000	0.00000	0.00019
	2	-0.00011	-0.00210	0.00000	0.00000	0.00000	-0.00003
	3	-0.00014	-0.00069	0.00000	0.00000	0.00000	0.00007
	4	2.33959	-0.02223	0.00000	0.00000	0.00000	-0.00283
	5	-0.00101	-0.00900	0.00000	0.00000	0.00000	0.00026
	6	-0.00134	-0.01180	0.00000	0.00000	0.00000	0.00036
	7	13.92055	-0.13228	0.00000	0.00000	0.00000	-0.01683
	8	-0.00104	-0.01033	0.00000	0.00000	0.00000	0.00022
	9	-0.00108	-0.00877	0.00000	0.00000	0.00000	0.00033
	10	2.33858	-0.03123	0.00000	0.00000	0.00000	-0.00256
	11	2.33914	-0.02598	0.00000	0.00000	0.00000	-0.00270
	12	7.95326	-0.08739	0.00000	0.00000	0.00000	-0.00926
	13	7.95450	-0.07643	0.00000	0.00000	0.00000	-0.00959

************* END OF LATEST ANALYSIS RESULT **************

75. FINISH

************** END OF STAAD-III **************

**** DATE= SEP 19,1999   TIME= 8:50:38 ****

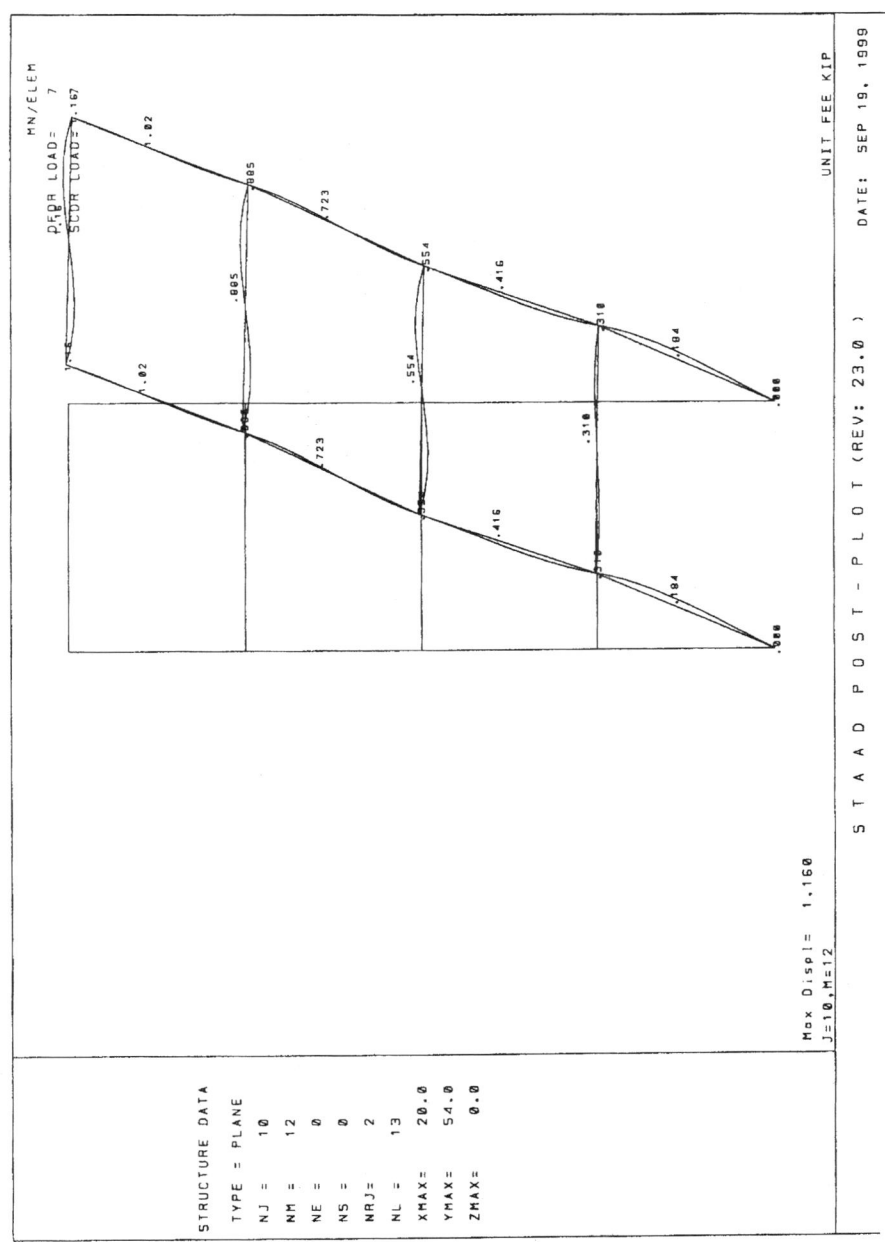

STRUCTURE DATA

TYPE = PLANE

NJ = 10
NM = 12
NE = 0
NS = 0
NRJ= 2
NL = 13

XMAX= 20.0
YMAX= 54.0
ZMAX= 0.0

Max Displ= 1.160
J=10,M=12

STAAD POST-PLOT (REV: 23.0)

UNIT FEE KIP   DATE: SEP 19, 1999

387

B. SMRF PROJECT PART II

```
******************************************************
*                                                    *
*             S T A A D - III                        *
*             Revision 23.0                          *
*             Proprietary Program of                 *
*             Research Engineers, Inc.               *
*             Date=    MAR 13, 1999                   *
*             Time=     8:52:27                       *
*             Build No.   1009.01.01                 *
*                                                    *
******************************************************
```

```
 1. STAAD PLANE ICBO FRAME
 2. *DRIFT & STRENGTH EVALUATION OF THE 94 PROTOTYPE 6 FRAME SMRF SYSTEM
 3. JOINT COORDINATES
 4. 1 0. 0.; 2 20. 0.; 3 0. 13.5; 4 20. 13.5; 5 0. 27.; 6 20. 27.
 5. 7 0. 40.5; 8 20. 40.5; 9 0. 54.; 10 20.0 54.
 6. MEMBER INCIDENCE
 7. 1 1 3; 2 2 4; 3 3 4; 4 4 6; 5 5 6; 6 3 5; 7 6 8; 8 7 8; 9 5 7
 8. 10 8 10; 11 9 10; 12 7 9
 9. MEMBER PROPERTY AMERICAN
10. 1 2 TABLE ST W14X193
11. 4 6 7 9 10 12 TABLE ST W14X176
12. 3 TABLE ST W36X135
13. 5 TABLE ST W36X135
14. 8 TABLE ST W36X135
15. 11 TABLE ST W36X135
16. UNIT INCHES
17. CONSTANTS
18. E 29000.0 ALL
19. MEMBER OFFSET
20. 3 5 8 11 START 7.0 0.0
21. 3 5 8 11 END  -7.0 0.0
22. PRINT MEMBER INFORMATION
```

'DRIFT & STRENGTH EVALUATION OF THE 94 PROTO

MEMBER INFORMATION

MEMBER	START JOINT	END JOINT	LENGTH (INCH)	BETA (DEG)	RELEASES
1	1	3	162.000	0.00	
2	2	4	162.000	0.00	
3	3	4	226.000	0.00	
4	4	6	162.000	0.00	
5	5	6	226.000	0.00	
6	3	5	162.000	0.00	
7	6	8	162.000	0.00	
8	7	8	226.000	0.00	
9	5	7	162.000	0.00	
10	8	10	162.000	0.00	
11	9	10	226.000	0.00	
12	7	9	162.000	0.00	

************ END OF DATA FROM INTERNAL STORAGE ************

23. SUPPORT
24. 1 2 FIXED
25. MEMBER RELEASE
26. 1 2 START MP 0.88
27. DRAW MEMBER JOINT PROPERTY SUPPORT

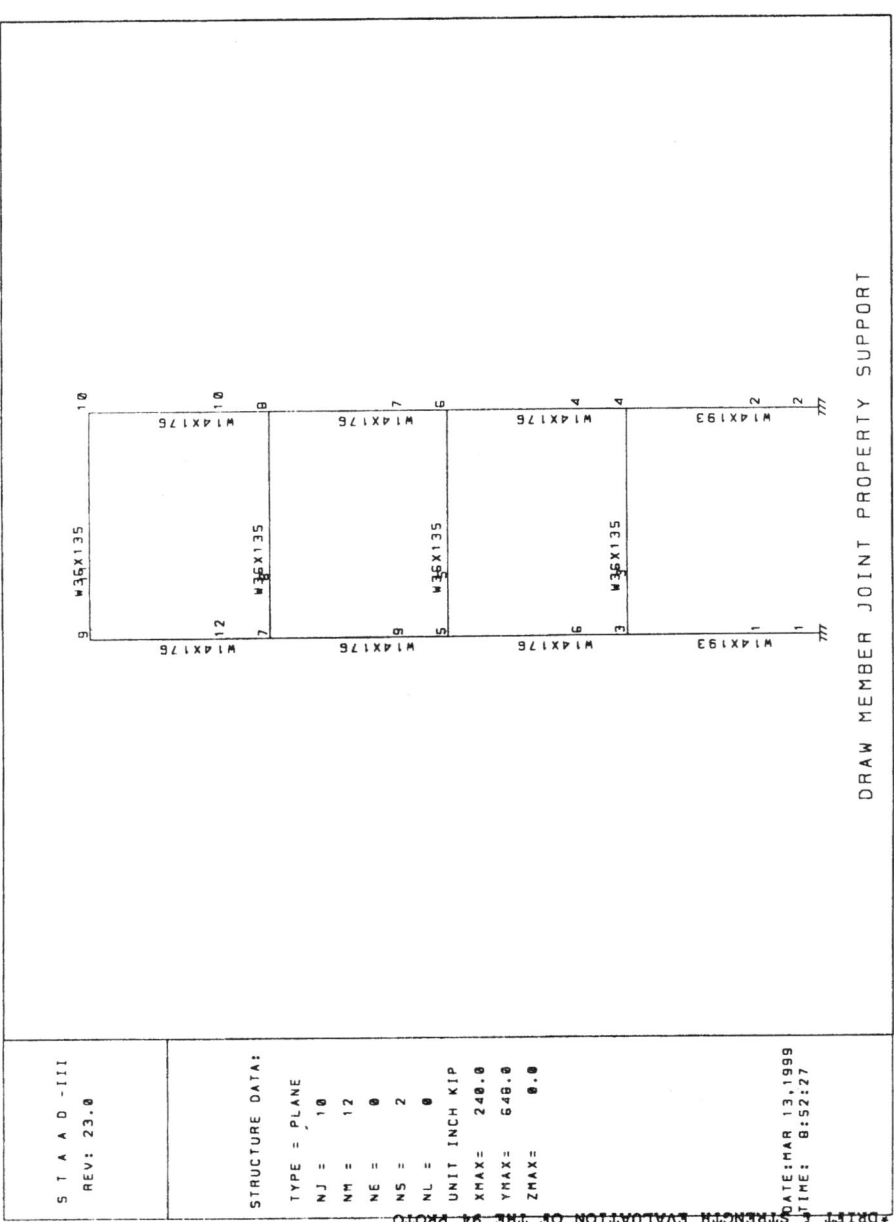

DRAW MEMBER JOINT PROPERTY SUPPORT

```
'DRIFT & STRENGTH EVALUATION OF THE 94 PROTO
28. UNIT FT
29. LOADING 1 DEAD
30. MEMBER LOAD
31. 3 UNI GY -0.6
32. 5 UNI GY -0.6
33. 8 UNI GY -0.6
34. 11 UNI GY -0.6
35. LOADING 2 LIVE
36. MEMBER LOAD
37. 3 UNI GY -0.4
38. 5 UNI GY -0.4
39. 8 UNI GY -0.4
40. 11 UNI GY -0.16
41. LOADING 6 EQ FROM LEFT AT *"STRENGTH LEVEL"
42. JOINT LOAD
43. 3 FX 7.8
44. 5 FX 15.6
45. 7 FX 23.5
46. 9 FX 25.2
47. LOAD COMBINATION 7 596 PERCENT (EQ)
48. 6 5.96
49. LOAD COMBINATION 8 DL + 70 PERCENT (LL) + 280 PERCENT (EQ)
50. 1 1.0 2 0.70 6 2.8
51. LOAD COMBINATION 9 85 PERCENT (DL) + 280 PERCENT (EQ)
52. 1 0.85 6 2.80
53. LOAD COMBINATION 10 120 PERC (DL)+ 50 PERC (LL) + 185 PERCENT (EQ)
54. 1 1.2 2 0.5 6 1.85
55. LOAD COMBINATION 11  120 PERCENT (DL) + (EQ)
56. 1 1.2 6 1.0
57. PERFORM ANALYSIS
```

P R O B L E M   S T A T I S T I C S

```
NUMBER OF JOINTS/MEMBER+ELEMENTS/SUPPORTS =    10/   12/    2
ORIGINAL/FINAL BAND-WIDTH =    2/    2
TOTAL PRIMARY LOAD CASES  =    3, TOTAL DEGREES OF FREEDOM =    24
SIZE OF STIFFNESS MATRIX =     216 DOUBLE PREC. WORDS
REQRD/AVAIL. DISK SPACE = 12.02/ 2047.7 MB,  EXMEM = 1968.7 MB
```

```
++ Processing Element Stiffness Matrix.           8:52:27
++ Processing Global Stiffness Matrix.            8:52:27
++ Processing Triangular Factorization.           8:52:27
++ Calculating Joint Displacements.               8:52:27
++ Calculating Member  Forces.                    8:52:27
```

```
58. PRINT MEMBER FORCES
```

MEMBER END FORCES   STRUCTURE TYPE = PLANE

ALL UNITS ARE -- KIP FEET

MEMBER	LOAD	JT	AXIAL	SHEAR-Y	SHEAR-Z	TORSION	MOM-Y	MOM-Z
1	1	1	22.60	-0.41	0.00	0.00	0.00	-0.34
		3	-22.60	0.41	0.00	0.00	0.00	-5.23
	2	1	12.81	-0.28	0.00	0.00	0.00	-0.23
		3	-12.81	0.28	0.00	0.00	0.00	-3.49
	6	1	-132.38	36.06	0.00	0.00	0.00	95.81
		3	132.38	-36.06	0.00	0.00	0.00	391.01
	7	1	-788.96	214.93	0.00	0.00	0.00	571.06
		3	788.96	-214.93	0.00	0.00	0.00	2330.44
	8	1	-339.09	100.37	0.00	0.00	0.00	267.78
		3	339.09	-100.37	0.00	0.00	0.00	1087.16
	9	1	-351.44	100.62	0.00	0.00	0.00	267.99
		3	351.44	-100.62	0.00	0.00	0.00	1090.39
	10	1	-211.37	66.08	0.00	0.00	0.00	176.73
		3	211.37	-66.08	0.00	0.00	0.00	715.35
	11	1	-105.26	35.57	0.00	0.00	0.00	95.40
		3	105.26	-35.57	0.00	0.00	0.00	384.73
2	1	2	22.60	0.41	0.00	0.00	0.00	0.34
		4	-22.60	-0.41	0.00	0.00	0.00	5.23
	2	2	12.81	0.28	0.00	0.00	0.00	0.23
		4	-12.81	-0.28	0.00	0.00	0.00	3.49
	6	2	132.38	36.04	0.00	0.00	0.00	95.73
		4	-132.38	-36.04	0.00	0.00	0.00	390.79
	7	2	788.96	214.79	0.00	0.00	0.00	570.55
		4	-788.96	-214.79	0.00	0.00	0.00	2329.12
	8	2	402.22	101.51	0.00	0.00	0.00	268.55
		4	-402.22	-101.51	0.00	0.00	0.00	1101.90
	9	2	389.86	101.26	0.00	0.00	0.00	268.33
		4	-389.86	-101.26	0.00	0.00	0.00	1098.67
	10	2	278.42	67.31	0.00	0.00	0.00	177.62
		4	-278.42	-67.31	0.00	0.00	0.00	730.99
	11	2	159.50	36.53	0.00	0.00	0.00	96.14
		4	-159.50	-36.53	0.00	0.00	0.00	397.07
3	1	3	-0.69	5.65	0.00	0.00	0.00	9.60
		4	0.69	5.65	0.00	0.00	0.00	-9.60
	2	3	-0.45	3.77	0.00	0.00	0.00	6.39
		4	0.45	3.77	0.00	0.00	0.00	-6.39
	6	3	3.91	-59.60	0.00	0.00	0.00	-561.50
		4	-3.91	59.60	0.00	0.00	0.00	-561.01
	7	3	23.29	-355.23	0.00	0.00	0.00	-3346.56
		4	-23.29	355.23	0.00	0.00	0.00	-3343.61
	8	3	9.94	-158.60	0.00	0.00	0.00	-1558.14
		4	-9.94	175.17	0.00	0.00	0.00	-1584.89

MEMBER END FORCES   STRUCTURE TYPE = PLANE

ALL UNITS ARE -- KIP FEET

MEMBER	LOAD	JT	AXIAL	SHEAR-Y	SHEAR-Z	TORSION	MOM-Y	MOM-Z
	9	3	10.36	-162.08	0.00	0.00	0.00	-1564.05
		4	-10.36	171.69	0.00	0.00	0.00	-1578.98
	10	3	6.18	-101.60	0.00	0.00	0.00	-1024.07
		4	-6.18	118.93	0.00	0.00	0.00	-1052.58
	11	3	3.08	-52.82	0.00	0.00	0.00	-549.99
		4	-3.08	66.38	0.00	0.00	0.00	-572.52
4	1	4	16.95	1.10	0.00	0.00	0.00	7.66
		6	-16.95	-1.10	0.00	0.00	0.00	7.21
	2	4	9.04	0.73	0.00	0.00	0.00	5.09
		6	-9.04	-0.73	0.00	0.00	0.00	4.77
	6	4	72.77	32.13	0.00	0.00	0.00	204.98
		6	-72.77	-32.13	0.00	0.00	0.00	228.78
	7	4	433.73	191.50	0.00	0.00	0.00	1221.71
		6	-433.73	-191.50	0.00	0.00	0.00	1363.52
	8	4	227.04	91.58	0.00	0.00	0.00	585.18
		6	-227.04	-91.58	0.00	0.00	0.00	651.13
	9	4	218.17	90.90	0.00	0.00	0.00	580.47
		6	-218.17	-90.90	0.00	0.00	0.00	646.71
	10	4	159.49	61.13	0.00	0.00	0.00	390.96
		6	-159.49	-61.13	0.00	0.00	0.00	434.28
	11	4	93.11	33.45	0.00	0.00	0.00	214.17
		6	-93.11	-33.45	0.00	0.00	0.00	237.43
5	1	5	0.10	5.65	0.00	0.00	0.00	10.76
		6	-0.10	5.65	0.00	0.00	0.00	-10.76
	2	5	0.03	3.77	0.00	0.00	0.00	7.24
		6	-0.03	3.77	0.00	0.00	0.00	-7.24
	6	5	7.81	-38.51	0.00	0.00	0.00	-362.95
		6	-7.81	38.51	0.00	0.00	0.00	-362.39
	7	5	46.52	-229.54	0.00	0.00	0.00	-2163.18
		6	-46.52	229.54	0.00	0.00	0.00	-2159.87
	8	5	21.97	-99.55	0.00	0.00	0.00	-1000.43
		6	-21.97	116.13	0.00	0.00	0.00	-1030.53
	9	5	21.94	-103.04	0.00	0.00	0.00	-1007.11
		6	-21.94	112.64	0.00	0.00	0.00	-1023.85
	10	5	14.57	-62.59	0.00	0.00	0.00	-654.92
		6	-14.57	79.91	0.00	0.00	0.00	-686.96
	11	5	7.93	-31.73	0.00	0.00	0.00	-350.04
		6	-7.93	45.29	0.00	0.00	0.00	-375.30
6	1	3	16.95	-1.10	0.00	0.00	0.00	-7.66
		5	-16.95	1.10	0.00	0.00	0.00	-7.21
	2	3	9.04	-0.73	0.00	0.00	0.00	-5.09
		5	-9.04	0.73	0.00	0.00	0.00	-4.77
	6	3	-72.77	32.17	0.00	0.00	0.00	205.26
		5	72.77	-32.17	0.00	0.00	0.00	229.03

*DRIFT & STRENGTH EVALUATION OF THE 94 PROTO

MEMBER END FORCES   **STRUCTURE TYPE = PLANE**

ALL UNITS ARE -- KIP FEET

MEMBER	LOAD	JT	AXIAL	SHEAR-Y	SHEAR-Z	TORSION	MOM-Y	MOM-Z
	7	3	-433.73	191.73	0.00	0.00	0.00	1223.34
		5	433.73	-191.73	0.00	0.00	0.00	1365.01
	8	3	-180.49	88.46	0.00	0.00	0.00	563.50
		5	180.49	-88.46	0.00	0.00	0.00	630.73
	9	3	-189.36	89.14	0.00	0.00	0.00	568.21
		5	189.36	-89.14	0.00	0.00	0.00	635.15
	10	3	-109.77	57.83	0.00	0.00	0.00	367.99
		5	109.77	-57.83	0.00	0.00	0.00	412.66
	11	3	-52.43	30.85	0.00	0.00	0.00	196.07
		5	52.43	-30.85	0.00	0.00	0.00	220.37
7	1	6	11.30	1.00	0.00	0.00	0.00	6.84
		8	-11.30	-1.00	0.00	0.00	0.00	6.67
	2	6	5.27	0.70	0.00	0.00	0.00	4.67
		8	-5.27	-0.70	0.00	0.00	0.00	4.84
	6	6	34.26	24.32	0.00	0.00	0.00	156.08
		8	-34.26	-24.32	0.00	0.00	0.00	172.29
	7	6	204.18	144.97	0.00	0.00	0.00	930.25
		8	-204.18	-144.97	0.00	0.00	0.00	1026.88
	8	6	110.92	69.60	0.00	0.00	0.00	447.14
		8	-110.92	-69.60	0.00	0.00	0.00	492.49
	9	6	105.53	68.96	0.00	0.00	0.00	442.84
		8	-105.53	-68.96	0.00	0.00	0.00	488.10
	10	6	79.58	46.55	0.00	0.00	0.00	299.30
		8	-79.58	-46.55	0.00	0.00	0.00	329.18
	11	6	47.82	25.53	0.00	0.00	0.00	164.29
		8	-47.82	-25.53	0.00	0.00	0.00	180.30
8	1	7	-0.29	5.65	0.00	0.00	0.00	11.20
		8	0.29	5.65	0.00	0.00	0.00	-11.20
	2	7	0.16	3.77	0.00	0.00	0.00	6.81
		8	-0.16	3.77	0.00	0.00	0.00	-6.81
	6	7	11.72	-24.77	0.00	0.00	0.00	-233.40
		8	-11.72	24.77	0.00	0.00	0.00	-233.07
	7	7	69.88	-147.62	0.00	0.00	0.00	-1391.07
		8	-69.88	147.62	0.00	0.00	0.00	-1389.12
	8	7	32.65	-61.07	0.00	0.00	0.00	-637.56
		8	-32.65	77.64	0.00	0.00	0.00	-668.56
	9	7	32.58	-64.55	0.00	0.00	0.00	-644.01
		8	-32.58	74.15	0.00	0.00	0.00	-662.12
	10	7	21.42	-37.16	0.00	0.00	0.00	-414.96
		8	-21.42	54.49	0.00	0.00	0.00	-448.02
	11	7	11.37	-17.99	0.00	0.00	0.00	-219.97
		8	-11.37	31.55	0.00	0.00°	0.00	-246.51
9	1	5	11.30	-1.00	0.00	0.00	0.00	-6.84
		7	-11.30	1.00	0.00	0.00	0.00	-6.67

MEMBER END FORCES   **STRUCTURE TYPE = PLANE**

ALL UNITS ARE -- KIP FEET

MEMBER	LOAD	JT	AXIAL	SHEAR-Y	SHEAR-Z	TORSION	MOM-Y	MOM-Z
	2	5	5.27	-0.70	0.00	0.00	0.00	-4.67
		7	-5.27	0.70	0.00	0.00	0.00	-4.84
	6	5	-34.26	24.38	0.00	0.00	0.00	156.39
		7	34.26	-24.38	0.00	0.00	0.00	172.69
	7	5	-204.18	145.28	0.00	0.00	0.00	932.07
		7	204.18	-145.28	0.00	0.00	0.00	1029.21
	8	5	-80.93	66.76	0.00	0.00	0.00	427.77
		7	80.93	-66.76	0.00	0.00	0.00	473.46
	9	5	-86.32	67.40	0.00	0.00	0.00	432.07
		7	86.32	-67.40	0.00	0.00	0.00	477.85
	10	5	-47.18	43.54	0.00	0.00	0.00	278.77
		7	47.18	-43.54	0.00	0.00	0.00	309.04
	11	5	-20.70	23.17	0.00	0.00	0.00	148.18
		7	20.70	-23.17	0.00	0.00	0.00	164.68
10	1	8	5.65	1.29	0.00	0.00	0.00	7.82
		10	-5.65	-1.29	0.00	0.00	0.00	9.66
	2	8	1.51	0.54	0.00	0.00	0.00	4.16
		10	-1.51	-0.54	0.00	0.00	0.00	3.17
	6	8	9.49	12.60	0.00	0.00	0.00	75.23
		10	-9.49	-12.60	0.00	0.00	0.00	94.88
	7	8	56.56	75.10	0.00	0.00	0.00	448.35
		10	-56.56	-75.10	0.00	0.00	0.00	565.50
	8	8	33.28	36.96	0.00	0.00	0.00	221.36
		10	-33.28	-36.96	0.00	0.00	0.00	277.56
	9	8	31.38	36.38	0.00	0.00	0.00	217.28
		10	-31.38	-36.38	0.00	0.00	0.00	273.89
	10	8	25.09	25.14	0.00	0.00	0.00	150.63
		10	-25.09	-25.14	0.00	0.00	0.00	188.72
	11	8	16.27	14.15	0.00	0.00	0.00	84.61
		10	-16.27	-14.15	0.00	0.00	0.00	106.48
11	1	9	1.29	5.65	0.00	0.00	0.00	6.37
		10	-1.29	5.65	0.00	0.00	0.00	-6.37
	2	9	0.54	1.51	0.00	0.00	0.00	2.29
		10	-0.54	1.51	0.00	0.00	0.00	-2.29
	6	9	12.60	-9.49	0.00	0.00	0.00	-89.39
		10	-12.60	9.49	0.00	0.00	0.00	-89.35
	7	9	75.10	-56.56	0.00	0.00	0.00	-532.77
		10	-75.10	56.56	0.00	0.00	0.00	-532.50
	8	9	36.96	-19.87	0.00	0.00	0.00	-242.32
		10	-36.96	33.28	0.00	0.00	0.00	-258.14
	9	9	36.38	-21.77	0.00	0.00	0.00	-244.88
		10	-36.38	31.38	0.00	0.00	0.00	-255.58
	10	9	25.14	-10.02	0.00	0.00	0.00	-156.58
		10	-25.14	25.09	0.00	0.00	0.00	-174.08
	11	9	14.15	-2.71	0.00	0.00	0.00	-81.75
		10	-14.15	16.27	0.00	0.00	0.00	-96.99

MEMBER END FORCES   STRUCTURE TYPE = PLANE

ALL UNITS ARE -- KIP FEET

MEMBER	LOAD	JT	AXIAL	SHEAR-Y	SHEAR-Z	TORSION	MOM-Y	MOM-Z
12	1	7	5.65	-1.29	0.00	0.00	0.00	-7.82
		9	-5.65	1.29	0.00	0.00	0.00	-9.66
	2	7	1.51	-0.54	0.00	0.00	0.00	-4.16
		9	-1.51	0.54	0.00	0.00	0.00	-3.17
	6	7	-9.49	12.60	0.00	0.00	0.00	75.16
		9	9.49	-12.60	0.00	0.00	0.00	94.93
	7	7	-56.56	75.09	0.00	0.00	0.00	447.97
		9	56.56	-75.09	0.00	0.00	0.00	565.77
	8	7	-19.87	33.60	0.00	0.00	0.00	199.73
		9	19.87	-33.60	0.00	0.00	0.00	253.91
	9	7	-21.77	34.18	0.00	0.00	0.00	203.81
		9	21.77	-34.18	0.00	0.00	0.00	257.58
	10	7	-10.02	21.48	0.00	0.00	0.00	127.59
		9	10.02	-21.48	0.00	0.00	0.00	162.43
	11	7	-2.71	11.05	0.00	0.00	0.00	65.78
		9	2.71	-11.05	0.00	0.00	0.00	83.33

************* END OF LATEST ANALYSIS RESULT *************

59. PRINT SUPPORT REACTIONS

SUPPORT REACTIONS -UNIT KIP FEET   STRUCTURE TYPE = PLANE

JOINT	LOAD	FORCE-X	FORCE-Y	FORCE-Z	MOM-X	MOM-Y	MOM Z
1	1	0.41	22.60	0.00	0.00	0.00	-0.34
	2	0.28	12.81	0.00	0.00	0.00	-0.23
	6	-36.06	-132.38	0.00	0.00	0.00	95.81
	7	-214.93	-788.96	0.00	0.00	0.00	571.06
	8	-100.37	-339.09	0.00	0.00	0.00	267.78
	9	-100.62	-351.44	0.00	0.00	0.00	267.99
	10	-66.08	-211.37	0.00	0.00	0.00	176.73
	11	-35.57	-105.26	0.00	0.00	0.00	95.40
2	1	-0.41	22.60	0.00	0.00	0.00	0.34
	2	-0.28	12.81	0.00	0.00	0.00	0.23
	6	-36.04	132.38	0.00	0.00	0.00	95.73
	7	-214.79	788.96	0.00	0.00	0.00	570.55
	8	-101.51	402.22	0.00	0.00	0.00	268.55
	9	-101.26	389.86	0.00	0.00	0.00	268.33
	10	-67.31	278.42	0.00	0.00	0.00	177.62
	11	-36.53	159.50	0.00	0.00	0.00	96.14

************* END OF LATEST ANALYSIS RESULT **************

60. FINISH

*************** END OF STAAD-III ***************

**** DATE= MAR 13,1999   TIME= 8:52:27 ****

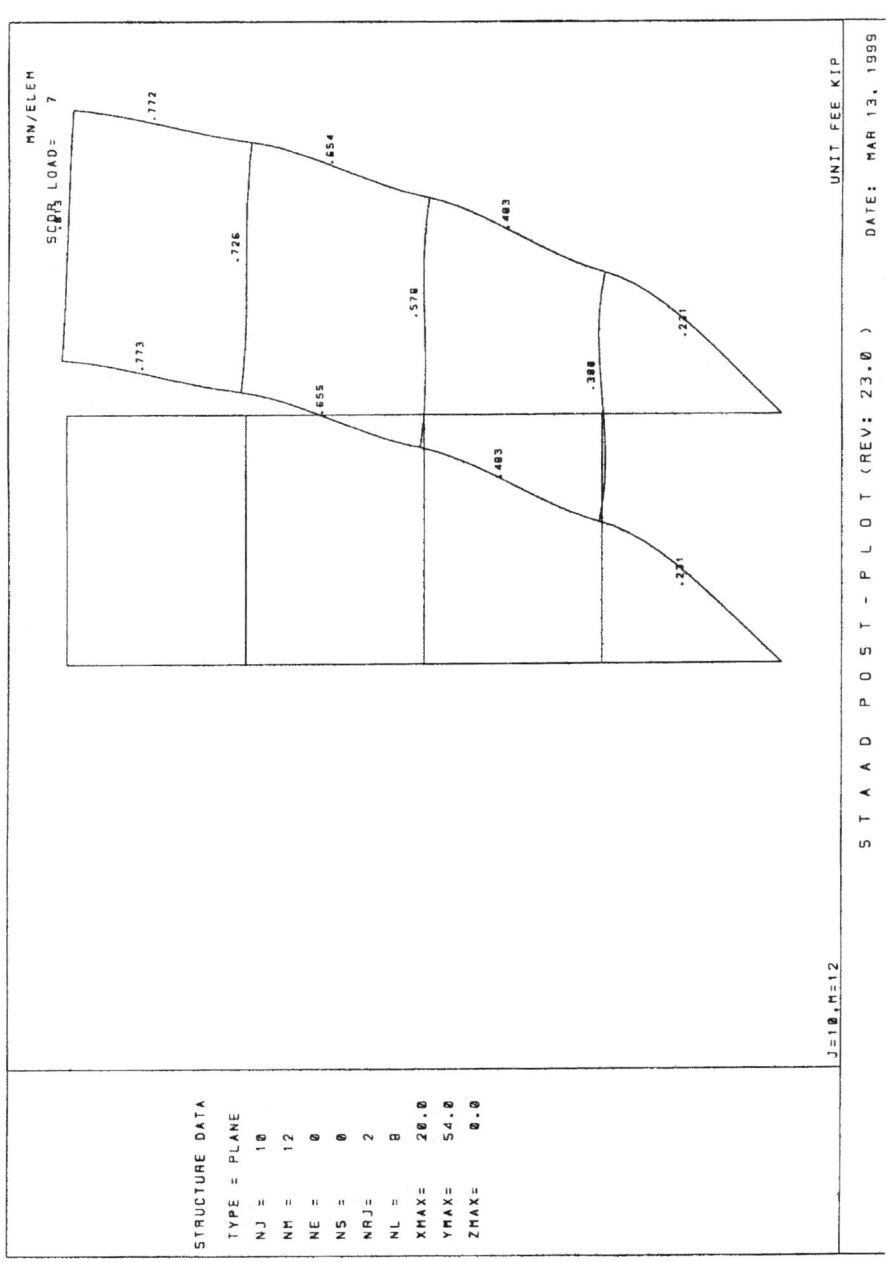

C. BRACED-FRAME PROJECT

```
**************************************************
*                                                *
*            S T A A D - III                     *
*            Revision 23.0                       *
*            Proprietary Program of              *
*            Research Engineers, Inc.            *
*            Date=    JUN 22, 1999               *
*            Time=     9:50:44                    *
*            Build No.    1009.01.01             *
*                                                *
***                                      *******
```

```
 1. STAAD PLANE ICBO FOUR STORY BRACED FRAME(SCBF)
 2. *SEISMIC ZONE 4 EARTHQUAKE, 85MPH WIND, LRFD DESIGN(AISC 2ND ED.)
 3. UNIT KIP FEET
 4. JOINT COORDINATES
 5. 1 0. 0.; 2 20. 0.; 3 0. 13.5; 4 20. 13.5; 5 0. 27.; 6 20. 27.
 6. 7 0. 40.5; 8 20 40.5; 9 0. 54; 10 20. 54.; 11 10 13.5; 12 10 40.5
 7. MEMBER INCIDENCE
 8. 1 1 3; 2 1 11; 3 2 11; 4 2 4; 5 3 11; 6 11 4; 7 11 6; 8 11 5
 9. 9 3 5; 10 4 6; 11 5 6; 12 5 12; 13 6 12; 14 6 8; 15 7 12
10. 16 12 8; 17 12 10; 18 12 9; 19 7 9; 20 8 10; 21 5 7; 22 9 10
11. MEMBER PROPERTY AMERICAN
12. 1 4 9 10 TABLE ST W12X96
13. 14 19 20 21 TABLE ST W12X40
14. 5 6 11 TABLE ST W10X54
15. 15 16 TABLE ST W10X39
16. 22 TABLE ST W10X33
17. 2 3 7 8  TABLE ST TUB70708
18. 12 13 17 18 TABLE ST TUB60606
19. MEMBER TRUSS
20. 2 3 7 8 12 13 17 18
21. PRINT MEMBER PROPERTY
```

*SEISMIC ZONE 4 EARTHQUAKE, 85MPH WIND, LRFD

MEMBER PROPERTIES. UNIT - INCH

MEMB	PROFILE	AX/ AY	IZ/ AZ	IY/ SZ	IX/ SY
1	ST W12 X96	28.20	833.00	270.00	6.86
		6.99	14.59	131.08	44.41
2	ST TUB 70708	12.40	84.60	84.60	141.00
		7.00	7.00	24.17	24.17
3	ST TUB 70708	12.40	84.60	84.60	141.00
		7.00	7.00	24.17	24.17
4	ST W12 X96	28.20	833.00	270.00	6.86
		6.99	14.59	131.08	44.41
5	ST W10 X54	15.80	303.00	103.00	1.82
		3.73	8.22	60.06	20.54
6	ST W10 X54	15.80	303.00	103.00	1.82
		3.73	8.22	60.06	20.54
7	ST TUB 70708	12.40	84.60	84.60	141.00
		7.00	7.00	24.17	24.17
8	ST TUB 70708	12.40	84.60	84.60	141.00
		7.00	7.00	24.17	24.17
9	ST W12 X96	28.20	833.00	270.00	6.86
		6.99	14.59	131.08	44.41
10	ST W12 X96	28.20	833.00	270.00	6.86
		6.99	14.59	131.08	44.41
11	ST W10 X54	15.80	303.00	103.00	1.82
		3.73	8.22	60.06	20.54
12	ST TUB 60606	8.08	41.60	41.60	68.50
		4.50	4.50	13.87	13.87
13	ST TUB 60606	8.08	41.60	41.60	68.50
		4.50	4.50	13.87	13.87
14	ST W12 X40	11.80	310.00	44.10	0.95
		3.52	5.50	51.93	11.02
15	ST W10 X39	11.50	209.00	45.00	0.98
		3.12	5.64	42.14	11.27
16	ST W10 X39	11.50	209.00	45.00	0.98
		3.12	5.64	42.14	11.27
17	ST TUB 60606	8.08	41.60	41.60	68.50
		4.50	4.50	13.87	13.87
18	ST TUB 60606	8.08	41.60	41.60	68.50
		4.50	4.50	13.87	13.87
19	ST W12 X40	11.80	310.00	44.10	0.95
		3.52	5.50	51.93	11.02
20	ST W12 X40	11.80	310.00	44.10	0.95
		3.52	5.50	51.93	11.02
21	ST W12 X40	11.80	310.00	44.10	0.95
		3.52	5.50	51.93	11.02
22	ST W10 X33	9.71	170.00	36.60	0.58
		2.82	4.62	34.94	9.20

*SEISMIC ZONE 4 EARTHQUAKE, 85MPH WIND, LRFD

*********** END OF DATA FROM INTERNAL STORAGE ************

22. UNIT INCHES
23. CONSTANTS
24. E 29000.0
25. MEMBER QFFSET
26. 5 15 START 6.5 0. 0.
27. 5 15 END -0. 0. 0.
28. 6 16 START 0. 0. 0.
29. 6 16 END -6.5 0. 0.
30. 11 22 START 6.5 0. 0.
31. 11 22 END -6.5 0. 0.
32. PRINT MEMBER INFORMATION

*SEISMIC ZONE 4 EARTHQUAKE, 85MPH WIND, LRFD

MEMBER INFORMATION

MEMBER	START JOINT	END JOINT	LENGTH (INCH)	BETA (DEG)	RELEASES
1	1	3	162.000	0.00	
2	1	11	201.604		TRUSS
3	2	11	201.604		TRUSS
4	2	4	162.000	0.00	
5	3	11	113.500	0.00	
6	11	4	113.500	0.00	
7	11	6	201.604		TRUSS
8	11	5	201.604		TRUSS
9	3	5	162.000	0.00	
10	4	6	162.000	0.00	
11	5	6	227.000	0.00	
12	5	12	201.604		TRUSS
13	6	12	201.604		TRUSS
14	6	8	162.000	0.00	
15	7	12	113.500	0.00	
16	12	8	113.500	0.00	
17	12	10	201.604		TRUSS
18	12	9	201.604		TRUSS
19	7	9	162.000	0.00	
20	8	10	162.000	0.00	
21	5	7	162.000	0.00	
22	9	10	227.000	0.00	

*********** END OF DATA FROM INTERNAL STORAGE ************

33. SUPPORT
34. 1 2 PINNED
35. DRAW MEMBER JOINT PROPERTY SUPPORT

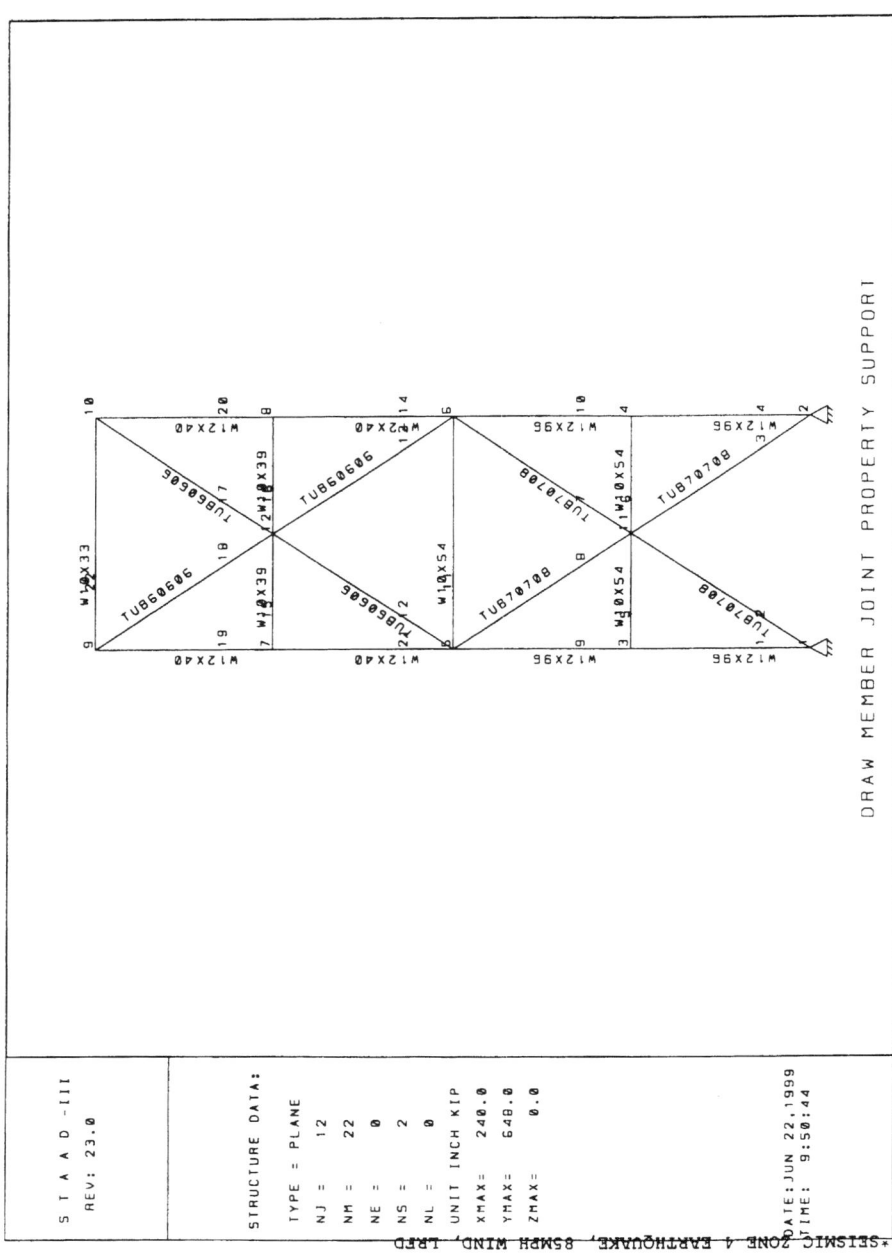

DRAW MEMBER JOINT PROPERTY SUPPORT

```
*SEISMIC ZONE 4 EARTHQUAKE, 85MPH WIND, LRFD
36. UNIT FT
37. LOADING 1 DEAD
38. MEMBER LOAD
39. 5 UNI GY -0.8
40. 6 UNI GY -0.8
41. 11 UNI GY -0.8
42. 15 UNI GY -0.8
43. 16 UNI GY -0.8
44. 22 UNI GY -0.6
45. LOADING `2 LIVE
46. MEMBER LOAD
47. 5 UNI GY -0.7
48. 6 UNI GY -0.7
49. 11 UNI GY -0.7
50. 15 UNI GY -0.7
51. 16 UNI GY -0.7
52. 22 UNI GY -0.2
53. LOADING 3 EQH FROM LEFT AT "STRENGTH LEVEL"
54. JOINT LOAD
55. 3 FX 11.65
56. 4 FX 11.65
57. 5 FX 23.3
58. 6 FX 23.3
59. 7 FX 35.0
60. 8 FX 35.0
61. 9 FX 31.0
62. 10 FX 31.0
63. LOADING 4 W FROM LEFT
64. JOINT LOAD
65. 3 FX 15.6
66. 5 FX 16.6
67. 7 FX 17.7
68. 9 FX 9.8
69. LOAD COMBINATION 5 120 PERCENT (DL) + 50 PERCENT (LL)+ 100 PERCENT (EQH)
70. 1 1.2 2 0.5 3 1.0
71. LOAD COMBINATION 6 90 PERCENT (DL) + 100 PERCENT (EQH)
72. 1 0.9 3 1.0
73. LOAD COMBINATION 7 120 PERCENT (DL) + 160 PERCENT (LL)
74. 1 1.2 2 1.6
75. LOAD COMBINATION 8 120 PERCENT (DL) + 130 PERCENT (W) + 50 PERCENT (LL)
76. 1 1.2 2 0.5 4 1.3
77. LOAD COMBINATION 9 180 PERCENT (DL) + 75 PERCENT LL + 150 PERCENT EQH
78. 1 1.8 2 0.75 3 1.5
79. LOAD COMBINATION 10 135 PERCENT (DL) + 150 PERCENT (EQH)
80. 1 1.35 3 1.5
81. LOAD COMBINATION 11 135 PERCENT (DL) + 195 PERCENT (W)
82. 1 1.35 4 1.95
83. LOAD COMBINATION 12 181 PERCENT (DL) + 50 PERCENT LL + 256 PERCENT EQH
84. 1 1.8144 2 0.5 3 2.56
85. LOAD COMBINATION 13 28.56 PERCENT (DL) + 256 PERCENT (EQH)
86. 1 0.2856 3 2.56
87. LOAD COMBINATION 14 120 PERCENT (DL) + 50 PERCENT LL + 256 PERCENT EQH
88. 1 1.2 2 0.5 3 2.56
89. LOAD COMBINATION 15 90 PERCENT (DL) + 256 PERCENT (EQH)
90. 1 0.9 3 2.56
91. LOAD COMBINATION 16 448 PERCENT (EQH) FOR M. INELASTIC R. DSPLACEMENT
```

```
*SEISMIC ZONE 4 EARTHQUAKE, 85MPH WIND, LRFD
92. 3 4.48
93. PERFORM ANALYSIS
```

P R O B L E M   S T A T I S T I C S

```
NUMBER OF JOINTS/MEMBER+ELEMENTS/SUPPORTS =    12/   22/    2
ORIGINAL/FINAL BAND-WIDTH =   10/    5
TOTAL PRIMARY LOAD CASES  =    4, TOTAL DEGREES OF FREEDOM =    32
SIZE OF STIFFNESS MATRIX  =    512 DOUBLE PREC. WORDS
REQRD/AVAIL. DISK SPACE = 12.04/ 2047.7 MB,  EXMEM = 1963.5 MB
```

```
++ Processing Element Stiffness Matrix.             9:50:44
++ Processing Global Stiffness Matrix.              9:50:45
++ Processing Triangular Factorization.             9:50:45
++ Calculating Joint Displacements.                 9:50:45
++ Calculating Member  Forces.                      9:50:45
```

```
94. PRINT MEMBER FORCES
```

*SEISMIC ZONE 4 EARTHQUAKE, 85MPH WIND, LRFD

MEMBER END FORCES   STRUCTURE TYPE = PLANE

ALL UNITS ARE -- KIP   FEET

MEMBER	LOAD	JT	AXIAL	SHEAR-Y	SHEAR-Z	TORSION	MOM-Y	MOM-Z
1	1	1	22.26	0.01	0.00	0.00	0.00	0.00
		3	-22.26	-0.01	0.00	0.00	0.00	0.08
	2	1	16.91	0.00	0.00	0.00	0.00	0.00
		3	-16.91	0.00	0.00	0.00	0.00	-0.03
	3	1	-250.90	-0.45	0.00	0.00	0.00	0.00
		3	250.90	0.45	0.00	0.00	0.00	-6.05
	4	1	-55.59	0.03	0.00	0.00	0.00	0.00
		3	55.59	-0.03	0.00	0.00	0.00	0.47
	5	1	-215.74	-0.44	0.00	0.00	0.00	0.00
		3	215.74	0.44	0.00	0.00	0.00	-5.97
	6	1	-230.87	-0.44	0.00	0.00	0.00	0.00
		3	230.87	0.44	0.00	0.00	0.00	-5.98
	7	1	53.76	0.00	0.00	0.00	0.00	0.00
		3	-53.76	0.00	0.00	0.00	0.00	0.05
	8	1	-37.11	0.05	0.00	0.00	0.00	0.00
		3	37.11	-0.05	0.00	0.00	0.00	0.69
	9	1	-323.61	-0.66	0.00	0.00	0.00	0.00
		3	323.61	0.66	0.00	0.00	0.00	-8.95
	10	1	-346.30	-0.66	0.00	0.00	0.00	0.00
		3	346.30	0.66	0.00	0.00	0.00	-8.97
	11	1	-78.36	0.08	0.00	0.00	0.00	0.00
		3	78.36	-0.08	0.00	0.00	0.00	1.02
	12	1	-593.47	-1.14	0.00	0.00	0.00	0.00
		3	593.47	1.14	0.00	0.00	0.00	-15.36
	13	1	-635.95	-1.15	0.00	0.00	0.00	0.00
		3	635.95	1.15	0.00	0.00	0.00	-15.46
	14	1	-607.14	-1.14	0.00	0.00	0.00	0.00
		3	607.14	1.14	0.00	0.00	0.00	-15.40
	15	1	-622.27	-1.14	0.00	0.00	0.00	0.00
		3	622.27	1.14	0.00	0.00	0.00	-15.41
	16	1	-1124.03	-2.01	0.00	0.00	0.00	0.00
		3	1124.03	2.01	0.00	0.00	0.00	-27.09
2	1	1	7.61	0.00	0.00	0.00	0.00	0.00
		11	-7.61	0.00	0.00	0.00	0.00	0.00
	2	1	6.03	0.00	0.00	0.00	0.00	0.00
		11	-6.03	0.00	0.00	0.00	0.00	0.00
	3	1	-170.35	0.00	0.00	0.00	0.00	0.00
		11	170.35	0.00	0.00	0.00	0.00	0.00
	4	1	-49.35	0.00	0.00	0.00	0.00	0.00
		11	49.35	0.00	0.00	0.00	0.00	0.00
	5	1	-158.20	0.00	0.00	0.00	0.00	0.00
		11	158.20	0.00	0.00	0.00	0.00	0.00
	6	1	-163.50	0.00	0.00	0.00	0.00	0.00
		11	163.50	0.00	0.00	0.00	0.00	0.00

*SEISMIC ZONE 4 EARTHQUAKE, 85MPH WIND, LRFD

MEMBER END FORCES   STRUCTURE TYPE = PLANE

ALL UNITS ARE -- KIP FEET

MEMBER	LOAD	JT	AXIAL	SHEAR-Y	SHEAR-Z	TORSION	MOM-Y	MOM-Z
	7	· 1	18.79	0.00	0.00	0.00	0.00	0.00
		11	-18.79	0.00	0.00	0.00	0.00	0.00
	8	1	-52.00	0.00	0.00	0.00	0.00	0.00
		11	52.00	0.00	0.00	0.00	0.00	0.00
	9	1	-237.30	0.00	0.00	0.00	0.00	0.00
		11	237.30	0.00	0.00	0.00	0.00	0.00
	10	1	-245.25	0.00	0.00	0.00	0.00	0.00
		11	245.25	0.00	0.00	0.00	0.00	0.00
	11	1	-85.94	0.00	0.00	0.00	0.00	0.00
		11	85.94	0.00	0.00	0.00	0.00	0.00
	12	1	-419.27	0.00	0.00	0.00	0.00	0.00
		11	419.27	0.00	0.00	0.00	0.00	0.00
	13	1	-433.93	0.00	0.00	0.00	0.00	0.00
		11	433.93	0.00	0.00	0.00	0.00	0.00
	14	1	-423.95	0.00	0.00	0.00	0.00	0.00
		11	423.95	0.00	0.00	0.00	0.00	0.00
	15	1	-429.25	0.00	0.00	0.00	0.00	0.00
		11	429.25	0.00	0.00	0.00	0.00	0.00
	16	1	-763.17	0.00	0.00	0.00	0.00	0.00
		11	763.17	0.00	0.00	0.00	0.00	0.00
3	1	2	7.61	0.00	0.00	0.00	0.00	0.00
		11	-7.61	0.00	0.00	0.00	0.00	0.00
	2	2	6.03	0.00	0.00	0.00	0.00	0.00
		11	-6.03	0.00	0.00	0.00	0.00	0.00
	3	2	170.35	0.00	0.00	0.00	0.00	0.00
		11	-170.35	0.00	0.00	0.00	0.00	0.00
	4	2	50.89	0.00	0.00	0.00	0.00	0.00
		11	-50.89	0.00	0.00	0.00	0.00	0.00
	5	2	182.50	0.00	0.00	0.00	0.00	0.00
		11	-182.50	0.00	0.00	0.00	0.00	0.00
	6	2	177.20	0.00	0.00	0.00	0.00	0.00
		11	-177.20	0.00	0.00	0.00	0.00	0.00
	7	2	18.79	0.00	0.00	0.00	0.00	0.00
		11	-18.79	0.00	0.00	0.00	0.00	0.00
	8	2	78.31	0.00	0.00	0.00	0.00	0.00
		11	-78.31	0.00	0.00	0.00	0.00	0.00
	9	2	273.76	0.00	0.00	0.00	0.00	0.00
		11	-273.76	0.00	0.00	0.00	0.00	0.00
	10	2	265.81	0.00	0.00	0.00	0.00	0.00
		11	-265.81	0.00	0.00	0.00	0.00	0.00
	11	2	109.52	0.00	0.00	0.00	0.00	0.00
		11	-109.52	0.00	0.00	0.00	0.00	0.00
	12	2	452.93	0.00	0.00	0.00	0.00	0.00
		11	-452.93	0.00	0.00	0.00	0.00	0.00
	13	2	438.27	0.00	0.00	0.00	0.00	0.00
		11	-438.27	0.00	0.00	0.00	0.00	0.00

*SEISMIC ZONE 4 EARTHQUAKE, 85MPH WIND, LRFD

MEMBER END FORCES   STRUCTURE TYPE = PLANE

ALL UNITS ARE -- KIP FEET

MEMBER	LOAD	JT	AXIAL	SHEAR-Y	SHEAR-Z	TORSION	MOM-Y	MOM-Z
	14	2	448.25	0.00	0.00	0.00	0.00	0.00
		11	-448.25	0.00	0.00	0.00	0.00	0.00
	15	2	442.95	0.00	0.00	0.00	0.00	0.00
		11	-442.95	0.00	0.00	0.00	0.00	0.00
	16	2	763.17	0.00	0.00	0.00	0.00	0.00
		11	-763.17	0.00	0.00	0.00	0.00	0.00
4	1	2	22.26	-0.01	0.00	0.00	0.00	0.00
		4	-22.26	0.01	0.00	0.00	0.00	-0.08
	2	2	16.91	0.00	0.00	0.00	0.00	0.00
		4	-16.91	0.00	0.00	0.00	0.00	0.03
	3	2	250.90	-0.45	0.00	0.00	0.00	0.00
		4	-250.90	0.45	0.00	0.00	0.00	-6.05
	4	2	54.35	0.00	0.00	0.00	0.00	0.00
		4	-54.35	0.00	0.00	0.00	0.00	0.01
	5	2	286.06	-0.45	0.00	0.00	0.00	0.00
		4	-286.06	0.45	0.00	0.00	0.00	-6.12
	6	2	270.93	-0.45	0.00	0.00	0.00	0.00
		4	-270.93	0.45	0.00	0.00	0.00	-6.12
	7	2	53.76	0.00	0.00	0.00	0.00	0.00
		4	-53.76	0.00	0.00	0.00	0.00	-0.05
	8	2	105.81	-0.01	0.00	0.00	0.00	0.00
		4	-105.81	0.01	0.00	0.00	0.00	-0.07
	9	2	429.09	-0.68	0.00	0.00	0.00	0.00
		4	-429.09	0.68	0.00	0.00	0.00	-9.19
	10	2	406.40	-0.68	0.00	0.00	0.00	0.00
		4	-406.40	0.68	0.00	0.00	0.00	-9.17
	11	2	136.02	-0.01	0.00	0.00	0.00	0.00
		4	-136.02	0.01	0.00	0.00	0.00	-0.09
	12	2	691.14	-1.16	0.00	0.00	0.00	0.00
		4	-691.14	1.16	0.00	0.00	0.00	-15.60
	13	2	648.66	-1.15	0.00	0.00	0.00	0.00
		4	-648.66	1.15	0.00	0.00	0.00	-15.50
	14	2	677.47	-1.15	0.00	0.00	0.00	0.00
		4	-677.47	1.15	0.00	0.00	0.00	-15.56
	15	2	662.34	-1.15	0.00	0.00	0.00	0.00
		4	-662.34	1.15	0.00	0.00	0.00	-15.55
	16	2	1124.03	-2.01	0.00	0.00	0.00	0.00
		4	-1124.03	2.01	0.00	0.00	0.00	-27.09
5	1	3	-2.03	3.86	0.00	0.00	0.00	6.36
		11	2.03	3.71	0.00	0.00	0.00	-5.63
	2	3	-1.72	3.37	0.00	0.00	0.00	5.52
		11	1.72	3.25	0.00	0.00	0.00	-4.94
	3	3	14.50	-1.59	0.00	0.00	0.00	-15.09
		11	-14.50	1.59	0.00	0.00	0.00	0.00
	4	3	16.01	-0.41	0.00	0.00	0.00	-4.02
		11	-16.01	0.41	0.00	0.00	0.00	0.13

*SEISMIC ZONE 4 EARTHQUAKE, 85MPH WIND, LRFD

MEMBER END FORCES   STRUCTURE TYPE = PLANE

ALL UNITS ARE -- KIP FEET

MEMBER	LOAD	JT	AXIAL	SHEAR-Y	SHEAR-Z	TORSION	MOM-Y	MOM-Z
	5	3	11.21	4.72	0.00	0.00	0.00	-4.69
		11	-11.21	7.67	0.00	0.00	0.00	-9.23
	6	3	12.68	1.88	0.00	0.00	0.00	-9.36
		11	-12.68	4.93	0.00	0.00	0.00	-5.07
	7	3	-5.19	10.03	0.00	0.00	0.00	16.47
		11	5.19	9.65	0.00	0.00	0.00	-14.67
	8	3	17.52	5.78	0.00	0.00	0.00	5.17
		11	-17.52	6.61	0.00	0.00	0.00	-9.06
	9	3	16.81	7.09	0.00	0.00	0.00	-7.04
		11	-16.81	11.50	0.00	0.00	0.00	-13.84
	10	3	19.02	2.82	0.00	0.00	0.00	-14.04
		11	-19.02	7.40	0.00	0.00	0.00	-7.60
	11	3	28.49	4.41	0.00	0.00	0.00	0.75
		11	-28.49	5.81	0.00	0.00	0.00	-7.35
	12	3	32.59	4.61	0.00	0.00	0.00	-24.31
		11	-32.59	12.43	0.00	0.00	0.00	-12.69
	13	3	36.55	-2.98	0.00	0.00	0.00	-36.80
		11	-36.55	5.14	0.00	0.00	0.00	-1.61
	14	3	33.84	2.24	0.00	0.00	0.00	-28.22
		11	-33.84	10.15	0.00	0.00	0.00	-9.23
	15	3	35.31	-0.61	0.00	0.00	0.00	-32.89
		11	-35.31	7.42	0.00	0.00	0.00	-5.07
	16	3	64.98	-7.15	0.00	0.00	0.00	-67.58
		11	-64.98	7.15	0.00	0.00	0.00	0.00
6	1	11	-2.03	3.71	0.00	0.00	0.00	5.63
		4	2.03	3.86	0.00	0.00	0.00	-6.36
	2	11	-1.72	3.25	0.00	0.00	0.00	4.94
		4	1.72	3.37	0.00	0.00	0.00	-5.52
	3	11	-14.50	-1.59	0.00	0.00	0.00	0.00
		4	14.50	1.59	0.00	0.00	0.00	-15.09
	4	11	-0.52	-0.45	0.00	0.00	0.00	-0.13
		4	0.52	0.45	0.00	0.00	0.00	-4.13
	5	11	-17.80	4.48	0.00	0.00	0.00	9.23
		4	17.80	7.91	0.00	0.00	0.00	-25.48
	6	11	-16.33	1.74	0.00	0.00	0.00	5.07
		4	16.33	5.07	0.00	0.00	0.00	-20.81
	7	11	-5.19	9.65	0.00	0.00	0.00	14.67
		4	5.19	10.03	0.00	0.00	0.00	-16.47
	8	11	-3.97	5.49	0.00	0.00	0.00	9.06
		4	3.97	6.90	0.00	0.00	0.00	-15.77
	9	11	-26.70	6.72	0.00	0.00	0.00	13.84
		4	26.70	11.87	0.00	0.00	0.00	-38.22
	10	11	-24.49	2.61	0.00	0.00	0.00	7.60
		4	24.49	7.60	0.00	0.00	0.00	-31.22
	11	11	-3.75	4.12	0.00	0.00	0.00	7.35
		4	3.75	6.09	0.00	0.00	0.00	-16.65

*SEISMIC ZONE 4 EARTHQUAKE, 85MPH WIND, LRFD

MEMBER END FORCES   STRUCTURE TYPE = PLANE

ALL UNITS ARE -- KIP   FEET

MEMBER	LOAD	JT	AXIAL	SHEAR-Y	SHEAR-Z	TORSION	MOM-Y	MOM-Z
	12	11	-41.67	4.27	0.00	0.00	0.00	12.69
		4	41.67	12.77	0.00	0.00	0.00	-52.92
	13	11	-37.71	-3.02	0.00	0.00	0.00	1.61
		4	37.71	5.19	0.00	0.00	0.00	-40.44
	14	11	-40.43	1.99	0.00	0.00	0.00	9.23
		4	40.43	10.40	0.00	0.00	0.00	-49.01
	15	11	-38.96	-0.75	0.00	0.00	0.00	5.07
		4	38.96	7.56	0.00	0.00	0.00	-44.34
	16	11	-64.98	-7.15	0.00	0.00	0.00	0.00
		4	64.98	7.15	0.00	0.00	0.00	-67.58
7	1	11	3.00	0.00	0.00	0.00	0.00	0.00
		6	-3.00	0.00	0.00	0.00	0.00	0.00
	2	11	1.99	0.00	0.00	0.00	0.00	0.00
		6	-1.99	0.00	0.00	0.00	0.00	0.00
	3	11	-145.98	0.00	0.00	0.00	0.00	0.00
		6	145.98	0.00	0.00	0.00	0.00	0.00
	4	11	-35.44	0.00	0.00	0.00	0.00	0.00
		6	35.44	0.00	0.00	0.00	0.00	0.00
	5	11	-141.39	0.00	0.00	0.00	0.00	0.00
		6	141.39	0.00	0.00	0.00	0.00	0.00
	6	11	-143.28	0.00	0.00	0.00	0.00	0.00
		6	143.28	0.00	0.00	0.00	0.00	0.00
	7	11	6.78	0.00	0.00	0.00	0.00	0.00
		6	-6.78	0.00	0.00	0.00	0.00	0.00
	8	11	-41.47	0.00	0.00	0.00	0.00	0.00
		6	41.47	0.00	0.00	0.00	0.00	0.00
	9	11	-212.08	0.00	0.00	0.00	0.00	0.00
		6	212.08	0.00	0.00	0.00	0.00	0.00
	10	11	-214.92	0.00	0.00	0.00	0.00	0.00
		6	214.92	0.00	0.00	0.00	0.00	0.00
	11	11	-65.05	0.00	0.00	0.00	0.00	0.00
		6	65.05	0.00	0.00	0.00	0.00	0.00
	12	11	-367.28	0.00	0.00	0.00	0.00	0.00
		6	367.28	0.00	0.00	0.00	0.00	0.00
	13	11	-372.86	0.00	0.00	0.00	0.00	0.00
		6	372.86	0.00	0.00	0.00	0.00	0.00
	14	11 ·	-369.12	0.00	0.00	0.00	0.00	0.00
		6	369.12	0.00	0.00	0.00	0.00	0.00
	15	11	-371.02	0.00	0.00	0.00	0.00	0.00
		6	371.02	0.00	0.00	0.00	0.00	0.00
	16	11	-654.01	0.00	0.00	0.00	0.00	0.00
		6	654.01	0.00	0.00	0.00	0.00	0.00
8	1	11	3.00	0.00	0.00	0.00	0.00	0.00
		5	-3.00	0.00	0.00	0.00	0.00	0.00
	2	11	1.99	0.00	0.00	0.00	0.00	0.00
		5	-1.99	0.00	0.00	0.00	0.00	0.00

*SEISMIC ZONE 4 EARTHQUAKE, 85MPH WIND, LRFD

MEMBER END FORCES   STRUCTURE TYPE = PLANE

ALL UNITS ARE -- KIP FEET

MEMBER	LOAD	JT	AXIAL	SHEAR-Y	SHEAR-Z	TORSION	MOM-Y	MOM-Z
	3	11	145.98	0.00	0.00	0.00	0.00	0.00
		5	-145.98	0.00	0.00	0.00	0.00	0.00
	4	11	37.03	0.00	0.00	0.00	0.00	0.00
		5	-37.03	0.00	0.00	0.00	0.00	0.00
	5	11	150.58	0.00	0.00	0.00	0.00	0.00
		5	-150.58	0.00	0.00	0.00	0.00	0.00
	6	11	148.69	0.00	0.00	0.00	0.00	0.00
		5	-148.69	0.00	0.00	0.00	0.00	0.00
	7	11	6.78	0.00	0.00	0.00	0.00	0.00
		5	-6.78	0.00	0.00	0.00	0.00	0.00
	8	11	52.74	0.00	0.00	0.00	0.00	0.00
		5	-52.74	0.00	0.00	0.00	0.00	0.00
	9	11	225.87	0.00	0.00	0.00	0.00	0.00
		5	-225.87	0.00	0.00	0.00	0.00	0.00
	10	11	223.03	0.00	0.00	0.00	0.00	0.00
		5	-223.03	0.00	0.00	0.00	0.00	0.00
	11	11	76.27	0.00	0.00	0.00	0.00	0.00
		5	-76.27	0.00	0.00	0.00	0.00	0.00
	12	11	380.16	0.00	0.00	0.00	0.00	0.00
		5	-380.16	0.00	0.00	0.00	0.00	0.00
	13	11	374.58	0.00	0.00	0.00	0.00	0.00
		5	-374.58	0.00	0.00	0.00	0.00	0.00
	14	11	378.31	0.00	0.00	0.00	0.00	0.00
		5	-378.31	0.00	0.00	0.00	0.00	0.00
	15	11	376.42	0.00	0.00	0.00	0.00	0.00
		5	-376.42	0.00	0.00	0.00	0.00	0.00
	16	11	654.01	0.00	0.00	0.00	0.00	0.00
		5	-654.01	0.00	0.00	0.00	0.00	0.00
9	1	3	18.40	-2.02	0.00	0.00	0.00	-8.53
		5	-18.40	2.02	0.00	0.00	0.00	-18.77
	2	3	13.54	-1.72	0.00	0.00	0.00	-7.32
		5	-13.54	1.72	0.00	0.00	0.00	-15.96
	3	3	-249.31	2.41	0.00	0.00	0.00	22.00
		5	249.31	-2.41	0.00	0.00	0.00	10.49
	4	3	-55.18	0.45	0.00	0.00	0.00	3.77
		5	55.18	-0.45	0.00	0.00	0.00	2.27
	5	3	-220.46	-0.88	0.00	0.00	0.00	8.10
		5	220.46	0.88	0.00	0.00	0.00	-20.01
	6	3	-232.75	0.59	0.00	0.00	0.00	14.32
		5	232.75	-0.59	0.00	0.00	0.00	-6.40
	7	3	43.73	-5.19	0.00	0.00	0.00	-21.94
		5	-43.73	5.19	0.00	0.00	0.00	-48.06
	8	3	-42.89	-2.71	0.00	0.00	0.00	-8.99
		5	42.89	2.71	0.00	0.00	0.00	-27.55
	9	3	-330.69	-1.32	0.00	0.00	0.00	12.15
		5	330.69	1.32	0.00	0.00	0.00	-30.02

*SEISMIC ZONE 4 EARTHQUAKE, 85MPH WIND, LRFD

MEMBER END FORCES   STRUCTURE TYPE = PLANE

ALL UNITS ARE -- KIP FEET

MEMBER	LOAD	JT	AXIAL	SHEAR-Y	SHEAR-Z	TORSION	MOM-Y	MOM-Z
	10	3	-349.12	0.88	0.00	0.00	0.00	21.48
		5	349.12	-0.88	0.00	0.00	0.00	-9.60
	11	3	-82.77	-1.86	0.00	0.00	0.00	-4.16
		5	82.77	1.86	0.00	0.00	0.00	-20.92
	12	3	-598.08	1.63	0.00	0.00	0.00	37.17
		5	598.08	-1.63	0.00	0.00	0.00	-15.18
	13	3	-632.97	5.58	0.00	0.00	0.00	53.87
		5	632.97	-5.58	0.00	0.00	0.00	21.49
	14	3	-609.38	2.87	0.00	0.00	0.00	42.41
		5	609.38	-2.87	0.00	0.00	0.00	-3.65
	15	3	-621.67	4.34	0.00	0.00	0.00	48.63
		5	621.67	-4.34	0.00	0.00	0.00	9.96
	16	3	-1116.89	10.78	0.00	0.00	0.00	98.54
		5	1116.89	-10.78	0.00	0.00	0.00	47.00
10	1	4	18.40	2.02	0.00	0.00	0.00	8.53
		6	-18.40	-2.02	0.00	0.00	0.00	18.77
	2	4	13.54	1.72	0.00	0.00	0.00	7.32
		6	-13.54	-1.72	0.00	0.00	0.00	15.96
	3	4	249.31	2.41	0.00	0.00	0.00	22.00
		6	-249.31	-2.41	0.00	0.00	0.00	10.49
	4	4	53.90	0.52	0.00	0.00	0.00	4.37
		6	-53.90	-0.52	0.00	0.00	0.00	2.61
	5	4	278.15	5.70	0.00	0.00	0.00	35.89
		6	-278.15	-5.70	0.00	0.00	0.00	40.99
	6	4	265.86	4.23	0.00	0.00	0.00	29.67
		6	-265.86	-4.23	0.00	0.00	0.00	27.38
	7	4	43.73	5.19	0.00	0.00	0.00	21.94
		6	-43.73	-5.19	0.00	0.00	0.00	48.06
	8	4	98.91	3.96	0.00	0.00	0.00	19.58
		6	-98.91	-3.96	0.00	0.00	0.00	33.89
	9	4	417.22	8.54	0.00	0.00	0.00	53.84
		6	-417.22	-8.54	0.00	0.00	0.00	61.49
	10	4	398.79	6.34	0.00	0.00	0.00	44.51
		6	-398.79	-6.34	0.00	0.00	0.00	41.07
	11	4	129.93	3.74	0.00	0.00	0.00	20.04
		6	-129.93	-3.74	0.00	0.00	0.00	30.42
	12	4	678.37	10.69	0.00	0.00	0.00	75.45
		6	-678.37	-10.69	0.00	0.00	0.00	68.89
	13	4	643.48	6.74	0.00	0.00	0.00	58.75
		6	-643.48	-6.74	0.00	0.00	0.00	32.21
	14	4	667.07	9.45	0.00	0.00	0.00	70.20
		6	-667.07	-9.45	0.00	0.00	0.00	57.36
	15	4	654.78	7.98	0.00	0.00	0.00	63.99
		6	-654.78	-7.98	0.00	0.00	0.00	43.75
	16	4	1116.89	10.78	0.00	0.00	0.00	98.54
		6	-1116.89	-10.78	0.00	0.00	0.00	47.00

*SEISMIC ZONE 4 EARTHQUAKE, 85MPH WIND, LRFD

MEMBER END FORCES   **STRUCTURE TYPE = PLANE**

ALL UNITS ARE -- KIP FEET

MEMBER	LOAD	JT	AXIAL	SHEAR-Y	SHEAR-Z	TORSION	MOM-Y	MOM-Z
11	1	5	-3.52	7.57	0.00	0.00	0.00	21.23
		6	3.52	7.57	0.00	0.00	0.00	-21.23
	2	5	-2.39	6.62	0.00	0.00	0.00	18.64
		6	2.39	6.62	0.00	0.00	0.00	-18.64
	3	5	0.00	-1.70	0.00	0.00	0.00	-16.10
		6	0.00	1.70	0.00	0.00	0.00	-16.10
	4	5	7.34	-0.38	0.00	0.00	0.00	-3.60
		6	-7.34	0.38	0.00	0.00	0.00	-3.60
	5	5	-5.42	10.69	0.00	0.00	0.00	18.70
		6	5.42	14.09	0.00	0.00	0.00	-50.90
	6	5	-3.17	5.11	0.00	0.00	0.00	3.01
		6	3.17	8.51	0.00	0.00	0.00	-35.21
	7	5	-8.05	19.67	0.00	0.00	0.00	55.31
		6	8.05	19.67	0.00	0.00	0.00	-55.31
	8	5	4.12	11.90	0.00	0.00	0.00	30.12
		6	-4.12	12.89	0.00	0.00	0.00	-39.48
	9	5	-8.12	16.03	0.00	0.00	0.00	28.05
		6	8.12	21.14	0.00	0.00	0.00	-76.35
	10	5	-4.75	7.66	0.00	0.00	0.00	4.52
		6	4.75	12.77	0.00	0.00	0.00	-52.81
	11	5	9.56	9.47	0.00	0.00	0.00	21.64
		6	-9.56	10.96	0.00	0.00	0.00	-35.68
	12	5	-7.58	12.68	0.00	0.00	0.00	6.63
		6	7.58	21.40	0.00	0.00	0.00	-89.06
	13	5	-1.00	-2.20	0.00	0.00	0.00	-35.15
		6	1.00	6.52	0.00	0.00	0.00	-47.28
	14	5	-5.42	8.03	0.00	0.00	0.00	-6.41
		6	5.42	16.75	0.00	0.00	0.00	-76.01
	15	5	-3.17	2.45	0.00	0.00	0.00	-22.10
		6	3.17	11.17	0.00	0.00	0.00	-60.32
	16	5	0.00	-7.63	0.00	0.00	0.00	-72.12
		6	0.00	7.63	0.00	0.00	0.00	-72.12
12	1	5	5.20	0.00	0.00	0.00	0.00	0.00
		12	-5.20	0.00	0.00	0.00	0.00	0.00
	2	5	3.72	0.00	0.00	0.00	0.00	0.00
		12	-3.72	0.00	0.00	0.00	0.00	0.00
	3	5	-109.17	0.00	0.00	0.00	0.00	0.00
		12	109.17	0.00	0.00	0.00	0.00	0.00
	4	5	-21.79	0.00	0.00	0.00	0.00	0.00
		12	21.79	0.00	0.00	0.00	0.00	0.00
	5	5	-101.06	0.00	0.00	0.00	0.00	0.00
		12	101.06	0.00	0.00	0.00	0.00	0.00
	6	5	-104.48	0.00	0.00	0.00	0.00	0.00
		12	104.48	0.00	0.00	0.00	0.00	0.00
	7	5	12.20	0.00	0.00	0.00	0.00	0.00
		12	-12.20	0.00	0.00	0.00	0.00	0.00

*SEISMIC ZONE 4 EARTHQUAKE, 85MPH WIND, LRFD

MEMBER END FORCES   STRUCTURE TYPE = PLANE

ALL UNITS ARE -- KIP FEET

MEMBER	LOAD	JT	AXIAL	SHEAR-Y	SHEAR-Z	TORSION	MOM-Y	MOM-Z
	8	5	-20.22	0.00	0.00	0.00	0.00	0.00
		12	20.22	0.00	0.00	0.00	0.00	0.00
	9	5	-151.59	0.00	0.00	0.00	0.00	0.00
		12	151.59	0.00	0.00	0.00	0.00	0.00
	10	5	-156.73	0.00	0.00	0.00	0.00	0.00
		12	156.73	0.00	0.00	0.00	0.00	0.00
	11	5	-35.46	0.00	0.00	0.00	0.00	0.00
		12	35.46	0.00	0.00	0.00	0.00	0.00
	12	5	-268.17	0.00	0.00	0.00	0.00	0.00
		12	268.17	0.00	0.00	0.00	0.00	0.00
	13	5	-277.98	0.00	0.00	0.00	0.00	0.00
		12	277.98	0.00	0.00	0.00	0.00	0.00
	14	5	-271.36	0.00	0.00	0.00	0.00	0.00
		12	271.36	0.00	0.00	0.00	0.00	0.00
	15	5	-274.79	0.00	0.00	0.00	0.00	0.00
		12	274.79	0.00	0.00	0.00	0.00	0.00
	16	5	-489.07	0.00	0.00	0.00	0.00	0.00
		12	489.07	0.00	0.00	0.00	0.00	0.00
13	1	6	5.20	0.00	0.00	0.00	0.00	0.00
		12	-5.20	0.00	0.00	0.00	0.00	0.00
	2	6	3.72	0.00	0.00	0.00	0.00	0.00
		12	-3.72	0.00	0.00	0.00	0.00	0.00
	3	6	109.17	0.00	0.00	0.00	0.00	0.00
		12	-109.17	0.00	0.00	0.00	0.00	0.00
	4	6	23.62	0.00	0.00	0.00	0.00	0.00
		12	-23.62	0.00	0.00	0.00	0.00	0.00
	5	6	117.27	0.00	0.00	0.00	0.00	0.00
		12	-117.27	0.00	0.00	0.00	0.00	0.00
	6	6	113.85	0.00	0.00	0.00	0.00	0.00
		12	-113.85	0.00	0.00	0.00	0.00	0.00
	7	6	12.20	0.00	0.00	0.00	0.00	0.00
		12	-12.20	0.00	0.00	0.00	0.00	0.00
	8	6	38.81	0.00	0.00	0.00	0.00	0.00
		12	-38.81	0.00	0.00	0.00	0.00	0.00
	9	6	175.91	0.00	0.00	0.00	0.00	0.00
		12	-175.91	0.00	0.00	0.00	0.00	0.00
	10	6	170.78	0.00	0.00	0.00	0.00	0.00
		12	-170.78	0.00	0.00	0.00	0.00	0.00
	11	6	53.09	0.00	0.00	0.00	0.00	0.00
		12	-53.09	0.00	0.00	0.00	0.00	0.00
	12	6	290.77	0.00	0.00	0.00	0.00	0.00
		12	-290.77	0.00	0.00	0.00	0.00	0.00
	13	6	280.95	0.00	0.00	0.00	0.00	0.00
		12	-280.95	0.00	0.00	0.00	0.00	0.00
	14	6	287.57	0.00	0.00	0.00	0.00	0.00
		12	-287.57	0.00	0.00	0.00	0.00	0.00

*SEISMIC ZONE 4 EARTHQUAKE, 85MPH WIND, LRFD

MEMBER END FORCES   STRUCTURE TYPE = PLANE

ALL UNITS ARE -- KIP   FEET

MEMBER	LOAD	JT	AXIAL	SHEAR-Y	SHEAR-Z	TORSION	MOM-Y	MOM-Z
	15	6	284.15	0.00	0.00	0.00	0.00	0.00
		12	-284.15	0.00	0.00	0.00	0.00	0.00
	16	6	489.07	0.00	0.00	0.00	0.00	0.00
		12	-489.07	0.00	0.00	0.00	0.00	0.00
14	1	6	9.06	0.65	0.00	0.00	0.00	6.56
		8	-9.06	-0.65	0.00	0.00	0.00	2.28
	2	6	5.52	0.72	0.00	0.00	0.00	6.27
		8	-5.52	-0.72	0.00	0.00	0.00	3.43
	3	6	42.57	1.02	0.00	0.00	0.00	6.53
		8	-42.57	-1.02	0.00	0.00	0.00	7.25
	4	6	6.06	0.21	0.00	0.00	0.00	1.19
		8	-6.06	-0.21	0.00	0.00	0.00	1.64
	5	6	56.21	2.17	0.00	0.00	0.00	17.54
		8	-56.21	-2.17	0.00	0.00	0.00	11.70
	6	6	50.73	1.61	0.00	0.00	0.00	12.44
		8	-50.73	-1.61	0.00	0.00	0.00	9.30
	7	6	19.71	1.94	0.00	0.00	0.00	17.90
		8	-19.71	-1.94	0.00	0.00	0.00	8.22
	8	6	21.51	1.42	0.00	0.00	0.00	12.56
		8	-21.51	-1.42	0.00	0.00	0.00	6.58
	9	6	84.31	3.25	0.00	0.00	0.00	26.31
		8	-84.31	-3.25	0.00	0.00	0.00	17.54
	10	6	76.09	2.41	0.00	0.00	0.00	18.66
		8	-76.09	-2.41	0.00	0.00	0.00	13.94
	11	6	24.05	1.29	0.00	0.00	0.00	11.19
		8	-24.05	-1.29	0.00	0.00	0.00	6.27
	12	6	128.19	4.16	0.00	0.00	0.00	31.76
		8	-128.19	-4.16	0.00	0.00	0.00	24.40
	13	6	111.58	2.80	0.00	0.00	0.00	18.59
		8	-111.58	-2.80	0.00	0.00	0.00	19.20
	14	6	122.62	3.76	0.00	0.00	0.00	27.73
		8	-122.62	-3.76	0.00	0.00	0.00	23.00
	15	6	117.15	3.20	0.00	0.00	0.00	22.62
		8	-117.15	-3.20	0.00	0.00	0.00	20.60
	16	6	190.73	4.57	0.00	0.00	0.00	29.26
		8	-190.73	-4.57	0.00	0.00	0.00	32.47
15	1	7	-1.16	4.03	0.00	0.00	0.00	7.37
		12	1.16	3.54	0.00	0.00	0.00	-5.06
	2	7	0.05	3.28	0.00	0.00	0.00	4.90
		12	-0.05	3.34	0.00	0.00	0.00	-5.19
	3	7	34.25	-0.83	0.00	0.00	0.00	-7.82
		12	-34.25	0.83	0.00	0.00	0.00	0.00
	4	7	17.43	-0.16	0.00	0.00	0.00	-1.51
		12	-17.43	0.16	0.00	0.00	0.00	-0.01
	5	7	32.88	5.65	0.00	0.00	0.00	3.48
		12	-32.88	6.74	0.00	0.00	0.00	-8.67

*SEISMIC ZONE 4 EARTHQUAKE, 85MPH WIND, LRFD

MEMBER END FORCES   STRUCTURE TYPE = PLANE

ALL UNITS ARE -- KIP FEET

MEMBER	LOAD	JT	AXIAL	SHEAR-Y	SHEAR-Z	TORSION	MOM-Y	MOM-Z
	6	7	33.21	2.80	0.00	0.00	0.00	-1.19
		12	-33.21	4.01	0.00	0.00	0.00	-4.56
	7	7	-1.32	10.08	0.00	0.00	0.00	16.69
		12	1.32	9.59	0.00	0.00	0.00	-14.37
	8	7	21.29	6.26	0.00	0.00	0.00	9.33
		12	-21.29	6.13	0.00	0.00	0.00	-8.69
	9	7	49.32	8.47	0.00	0.00	0.00	5.21
		12	-49.32	10.12	0.00	0.00	0.00	-13.00
	10	7	49.81	4.20	0.00	0.00	0.00	-1.78
		12	-49.81	6.02	0.00	0.00	0.00	-6.84
	11	7	32.42	5.12	0.00	0.00	0.00	7.00
		12	-32.42	5.09	0.00	0.00	0.00	-6.86
	12	7	85.60	6.83	0.00	0.00	0.00	-4.19
		12	-85.60	10.21	0.00	0.00	0.00	-11.78
	13	7	87.36	-0.97	0.00	0.00	0.00	-17.91
		12	-87.36	3.13	0.00	0.00	0.00	-1.45
	14	7	86.32	4.36	0.00	0.00	0.00	-8.72
		12	-86.32	8.03	0.00	0.00	0.00	-8.67
	15	7	86.64	1.51	0.00	0.00	0.00	-13.38
		12	-86.64	5.30	0.00	0.00	0.00	-4.56
	16	7	153.45	-3.70	0.00	0.00	0.00	-35.02
		12	-153.45	3.70	0.00	0.00	0.00	0.00
16	1	12	-1.16	3.54	0.00	0.00	0.00	5.06
		8	1.16	4.03	0.00	0.00	0.00	-7.37
	2	12	0.05	3.34	0.00	0.00	0.00	5.19
		8	-0.05	3.28	0.00	0.00	0.00	-4.90
	3	12	-34.25	-0.83	0.00	0.00	0.00	0.00
		8	34.25	0.83	0.00	0.00	0.00	-7.82
	4	12	0.19	-0.15	0.00	0.00	0.00	0.01
		8	-0.19	0.15	0.00	0.00	0.00	-1.46
	5	12	-35.63	5.09	0.00	0.00	0.00	8.67
		8	35.63	7.30	0.00	0.00	0.00	-19.11
	6	12	-35.30	2.36	0.00	0.00	0.00	4.56
		8	35.30	4.45	0.00	0.00	0.00	-14.45
	7	12	-1.32	9.59	0.00	0.00	0.00	14.37
		8	1.32	10.08	0.00	0.00	0.00	-16.69
	8	12	-1.12	5.72	0.00	0.00	0.00	8.69
		8	1.12	6.67	0.00	0.00	0.00	-13.19
	9	12	-53.44	7.64	0.00	0.00	0.00	13.00
		8	53.44	10.95	0.00	0.00	0.00	-28.67
	10	12	-52.95	3.54	0.00	0.00	0.00	6.84
		8	52.95	6.68	0.00	0.00	0.00	-21.67
	11	12	-1.20	4.48	0.00	0.00	0.00	6.86
		8	1.20	5.74	0.00	0.00	0.00	-12.80
	12	12	-89.77	5.98	0.00	0.00	0.00	11.78
		8	89.77	11.06	0.00	0.00	0.00	-35.83

*SEISMIC ZONE 4 EARTHQUAKE, 85MPH WIND, LRFD

MEMBER END FORCES   STRUCTURE TYPE = PLANE

ALL UNITS ARE -- KIP   FEET

MEMBER	LOAD	JT	AXIAL	SHEAR-Y	SHEAR-Z	TORSION	MOM-Y	MOM-Z
	13	12	-88.02	-1.10	0.00	0.00	0.00	1.45
		8	88.02	3.27	0.00	0.00	0.00	-22.12
	14	12	-89.06	3.80	0.00	0.00	0.00	8.67
		8	89.06	8.59	0.00	0.00	0.00	-31.31
	15	12	-88.73	1.07	0.00	0.00	0.00	4.56
		8	88.73	5.74	0.00	0.00	0.00	-26.64
	16	12	-153.45	-3.70	0.00	0.00	0.00	0.00
		8	153.45	3.70	0.00	0.00	0.00	-35.02
17	1	12	0.80	0.00	0.00	0.00	0.00	0.00
		10	-0.80	0.00	0.00	0.00	0.00	0.00
	2	12	-0.44	0.00	0.00	0.00	0.00	0.00
		10	0.44	0.00	0.00	0.00	0.00	0.00
	3	12	-51.62	0.00	0.00	0.00	0.00	0.00
		10	51.62	0.00	0.00	0.00	0.00	0.00
	4	12	-7.31	0.00	0.00	0.00	0.00	0.00
		10	7.31	0.00	0.00	0.00	0.00	0.00
	5	12	-50.88	0.00	0.00	0.00	0.00	0.00
		10	50.88	0.00	0.00	0.00	0.00	0.00
	6	12	-50.90	0.00	0.00	0.00	0.00	0.00
		10	50.90	0.00	0.00	0.00	0.00	0.00
	7	12	0.26	0.00	0.00	0.00	0.00	0.00
		10	-0.26	0.00	0.00	0.00	0.00	0.00
	8	12	-8.77	0.00	0.00	0.00	0.00	0.00
		10	8.77	0.00	0.00	0.00	0.00	0.00
	9	12	-76.32	0.00	0.00	0.00	0.00	0.00
		10	76.32	0.00	0.00	0.00	0.00	0.00
	10	12	-76.35	0.00	0.00	0.00	0.00	0.00
		10	76.35	0.00	0.00	0.00	0.00	0.00
	11	12	-13.18	0.00	0.00	0.00	0.00	0.00
		10	13.18	0.00	0.00	0.00	0.00	0.00
	12	12	-130.92	0.00	0.00	0.00	0.00	0.00
		10	130.92	0.00	0.00	0.00	0.00	0.00
	13	12	-131.92	0.00	0.00	0.00	0.00	0.00
		10	131.92	0.00	0.00	0.00	0.00	0.00
	14	12	-131.41	0.00	0.00	0.00	0.00	0.00
		10	131.41	0.00	0.00	0.00	0.00	0.00
	15	12	-131.43	0.00	0.00	0.00	0.00	0.00
		10	131.43	0.00	0.00	0.00	0.00	0.00
	16	12	-231.26	0.00	0.00	0.00	0.00	0.00
		10	231.26	0.00	0.00	0.00	0.00	0.00
18	1	12	0.80	0.00	0.00	0.00	0.00	0.00
		9	-0.80	0.00	0.00	0.00	0.00	0.00
	2	12	-0.44	0.00	0.00	0.00	0.00	0.00
		9	0.44	0.00	0.00	0.00	0.00	0.00
	3	12	51.62	0.00	0.00	0.00	0.00	0.00
		9	-51.62	0.00	0.00	0.00	0.00	0.00

*SEISMIC ZONE 4 EARTHQUAKE, 85MPH WIND, LRFD

MEMBER END FORCES   STRUCTURE TYPE = PLANE

ALL UNITS ARE -- KIP FEET

MEMBER	LOAD	JT	AXIAL	SHEAR-Y	SHEAR-Z	TORSION	MOM-Y	MOM-Z
	4	12	9.13	0.00	0.00	0.00	0.00	0.00
		9	-9.13	0.00	0.00	0.00	0.00	0.00
	5	12	52.36	0.00	0.00	0.00	0.00	0.00
		9	-52.36	0.00	0.00	0.00	0.00	0.00
	6	12	52.34	0.00	0.00	0.00	0.00	0.00
		9	-52.34	0.00	0.00	0.00	0.00	0.00
	7	12	0.26	0.00	0.00	0.00	0.00	0.00
		9	-0.26	0.00	0.00	0.00	0.00	0.00
	8	12	12.62	0.00	0.00	0.00	0.00	0.00
		9	-12.62	0.00	0.00	0.00	0.00	0.00
	9	12	78.54	0.00	0.00	0.00	0.00	0.00
		9	-78.54	0.00	0.00	0.00	0.00	0.00
	10	12	78.51	0.00	0.00	0.00	0.00	0.00
		9	-78.51	0.00	0.00	0.00	0.00	0.00
	11	12	18.89	0.00	0.00	0.00	0.00	0.00
		9	-18.89	0.00	0.00	0.00	0.00	0.00
	12	12	133.38	0.00	0.00	0.00	0.00	0.00
		9	-133.38	0.00	0.00	0.00	0.00	0.00
	13	12	132.38	0.00	0.00	0.00	0.00	0.00
		9	-132.38	0.00	0.00	0.00	0.00	0.00
	14	12	132.89	0.00	0.00	0.00	0.00	0.00
		9	-132.89	0.00	0.00	0.00	0.00	0.00
	15	12	132.87	0.00	0.00	0.00	0.00	0.00
		9	-132.87	0.00	0.00	0.00	0.00	0.00
	16	12	231.26	0.00	0.00	0.00	0.00	0.00
		9	-231.26	0.00	0.00	0.00	0.00	0.00
19	1	7	5.03	-1.82	0.00	0.00	0.00	-7.27
		9	-5.03	1.82	0.00	0.00	0.00	-17.26
	2	7	2.24	-0.67	0.00	0.00	0.00	-3.25
		9	-2.24	0.67	0.00	0.00	0.00	-5.83
	3	7	-41.75	0.27	0.00	0.00	0.00	1.02
		9	41.75	-0.27	0.00	0.00	0.00	2.68
	4	7	-7.37	-0.01	0.00	0.00	0.00	-0.37
		9	7.37	0.01	0.00	0.00	0.00	0.26
	5	7	-34.59	-2.24	0.00	0.00	0.00	-9.33
		9	34.59	2.24	0.00	0.00	0.00	-20.95
	6	7	-37.22	-1.36	0.00	0.00	0.00	-5.53
		9	37.22	1.36	0.00	0.00	0.00	-12.86
	7	7	9.63	-3.26	0.00	0.00	0.00	-13.92
		9	-9.63	3.26	0.00	0.00	0.00	-30.05
	8	7	-2.42	-2.53	0.00	0.00	0.00	-10.83
		9	2.42	2.53	0.00	0.00	0.00	-23.29
	9	7	-51.88	-3.36	0.00	0.00	0.00	-14.00
		9	51.88	3.36	0.00	0.00	0.00	-31.43
	10	7	-55.83	-2.04	0.00	0.00	0.00	-8.29
		9	55.83	2.04	0.00	0.00	0.00	-19.28

*SEISMIC ZONE 4 EARTHQUAKE, 85MPH WIND, LRFD

MEMBER END FORCES   STRUCTURE TYPE = PLANE

ALL UNITS ARE -- KIP FEET

MEMBER	LOAD	JT	AXIAL	SHEAR-Y	SHEAR-Z	TORSION	MOM-Y	MOM-Z
	11	7	-7.58	-2.47	0.00	0.00	0.00	-10.53
		9	7.58	2.47	0.00	0.00	0.00	-22.79
	12	7	-96.62	-2.93	0.00	0.00	0.00	-12.21
		9	96.62	2.93	0.00	0.00	0.00	-27.38
	13	7	-105.44	0.18	0.00	0.00	0.00	0.53
		9	105.44	-0.18	0.00	0.00	0.00	1.93
	14	7	-99.71	-1.82	0.00	0.00	0.00	-7.74
		9	99.71	1.82	0.00	0.00	0.00	-16.77
	15	7	-102.35	-0.93	0.00	0.00	0.00	-3.94
		9	102.35	0.93	0.00	0.00	0.00	-8.68
	16	7	-187.03	1.23	0.00	0.00	0.00	4.56
		9	187.03	-1.23	0.00	0.00	0.00	12.00
20	1	8	5.03	1.82	0.00	0.00	0.00	7.27
		10	-5.03	-1.82	0.00	0.00	0.00	17.26
	2	8	2.24	0.67	0.00	0.00	0.00	3.25
		10	-2.24	-0.67	0.00	0.00	0.00	5.83
	3	8	41.75	0.27	0.00	0.00	0.00	1.02
		10	-41.75	-0.27	0..00	0.00	0.00	2.68
	4	8	5.91	0.02	0.00	0.00	0.00	-0.09
		10	-5.91	-0.02	0.00	0.00	0.00	0.33
	5	8	48.91	2.79	0.00	0.00	0.00	11.37
		10	-48.91	-2.79	0.00	0.00	0.00	26.31
	6	8	46.28	1.91	0.00	0.00	0.00	7.56
		10	-46.28	-1.91	0.00	0.00	0.00	18.21
	7	8	9.63	3.26	0.00	0.00	0.00	13.92
		10	-9.63	-3.26	0.00	0.00	0.00	30.05
	8	8	14.84	2.54	0.00	0.00	0.00	10.23
		10	-14.84	-2.54	0.00	0.00	0.00	24.07
	9	8	73.36	4.19	0.00	0.00	0.00	17.05
		10	-73.36	-4.19	0.00	0.00	0.00	39.46
	10	8	69.42	2.86	0.00	0.00	0.00	11.34
		10	-69.42	-2.86	0.00	0.00	0.00	27.32
	11	8	18.31	2.49	0.00	0.00	0.00	9.64
		10	-18.31	-2.49	0.00	0.00	0.00	23.95
	12	8	117.13	4.33	0.00	0.00	0.00	17.43
		10	-117.13	-4.33	0.00	0.00	0.00	41.09
	13	8	108.31	1.22	0.00	0.00	0.00	4.68
		10	-108.31	-1.22	0.00	0.00	0.00	11.79
	14	8	114.04	3.22	0.00	0.00	0.00	12.96
		10	-114.04	-3.22	0.00	0.00	0.00	30.49
	15	8	111.41	2.34	0.00	0.00	0.00	9.15
		10	-111.41	-2.34	0.00	0.00	0.00	22.39
	16	8	187.03	1.23	0.00	0.00	0.00	4.56
		10	-187.03	-1.23	0.00	0.00	0.00	12.00
21	1	5	9.06	-0.65	0.00	0.00	0.00	-6.56
		7	-9.06	0.65	0.00	0.00	0.00	-2.28

*SEISMIC ZONE 4 EARTHQUAKE, 85MPH WIND, LRFD

MEMBER END FORCES   STRUCTURE TYPE = PLANE

ALL UNITS ARE -- KIP FEET

MEMBER	LOAD	JT	AXIAL	SHEAR-Y	SHEAR-Z	TORSION	MOM-Y	MOM-Z
	2	5	5.52	-0.72	0.00	0.00	0.00	-6.27
		7	-5.52	0.72	0.00	0.00	0.00	-3.43
	3	5	-42.57	1.02	0.00	0.00	0.00	6.53
		7	42.57	-1.02	0.00	0.00	0.00	7.25
	4	5	-7.53	0.26	0.00	0.00	0.00	1.54
		7	7.53	-0.26	0.00	0.00	0.00	1.97
	5	5	-28.94	-0.12	0.00	0.00	0.00	-4.48
		7	28.94	0.12	0.00	0.00	0.00	2.80
	6	5	-34.42	0.43	0.00	0.00	0.00	0.62
		7	34.42	-0.43	0.00	0.00	0.00	5.20
	7	5	19.71	-1.94	0.00	0.00	0.00	-17.90
		7	-19.71	1.94	0.00	0.00	0.00	-8.22
	8	5	3.84	-0.81	0.00	0.00	0.00	-9.01
		7	-3.84	0.81	0.00	0.00	0.00	-1.89
	9	5	-43.41	-0.19	0.00	0.00	0.00	-6.72
		7	43.41	0.19	0.00	0.00	0.00	4.20
	10	5	-51.63	0.65	0.00	0.00	0.00	0.94
		7	51.63	-0.65	0.00	0.00	0.00	7.80
	11	5	-2.46	-0.38	0.00	0.00	0.00	-5.86
		7	2.46	0.38	0.00	0.00	0.00	0.76
	12	5	-89.79	1.07	0.00	0.00	0.00	1.68
		7	89.79	-1.07	0.00	0.00	0.00	12.70
	13	5	-106.40	2.43	0.00	0.00	0.00	14.84
		7	106.40	-2.43	0.00	0.00	0.00	17.90
	14	5	-95.36	1.47	0.00	0.00	0.00	5.71
		7	95.36	-1.47	0.00	0.00	0.00	14.10
	15	5	-100.84	2.02	0.00	0.00	0.00	10.81
		7	100.84	-2.02	0.00	0.00	0.00	16.50
	16	5	-190.73	4.57	0.00	0.00	0.00	29.26
		7	190.73	-4.57	0.00	0.00	0.00	32.47
22	1	9	1.34	5.68	0.00	0.00	0.00	14.19
		10	-1.34	5.68	0.00	0.00	0.00	-14.19
	2	9	0.93	1.89	0.00	0.00	0.00	4.81
		10	-0.93	1.89	0.00	0.00	0.00	-4.81
	3	9	0.00	-0.27	0.00	0.00	0.00	-2.53
		10	0.00	0.27	0.00	0.00	0.00	-2.53
	4	9	4.37	-0.03	0.00	0.00	0.00	-0.25
		10	-4.37	0.03	0.00	0.00	0.00	-0.32
	5	9	2.08	7.49	0.00	0.00	0.00	16.90
		10	-2.08	8.02	0.00	0.00	0.00	-21.96
	6	9	1.21	4.84	0.00	0.00	0.00	10.23
		10	-1.21	5.38	0.00	0.00	0.00	-15.30
	7	9	3.10	9.84	0.00	0.00	0.00	24.72
		10	-3.10	9.84	0.00	0.00	0.00	-24.72
	8	9	7.76	7.72	0.00	0.00	0.00	19.11
		10	-7.76	7.79	0.00	0.00	0.00	-19.84

*SEISMIC ZONE 4 EARTHQUAKE, 85MPH WIND, LRFD

MEMBER END FORCES   STRUCTURE TYPE = PLANE

ALL UNITS ARE -- KIP FEET

MEMBER	LOAD	JT	AXIAL	SHEAR-Y	SHEAR-Z	TORSION	MOM-Y	MOM-Z
	9	9	3.11	11.23	0.00	0.00	0.00	25.34
		10	-3.11	12.04	0.00	0.00	0.00	-32.95
	10	9	1.81	7.26	0.00	0.00	0.00	15.35
		10	-1.81	8.06	0.00	0.00	0.00	-22.95
	11	9	10.33	7.60	0.00	0.00	0.00	18.67
		10	-10.33	7.72	0.00	0.00	0.00	-19.77
	12	9	2.90	10.56	0.00	0.00	0.00	21.66
		10	-2.90	11.93	0.00	0.00	0.00	-34.63
	13	9	0.38	0.93	0.00	0.00	0.00	-2.43
		10	-0.38	2.31	0.00	0.00	0.00	-10.54
	14	9	2.08	7.07	0.00	0.00	0.00	12.94
		10	-2.08	8.44	0.00	0.00	0.00	-25.92
	15	9	1.21	4.42	0.00	0.00	0.00	6.28
		10	-1.21	5.79	0.00	0.00	0.00	-19.25
	16	9	0.00	-1.20	0.00	0.00	0.00	-11.35
		10	0.00	1.20	0.00	0.00	0.00	-11.35

************* END OF LATEST ANALYSIS RESULT **************

95. PRINT SUPPORT REACTIONS

*SEISMIC ZONE 4 EARTHQUAKE, 85MPH WIND, LRFD

SUPPORT REACTIONS -UNIT KIP   FEET   STRUCTURE TYPE = PLANE

JOINT	LOAD	FORCE-X	FORCE-Y	FORCE-Z	MOM-X	MOM-Y	MOM Z
1	1	4.53	28.38	0.00	0.00	0.00	0.00
	2	3.59	21.75	0.00	0.00	0.00	0.00
	3	-100.95	-387.79	0.00	0.00	0.00	0.00
	4	-29.41	-95.24	0.00	0.00	0.00	0.00
	5	-93.72	-342.86	0.00	0.00	0.00	0.00
	6	-96.88	-362.25	0.00	0.00	0.00	0.00
	7	11.18	68.86	0.00	0.00	0.00	0.00
	8	-31.00	-78.89	0.00	0.00	0.00	0.00
	9	-140.58	-514.29	0.00	0.00	0.00	0.00
	10	-145.31	-543.37	0.00	0.00	0.00	0.00
	11	-51.23	-147.42	0.00	0.00	0.00	0.00
	12	-248.42	-930.38	0.00	0.00	0.00	0.00
	13	-257.14	-984.63	0.00	0.00	0.00	0.00
	14	-251.20	-947.81	0.00	0.00	0.00	0.00
	15	-254.36	-967.20	0.00	0.00	0.00	0.00
	16	-452.26	-1737.29	0.00	0.00	0.00	0.00
2	1	-4.53	28.38	0.00	0.00	0.00	0.00
	2	-3.59	21.75	0.00	0.00	0.00	0.00
	3	-100.95	387.79	0.00	0.00	0.00	0.00
	4	-30.29	95.24	0.00	0.00	0.00	0.00
	5	-108.18	432.71	0.00	0.00	0.00	0.00
	6	-105.02	413.32	0.00	0.00	0.00	0.00
	7	-11.18	68.86	0.00	0.00	0.00	0.00
	8	-46.61	168.74	0.00	0.00	0.00	0.00
	9	-162.27	649.07	0.00	0.00	0.00	0.00
	10	-157.54	619.99	0.00	0.00	0.00	0.00
	11	-65.18	224.03	0.00	0.00	0.00	0.00
	12	-268.44	1055.10	0.00	0.00	0.00	0.00
	13	-259.72	1000.84	0.00	0.00	0.00	0.00
	14	-265.66	1037.66	0.00	0.00	0.00	0.00
	15	-262.51	1018.27	0.00	0.00	0.00	0.00
	16	-452.26	1737.29	0.00	0.00	0.00	0.00

************* END OF LATEST ANALYSIS RESULT **************

96. FINISH

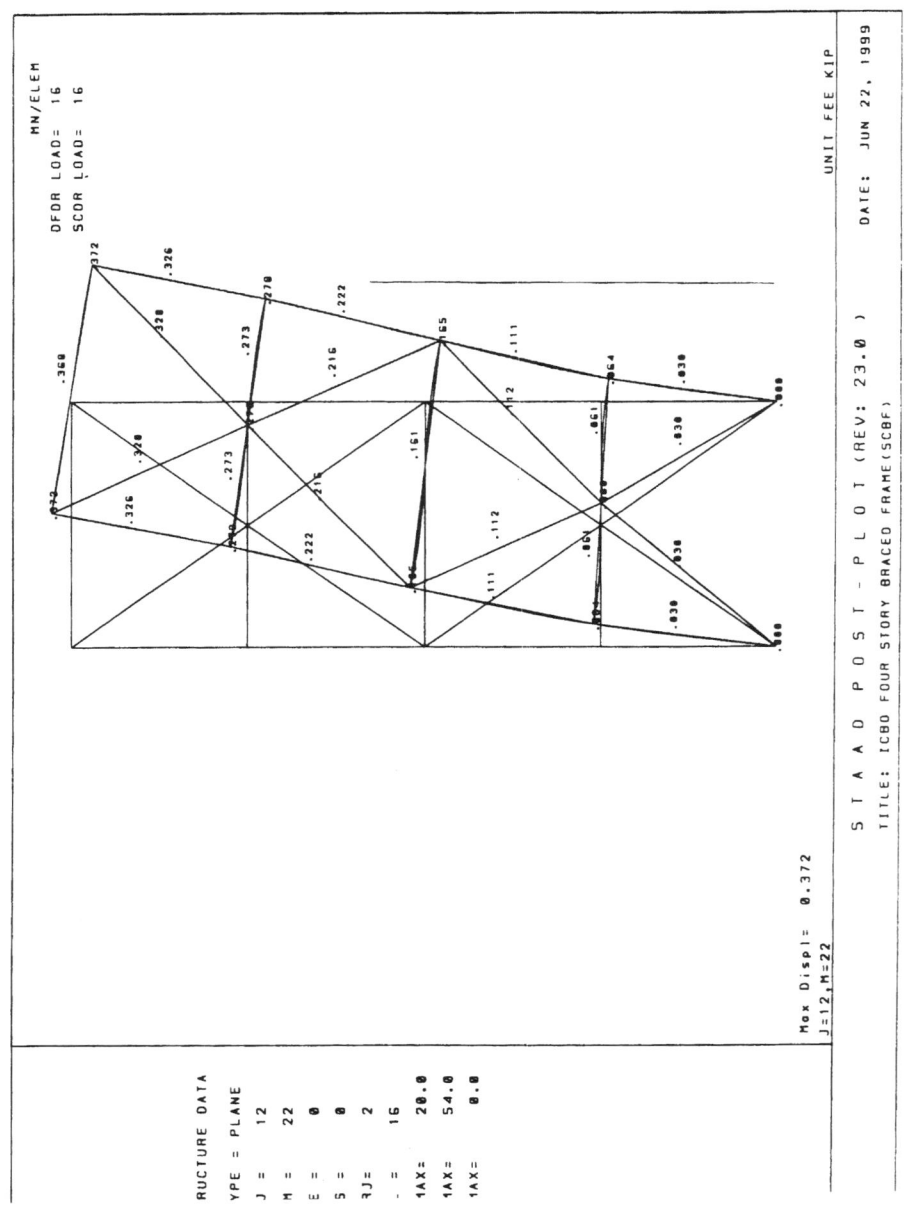

INDEX